CISM Courses and Lectures

The series presents lecture notes, monographs, edited works and proceedings in the field of Mechanics, Engineering, Computer Science and Applied Mathematics.
Purpose of the series is to make known in the international scientific and technical community results obtained in some of the activities organized by CISM, the International Centre for Mechanical Sciences.

International Centre for Mechanical Sciences

Courses and Lectures Vol. 558

For further volumes:
www.springer.com/series/76

Peter Hagedorn · Gottfried Spelsberg-Korspeter
Editors

Active and Passive Vibration Control of Structures

Springer

Editors

Peter Hagedorn
TU Darmstadt, Darmstadt, Germany

Gottfried Spelsberg-Korspeter
TU Darmstadt, Darmstadt, Germany

ISSN 0254-1971
ISBN 978-3-7091-1820-7 ISBN 978-3-7091-1821-4 (eBook)
DOI 10.1007/ 978-3-7091-1821-4
Springer Wien Heidelberg New York Dordrecht London

All contributions have been typeset by the authors

Printed on acid-free paper

Springer is part of Springer Science+Business Media (www.springer.com)

PREFACE

Active and passive vibration control of structures form a topic of very actual interest in many different fields of engineering, for example in the automotive industry, in aerospace engineering (e.g. in large telescopes) and also in civil engineering. The material presented in this book tries to fill gaps between structural mechanics, vibrations and modern control theory. It also establishes links between the different applications in structural control. In vibration control it is still common practice to a large extent to regard the design of mechanical structures and their damping and control as different topics, which often are treated sequentially, with the design of the structure to be carried out first, followed later by designing the active and passive vibration control. Similarly, in the standard curricula at most universities, the mechanical modelling and the design of structures as well as their control are treated as separate subjects. This book intends to fill some of the gaps between these different issues. The intention is to give a solid foundation of the mechanical modelling and the vibration control for discrete and continuous structures, with an emphasis on the interfaces of the different disciplines.

This book was written accompanying the CISM Course no. 418 entitled 'Active and Passive Vibration Control of Structures' *held in Udine from May 27 to 31, 2013. Therefore it is directed to young researchers, to doctoral students and also to engineers working in fields related to structures, vibrations and control.*

A thorough introduction into the relevant theory both of the mechanical modelling as well as of the vibration control theory are presented and the most important design goals are discussed. Various strategies for modelling complex mechanical structures are given and an introduction to active, passive and semi-active strategies for vibration control are discussed. In a number of examples from different areas it is shown that a comprehensive approach, in which both the mechanical design problem and the development of suitable controls are considered simultaneously, can present substantial advantages.

The organization of the book is as follows.

Chapter I, by PETER HAGEDORN *(Technische Universität Darmstadt, Germany), treats equations of motion for discrete and con-*

tinuous mechanical systems laying the foundation for the creation of control models.

Chapter II, by GOTTFRIED SPELSBERG-KORSPETER *(Technische Universität Darmstadt, Germany), gives an introduction to variational principles in mechanics and control relating to mechanical modelling and the development of control strategies.*

Chapter III, by ANDRE PREUMONT, DAVID ALALUF *and* RENAUD BASTAITS *(Université Libre de Bruxelles, Belgium), treat hybrid mass dampers for mitigating the dynamic response of buildings.*

Chapter IV, by ANDRE PREUMONT *and* BILAL MOKRANI *(Université Libre de Bruxelles, Belgium), discuss the theory of electromagnetic and piezoelectric transducers and presents some applications in structural control.*

Chapter V, by ROBERT SKELTON *(University of California San Diego, United States), focuses on structure and design of control systems with an emphasis on the advantages of using matrix inequalities.*

Chapter VI, by ADNAN AKAY *(Bilkent University, Turkey) and* ANTONIO CARCATERRA *(University of Rome, Italy), address the physics and modelling of structural damping which is extremely important in almost every structural control problem.*

Chapter VII, by RAINER NORDMANN *(Technische Universität Darmstadt, Germany), deals with active magnetic bearings, which can be used for control purposes in rotating machinery.*

Summarizing, the material presented in this book will offer a unified view on active and passive control and the mechanical modelling of structures presented from the point of view of experienced researchers with quite different perspectives.

The authors want to express their gratitude for the support of the CISM organization, in particular to Professor Friedrich Pfeiffer for chairing the course in Udine, and to Carla Toros for her tremendous support in organizing it.

The authors sincerely thank Manuel Eckstein, who carried the main burden of editing the manuscripts and coordinating the different chapters, as well to Eduard Heffel, Matthias Heymanns, Henning Spiegelberg and Andreas Wagner.

Peter Hagedorn
Gottfried Spelsberg-Korspeter

CONTENTS

Mechanical Systems: Equations of Motion and Stability
by P. Hagedorn 1

Variational Principles in Mechanics and Control
by G. Spelsberg-Korspeter 133

Hybrid Mass Damper: A Tutorial Example
by A. Preumont, D. Alaluf and R. Bastaits 179

Electromagnetic and Piezoelectric Transducers
by A. Preumont and B. Mokrani 213

LMIs in Control Optimization
by R.E. Skelton 249

Damping Mechanisms
by A. Akay and A. Carcaterra 259

Vibration Control and Failure Diagnosis in Rotating Machinery by means of Active Magnetic Bearings
by R. Nordmann 301

Mechanical Systems: Equations of Motion and Stability

Peter Hagedorn
TU Darmstadt, Darmstadt, Germany

Abstract The Chapter 'Mechanical Systems: Equations of Motion and Stability' corresponds to the material presented in five lectures given at the CISM Course no. 418. The first parts deal with the form of the equations of motion of mechanical systems, in particular the linearized equations and the influence and importance of the different terms (inertia terms, damping, gyroscopic terms, restoring terms and circulatory terms as well as with their physical origin). This is done both for discrete systems, and the corresponding material is part of the recent book *Hagedorn & Hochlenert, Technische Schwingungslehre, Verlag Harri Deutsch, Frankfurt, 2012*, as well as for continuous systems, the material being adapted from *Hagedorn & DasGupta, Vibrations and Waves in Continuous Mechanical Systems, Wiley, Chichester, 2007.* Almost all the material is presented in typical elementary vibration courses, but here certain aspects will be highlighted, which are not always stressed in basic vibration courses. The third part deals with LIAPOUNOV stability, the material is from the author's earlier book *Hagedorn, Non-Linear Oscillations, 2nd edition, Oxford Science Publications, 1988.* The material of these five lectures is used in the other lectures of the course.

The author prepared most of the material in 2012 and 2013, while staying at the University of Canterbury in Christchurch, New Zealand. The author thanks the Department of Mechanical Engineering of the UC for providing the infrastructure and assistance.

P. Hagedorn, G. Spelsberg-Korspeter (Eds.), *Active and Passive Vibration Control of Structures*, CISM International Centre for Mechanical Sciences
DOI 10.1007/ 978-3-7091-1821-4_1 © CISM Udine 2014

1 Equations of Motion of Discrete Mechanical Systems

In this lecture we will shortly recapitulate the form of the equations of motion of discrete mechanical systems (which may of course be an approximation of continuous systems). We will highlight certain aspects which, although elementary, are not always stressed in basic vibration courses.

For a holonomic system of n degrees of freedom and generalized coordinates

$$\boldsymbol{q} = (q_1, q_2, \cdots, q_n)^{\mathsf{T}} \tag{1}$$

the equations of motion can be obtained from the LAGRANGE equations of the second type, based on the Lagrangian

$$L = T - U, \tag{2}$$

where T is the kinetic energy and U the potential energy function (which we assume may also depend on the time t). LAGRANGE's equations then read

$$\frac{\partial}{\partial t}\frac{\partial L}{\dot{q}_s} - \frac{\partial L}{\partial q_s} = Q_s, \tag{3}$$

where the Q_s are the generalized forces not represented by the potential U. For a large class of systems these equations can be written as

$$\boldsymbol{M}\ddot{\boldsymbol{q}} + \boldsymbol{G}\dot{\boldsymbol{q}} + \boldsymbol{K}\boldsymbol{q} = \boldsymbol{f}(\boldsymbol{q}, \dot{\boldsymbol{q}}, t). \tag{4}$$

The term $\boldsymbol{f}(\boldsymbol{q}, \dot{\boldsymbol{q}}, t)$ contains for example the damping and other nonconservative terms, as well as for example control forces. The linearized equations (linearized about an equilibrium of the unforced autonomous system) can then be written as

$$\boldsymbol{M}\ddot{\boldsymbol{q}} + (\boldsymbol{D} + \boldsymbol{G})\dot{\boldsymbol{q}} + (\boldsymbol{K} + \boldsymbol{N})\boldsymbol{q} = \boldsymbol{f}(t). \tag{5}$$

This linearized form of the equations of motion is usually employed to develop appropriate active or passive vibration control. The control strategies based on these linear models may then later be tested for the nonlinear model.

In many cases, setting up the equations of motion using LAGRANGE's equations is not a practical approach and other methods may be more efficient. The form of the equations will however be the same as above. Unless stated otherwise, we will assume the following properties for the matrices:

Mass matrix:	$\boldsymbol{M} = \boldsymbol{M}^{\mathsf{T}}$,	$\boldsymbol{M} > 0$	(symmetric, positive definite)
Damping matrix:	$\boldsymbol{D} = \boldsymbol{D}^{\mathsf{T}}$,	$\boldsymbol{D} \geq 0$	(symmetric, positive semidef.)
Stiffness matrix:	$\boldsymbol{K} = \boldsymbol{K}^{\mathsf{T}}$,	$\boldsymbol{K} \geq 0$	(symmetric, positive semidef.)
Gyroscopic matrix:	$\boldsymbol{G} = -\boldsymbol{G}^{\mathsf{T}}$,		(skew symmetric)
Circulatory matrix:	$\boldsymbol{N} = -\boldsymbol{N}^{\mathsf{T}}$,		(skew symmetric)

In this lecture and in the next one, we will discuss in some more detail the significance of the different matrices for the behavior of the mechanical systems.

We will first consider the free vibrations, i.e. the case $\boldsymbol{f}(t) = 0$:

$$\boldsymbol{M}\ddot{\boldsymbol{q}} + (\boldsymbol{D} + \boldsymbol{G})\dot{\boldsymbol{q}} + (\boldsymbol{K} + \boldsymbol{N})\boldsymbol{q} = \boldsymbol{0}. \tag{6}$$

1.1 The Eigenvalue Problem

Since (5) is a system of ode's with constant coefficients, the exponential *ansatz*

$$\boldsymbol{q}(t) = \boldsymbol{r}e^{\lambda t} \tag{7}$$

is successful, leading to

$$\Big[\lambda^2\boldsymbol{M} + \lambda(\boldsymbol{D} + \boldsymbol{G}) + \boldsymbol{K} + \boldsymbol{N}\Big]\boldsymbol{r}e^{\lambda t} = \boldsymbol{0}. \tag{8}$$

In order for (8) to be valid for all times, the condition

$$\Big[\lambda^2\boldsymbol{M} + \lambda(\boldsymbol{D} + \boldsymbol{G}) + \boldsymbol{K} + \boldsymbol{N}\Big]\boldsymbol{r} = \boldsymbol{0} \tag{9}$$

must be fulfilled. Equation (9) is the eigenvalue problem, and the sought values of λ and $\boldsymbol{r}$ are respectively the *eigenvalues* and the *eigenvectors.*

The eigenvalue problem is a linear homogeneous algebraic system in the unknown vector $\boldsymbol{r}$. A necessary condition for the existence of non-trivial solutions in $\boldsymbol{r}$, is that the determinant of the coefficient matrix vanishes, and this leads to the characteristic equation

$$\det\Big(\lambda^2\boldsymbol{M} + \lambda(\boldsymbol{D} + \boldsymbol{G}) + \boldsymbol{K} + \boldsymbol{N}\Big) = 0. \tag{10}$$

The left-hand side is a polynomial of degree $2n$ in λ and can be written as

$$a_{2n}\lambda^{2n} + \ldots + a_2\lambda^2 + a_1\lambda + a_0 = 0. \tag{11}$$

The $2n$ solutions λ_i $(i = 1, 2, \ldots, 2n)$ are real or appear in complex conjugate pairs, since all the coefficients a_k are real. For each eigenvalue λ_i the corresponding nontrivial eigenvector $\boldsymbol{r}_i$ can then be calculated from

$$\left[\lambda_i^2 \boldsymbol{M} + \lambda_i(\boldsymbol{D} + \boldsymbol{G}) + \boldsymbol{K} + \boldsymbol{N}\right]\boldsymbol{r}_i = \boldsymbol{0}, \qquad i = 1, \ldots, 2n. \tag{12}$$

We will not deal with the case of multiple eigenvalues with multiplicity larger than the number of independent eigenvectors; the case of 'semi-simple' multiple eigenvalues will however play an important role in a later lecture. Obviously the eigenvectors resulting from (12) can at most be determined up to a multiplicative (complex) factor, since we found the eigenvalues by setting the determinant of the coefficient matrix equal to zero and the rows and columns of the coefficient matrix are therefore linearly dependent.

Only in very particular cases, matrix eigenvalue problems can be solved analytically. As a rule, eigenvalues and eigenvectors will be numerically calculated using appropriate numerical tools. For example MATLAB immediately calculates the eigenvalues and eigenvectors with the command `polyeig(K+N,D+G,M)`, if the corresponding numerical values are attributed to the matrices `M`, `D`, `G`, `K` and `N`.

Normalization of Eigenvectors It is often important to normalize the eigenvectors, if only for example to compare numerical or experimental results. From (12) it is obvious that a complex eigenvalue will in general imply a complex eigenvector $\boldsymbol{r}_i$, and that a real eigenvector can be found for each real eigenvalue. Real eigenvectors can be normalized e.g. according to

$$\boldsymbol{r}_i^{\mathsf{T}}\boldsymbol{r}_i = 1 \qquad \text{or} \qquad \boldsymbol{r}_i^{\mathsf{T}}\boldsymbol{M}\boldsymbol{r}_i = 1 \tag{13a}$$

with respect to their magnitude or with respect to the mass matrix. Let $\underset{\sim}{\boldsymbol{r}}_i$ be a non-normalized eigenvector, then the corresponding normalized eigenvector $\boldsymbol{r}_i$ is

$$\boldsymbol{r}_i = \frac{\underset{\sim}{\boldsymbol{r}}_i}{\sqrt{\underset{\sim}{\boldsymbol{r}}_i^{\mathsf{T}}\underset{\sim}{\boldsymbol{r}}_i}} \qquad \text{or} \qquad \boldsymbol{r}_i = \frac{\underset{\sim}{\boldsymbol{r}}_i}{\sqrt{\underset{\sim}{\boldsymbol{r}}_i^{\mathsf{T}}\boldsymbol{M}\underset{\sim}{\boldsymbol{r}}_i}}. \tag{13b}$$

This eigenvector is still not uniquely determined, as is obvious that the real eigenvector $\tilde{\boldsymbol{r}}_i = -\boldsymbol{r}_i$ also fulfills the normalization. This is more involved for complex eigenvectors, whose absolute value can be normalized in analogy to (13a) with

$$\boldsymbol{r}_i^{*}\boldsymbol{r}_i = 1 \qquad \text{or} \qquad \boldsymbol{r}_i^{*}\boldsymbol{M}\boldsymbol{r}_i = 1. \tag{13c}$$

Here $\boldsymbol{r}_i^*$ is the complex conjugate transposed vector to $\boldsymbol{r}_i$. Since the normalization of complex eigenvalues is less commonly discussed in the literature, we will describe it in more detail. The normalization (13c) is here also fulfilled by the complex eigenvector $\tilde{\boldsymbol{r}}_i = e^{j\beta_i}\boldsymbol{r}_i$ for an arbitrary β_i, i.e. each component of $\boldsymbol{r}_i$ can still be rotated in the complex plane by an equal but arbitrary angle β_i. The components of the real eigenvectors can in the same fashion be rotated by the angle π, remaining real, or by an arbitrary angle, so that the eigenvectors become complex.

In order to obtain uniquely determined eigenvectors, so that the results of different computations can be compared with each other, it is convenient to fix the angle β_i conveniently. This can be done by choosing β_i so that a complex eigenvector is reduced to a real form, if this is possible. As an example, consider the eigenvector

$$\boldsymbol{r} = \boldsymbol{a} + j\boldsymbol{b} \tag{14}$$

with real part $\boldsymbol{a}$ and imaginary part $\boldsymbol{b}$, as well as its representation rotated by an angle β

$$\begin{aligned}\tilde{\boldsymbol{r}} &= e^{j\beta}(\boldsymbol{a} + j\boldsymbol{b}) = (\boldsymbol{a}\cos\beta - \boldsymbol{b}\sin\beta) + j(\boldsymbol{a}\sin\beta + \boldsymbol{b}\cos\beta) \\ &= \tilde{\boldsymbol{a}} + j\tilde{\boldsymbol{b}}.\end{aligned} \tag{15}$$

The angle β will now be chosen in such a way that the real part of $\tilde{\boldsymbol{r}}$ becomes 'as large as possible'. We formulate this requirement by

$$\begin{aligned}\max_\beta \tilde{\boldsymbol{a}}^\mathsf{T}\tilde{\boldsymbol{a}} &= \max_\beta\, (\boldsymbol{a}\cos\beta - \boldsymbol{b}\sin\beta)^\mathsf{T}(\boldsymbol{a}\cos\beta - \boldsymbol{b}\sin\beta) \\ &= \max_\beta\, \boldsymbol{a}^\mathsf{T}\boldsymbol{a}\, \cos^2\beta - 2\boldsymbol{a}^\mathsf{T}\boldsymbol{b}\, \cos\beta\sin\beta + \boldsymbol{b}^\mathsf{T}\boldsymbol{b}\, \sin^2\beta \\ &= \max_\beta \begin{bmatrix}\cos\beta & \sin\beta\end{bmatrix}\begin{bmatrix}\boldsymbol{a}^\mathsf{T}\boldsymbol{a} & -\boldsymbol{a}^\mathsf{T}\boldsymbol{b} \\ -\boldsymbol{a}^\mathsf{T}\boldsymbol{b} & \boldsymbol{b}^\mathsf{T}\boldsymbol{b}\end{bmatrix}\begin{bmatrix}\cos\beta \\ \sin\beta\end{bmatrix},\end{aligned} \tag{16}$$

i.e. by the maximization of a positive definite quadratic form with the corresponding eigenvalue problem in μ

$$\begin{bmatrix}\boldsymbol{a}^\mathsf{T}\boldsymbol{a} - \mu & -\boldsymbol{a}^\mathsf{T}\boldsymbol{b} \\ -\boldsymbol{a}^\mathsf{T}\boldsymbol{b} & \boldsymbol{b}^\mathsf{T}\boldsymbol{b} - \mu\end{bmatrix}\begin{bmatrix}\cos\beta \\ \sin\beta\end{bmatrix} = \begin{bmatrix}0 \\ 0\end{bmatrix}. \tag{17}$$

The two eigenvalues for (17) are

$$\mu_{1,2} = \frac{\boldsymbol{a}^\mathsf{T}\boldsymbol{a} + \boldsymbol{b}^\mathsf{T}\boldsymbol{b}}{2} \pm \sqrt{\frac{\left(\boldsymbol{a}^\mathsf{T}\boldsymbol{a} - \boldsymbol{b}^\mathsf{T}\boldsymbol{b}\right)^2}{4} - \left(\boldsymbol{a}^\mathsf{T}\boldsymbol{b}\right)^2}\,. \tag{18}$$

Substituting the larger of the two eigenvalues μ_2 in (17) leads to the two equations

$$\left(\boldsymbol{b}^{\mathsf{T}}\boldsymbol{b}-\boldsymbol{a}^{\mathsf{T}}\boldsymbol{a}+\sqrt{(\boldsymbol{b}^{\mathsf{T}}\boldsymbol{b}-\boldsymbol{a}^{\mathsf{T}}\boldsymbol{a})^2-4(\boldsymbol{a}^{\mathsf{T}}\boldsymbol{b})^2}\right)\cos\beta+2\boldsymbol{a}^{\mathsf{T}}\boldsymbol{b}\sin\beta=0, \qquad (19a)$$

$$2\boldsymbol{a}^{\mathsf{T}}\boldsymbol{b}\cos\beta+\left(\boldsymbol{a}^{\mathsf{T}}\boldsymbol{a}-\boldsymbol{b}^{\mathsf{T}}\boldsymbol{b}+\sqrt{(\boldsymbol{a}^{\mathsf{T}}\boldsymbol{a}-\boldsymbol{b}^{\mathsf{T}}\boldsymbol{b})^2-4(\boldsymbol{a}^{\mathsf{T}}\boldsymbol{b})^2}\right)\sin\beta=0 \qquad (19b)$$

for the determination of $\cos\beta$ and $\sin\beta$. The two equations are of course linearly dependent, but one should be careful in selecting one of the equations. For $\boldsymbol{a}^{\mathsf{T}}\boldsymbol{b}=0$ and $\boldsymbol{a}^{\mathsf{T}}\boldsymbol{a}=\boldsymbol{b}^{\mathsf{T}}\boldsymbol{b}$ both equations are identically fulfilled, independently of the value of β. In this case we choose $\beta=0$. For $\boldsymbol{a}^{\mathsf{T}}\boldsymbol{b}=0$ and $\boldsymbol{a}^{\mathsf{T}}\boldsymbol{a}>\boldsymbol{b}^{\mathsf{T}}\boldsymbol{b}$ both coefficients of the first equation vanish and the second equation gives $\sin\beta=0$. For $\boldsymbol{a}^{\mathsf{T}}\boldsymbol{b}=0$ and $\boldsymbol{a}^{\mathsf{T}}\boldsymbol{a}<\boldsymbol{b}^{\mathsf{T}}\boldsymbol{b}$ the first equation gives $\cos\beta=0$, and both coefficients of the second equation vanish. For $\boldsymbol{a}^{\mathsf{T}}\boldsymbol{b}\neq 0$ either the first or the second equation can be used. Summarizing, one has

$$\tan\beta=\begin{cases}0 & \text{for}\quad \boldsymbol{a}^{\mathsf{T}}\boldsymbol{b}=0,\ \boldsymbol{a}^{\mathsf{T}}\boldsymbol{a}\geq\boldsymbol{b}^{\mathsf{T}}\boldsymbol{b}\\ \pm\infty & \text{for}\quad \boldsymbol{a}^{\mathsf{T}}\boldsymbol{b}=0,\ \boldsymbol{a}^{\mathsf{T}}\boldsymbol{a}<\boldsymbol{b}^{\mathsf{T}}\boldsymbol{b}\\ \dfrac{\boldsymbol{a}^{\mathsf{T}}\boldsymbol{a}-\boldsymbol{b}^{\mathsf{T}}\boldsymbol{b}}{2\boldsymbol{a}^{\mathsf{T}}\boldsymbol{b}}-\sqrt{\dfrac{(\boldsymbol{a}^{\mathsf{T}}\boldsymbol{a}-\boldsymbol{b}^{\mathsf{T}}\boldsymbol{b})^2}{4(\boldsymbol{a}^{\mathsf{T}}\boldsymbol{b})^2}-1} & \text{for}\quad \boldsymbol{a}^{\mathsf{T}}\boldsymbol{b}\neq 0\end{cases}. \qquad (20)$$

In case one desires to maximize the real part with respect to the mass matrix, according to $\max\tilde{\boldsymbol{a}}^{\mathsf{T}}\boldsymbol{M}\tilde{\boldsymbol{a}}$, the matrix $\boldsymbol{M}$ is introduced in the scalar products. A normalization of the eigenvectors with β according to (20) subject to (13c), will therefore always result in a real representation of the eigenvector, if this is at all possible. In order to make the normalization completely unique, since (20) still permits a factor -1, one may for example prescribe that the first non vanishing component of $\boldsymbol{a}$ be positive.

For many applications, the normalization according to (13) is completely sufficient. Yet, for the comparison of different numerical results the determination of β will in some cases be necessary and a completely unique normalization of the eigenvectors may be in order.

General Solution for Free Vibrations For distinct eigenvalues ($\lambda_i\neq\lambda_k$) (12) gives $2n$ linear independent solutions of the form $\boldsymbol{r}_i\,e^{\lambda_i t}$. With complex integration constants, adding the individual solutions leads to the general solution of (6) in the form

$$\boldsymbol{q}(t) = \sum_{i=1}^{2n} \underset{\sim}{K}_i \, \boldsymbol{r}_i \, e^{\lambda_i t}. \tag{21}$$

The integration constants $\underset{\sim}{K}_i$ $(i = 1, \ldots, 2n)$ can be determined from the initial conditions

$$\boldsymbol{q}(0) = \boldsymbol{q}_0, \qquad \dot{\boldsymbol{q}}(0) = \dot{\boldsymbol{q}}_0. \tag{22}$$

For real initial conditions, $\boldsymbol{q}(t)$ will then also be real.

It may however be convenient to directly write (21) in real form. In doing this we assume first that the eigenvalues are all complex. From (12) one recognizes that for each pair of complex conjugate eigenvalues there is a pair of complex conjugate eigenvectors, so that

$$\lambda_{i,n+i} = -\delta_i \pm j\omega_{\mathrm{d}i}, \qquad \boldsymbol{r}_{i,n+i} = \boldsymbol{a}_i \pm j\boldsymbol{b}_i \tag{23}$$

holds. Next, we choose the integration constants in complex conjugate pairs $\underset{\sim}{K}_i = \underset{\sim}{K}^*_{n+i}$ and write them as

$$\underset{\sim}{K}_i = \frac{1}{2} K_i \, e^{j\gamma_i}, \qquad \underset{\sim}{K}_{n+i} = \frac{1}{2} K_i \, e^{-j\gamma_i} \tag{24}$$

with the new integration constants K_i and γ_i. With this substitution, (21) assumes the real form

$$\begin{aligned}
\boldsymbol{q}(t) &= \sum_{i=1}^{n} \frac{1}{2} K_i \, (\boldsymbol{a}_i + j\boldsymbol{b}_i) \, e^{-\delta_i t + j(\omega_{\mathrm{d}i} t + \gamma_i)} + \frac{1}{2} K_i \, (\boldsymbol{a}_i - j\boldsymbol{b}_i) \, e^{-\delta_i t - j(\omega_{\mathrm{d}i} t + \gamma_i)} \\
&= \sum_{i=1}^{n} K_i \, e^{-\delta_i t} \Big[\boldsymbol{a}_i \Big(e^{j(\omega_{\mathrm{d}i} t + \gamma_i)} + e^{-j(\omega_{\mathrm{d}i} t + \gamma_i)} \Big) \\
&\qquad\qquad + j\boldsymbol{b}_i \Big(e^{j(\omega_{\mathrm{d}i} t + \gamma_i)} - e^{-j(\omega_{\mathrm{d}i} t + \gamma_i)} \Big) \Big] \\
&= \sum_{i=1}^{n} K_i \, e^{-\delta_i t} \Big[\boldsymbol{a}_i \cos(\omega_{\mathrm{d}i} t + \gamma_i) - \boldsymbol{b}_i \sin(\omega_{\mathrm{d}i} t + \gamma_i) \Big].
\end{aligned} \tag{25a}$$

If the system does not have $2n$ complex eigenvalues, but only $2s$ complex eigenvalues $\lambda_{i,s+i}$ $(i = 1, \ldots, s)$ and $2(n-s)$ real eigenvalues λ_i $(i = 2s+1, \ldots, 2n)$ with the corresponding real eigenvector, one has

$$\boldsymbol{q}(t) = \sum_{i=1}^{s} K_i \, e^{-\delta_i t} \Big[\boldsymbol{a}_i \cos(\omega_{\mathrm{d}i} t + \gamma_i) - \boldsymbol{b}_i \sin(\omega_{\mathrm{d}i} t + \gamma_i) \Big] + \sum_{i=2s+1}^{2n} K_i \, \boldsymbol{r}_i \, e^{\lambda_i t}. \tag{25b}$$

The behavior of the different particular solutions corresponding to the different eigenvalues strongly depends in particular on the real parts of the eigenvalues, as is well known. They determine the stability of the solutions and of the system. An important aspect of (25) is that the eigensolutions in general (when all the γ are different from zero) are such that their different coordinates vanish at different times. This is quite different in the particular case which we examine next.

1.2 $\boldsymbol{M}$-$\boldsymbol{K}$-Systems

An important particular case is that in which the matrices $\boldsymbol{D}$, $\boldsymbol{G}$ and $\boldsymbol{N}$ vanish. The equations of motion are then of the form

$$\boldsymbol{M}\ddot{\boldsymbol{q}} + \boldsymbol{K}\boldsymbol{q} = \boldsymbol{0}. \tag{26}$$

We will recapitulate their main properties and then later examine the way in which the other terms in the equations of motion influence the solutions.

Properties of the Eigenvalues For $\boldsymbol{M}$-$\boldsymbol{K}$-systems the characteristic equation (10) simplifies to

$$\det(\lambda^2\boldsymbol{M} + \boldsymbol{K}) = 0. \tag{27}$$

This is a polynomial of degree n in λ^2. The roots λ_i^2 $(i = 1, \dots, n)$ not necessarily are single. If they are, then for each λ_i^2 there is exactly one nontrivial possibly complex eigenvector $\boldsymbol{r}_i$, satisfying

$$(\lambda_i^2\boldsymbol{M} + \boldsymbol{K})\boldsymbol{r}_i = \boldsymbol{0}. \tag{28}$$

Multiplying (28) from the left with the vector $\boldsymbol{r}_i^*$ leads to

$$\lambda_i^2\, \boldsymbol{r}_i^*\boldsymbol{M}\boldsymbol{r}_i + \boldsymbol{r}_i^*\boldsymbol{K}\boldsymbol{r}_i = 0. \tag{29}$$

In doing this we have in a way ‘projected’ the eigenvalue problem (28) on $\boldsymbol{r}_i^*$, which of course leads to a loss of information. It would not be correct to assume that all the solutions of (29) also fulfill (28). The inverse is however true.

Since $\boldsymbol{M}$ and $\boldsymbol{K}$ were assumed symmetric, the expressions $\boldsymbol{r}_i^*\boldsymbol{M}\boldsymbol{r}_i$ and $\boldsymbol{r}_i^*\boldsymbol{K}\boldsymbol{r}_i$ are real for arbitrary vectors $\boldsymbol{r}_i$. For a positive definite matrix one has $\boldsymbol{r}_i^*\boldsymbol{M}\boldsymbol{r}_i > 0$ and (29) can be transformed into

$$\lambda_i^2 = -\frac{\boldsymbol{r}_i^*\boldsymbol{K}\boldsymbol{r}_i}{\boldsymbol{r}_i^*\boldsymbol{M}\boldsymbol{r}_i}. \tag{30}$$

If also $\boldsymbol{K}$ is positive definite, we can conclude that λ^2 is a negative real number. Since we obtained this result from the 'projected' equation (29), at this point we only can be sure that one of the two solutions of (30) will also fulfill (28). We do however know that the eigenvalues appear in complex conjugate pairs

$$\lambda_i = \lambda_{n+i}^* \,, \qquad i = 1, \ldots, n, \tag{31}$$

so that in fact both solutions of (30) fulfill (28). The $2n$ eigenvalus

$$\lambda_{i,n+i} = \pm j \sqrt{\frac{\boldsymbol{r}_i^* \boldsymbol{K} \boldsymbol{r}_i}{\boldsymbol{r}_i^* \boldsymbol{M} \boldsymbol{r}_i}} \,, \qquad i = 1, \ldots, n \tag{32}$$

of any $\boldsymbol{M}$-$\boldsymbol{K}$-system (26) are therefore purely imaginary (if $\boldsymbol{K}$ is positive definite). This according to (28) implies that the eigenvectors can be chosen real, where $\boldsymbol{r}_i$ is the eigenvector associated to λ_i and λ_i^*. If the stiffness matrix is not positive definite but only positive semidefinite ($\boldsymbol{r}_i^\mathsf{T} \boldsymbol{K} \boldsymbol{r}_i \geq 0$), then there are pairs of zero eigenvaluse $\lambda_i = \lambda_{n+i} = 0$; with positive definite stiffness matrix ($\boldsymbol{r}_i^\mathsf{T} \boldsymbol{K} \boldsymbol{r}_i > 0$) all the eigenvalues are complex conjugate in strict sense.

Orthogonality Relations of Eigenvectors If $(\lambda_i, \boldsymbol{r}_i)$ and $(\lambda_k, \boldsymbol{r}_k)$ are two eigenpairs, i.e. pairs of eigenvalues with the corresponding eigenvectors, then (29) impies

$$\lambda_i^2 \boldsymbol{M} \boldsymbol{r}_i + \boldsymbol{K} \boldsymbol{r}_i = \boldsymbol{0}, \tag{33a}$$

$$\lambda_k^2 \boldsymbol{M} \boldsymbol{r}_k + \boldsymbol{K} \boldsymbol{r}_k = \boldsymbol{0}. \tag{33b}$$

Multiplying (33a), respectively (33b) from the left with $\boldsymbol{r}_k^\mathsf{T}$, respectively with $\boldsymbol{r}_i^\mathsf{T}$, leads to

$$\lambda_i^2 \, \boldsymbol{r}_k^\mathsf{T} \boldsymbol{M} \boldsymbol{r}_i + \boldsymbol{r}_k^\mathsf{T} \boldsymbol{K} \boldsymbol{r}_i = 0, \tag{34a}$$

$$\lambda_k^2 \, \boldsymbol{r}_i^\mathsf{T} \boldsymbol{M} \boldsymbol{r}_k + \boldsymbol{r}_i^\mathsf{T} \boldsymbol{K} \boldsymbol{r}_k = 0. \tag{34b}$$

Due to the symmetry of $\boldsymbol{M}$ and $\boldsymbol{K}$ we have $\boldsymbol{r}_k^\mathsf{T} \boldsymbol{M} \boldsymbol{r}_i = \boldsymbol{r}_i^\mathsf{T} \boldsymbol{M} \boldsymbol{r}_k$ and $\boldsymbol{r}_k^\mathsf{T} \boldsymbol{K} \boldsymbol{r}_i = \boldsymbol{r}_i^\mathsf{T} \boldsymbol{K} \boldsymbol{r}_k$, so that the the difference of (34a) and (34b) gives

$$(\lambda_i^2 - \lambda_k^2) \, \boldsymbol{r}_i^\mathsf{T} \boldsymbol{M} \boldsymbol{r}_k = 0. \tag{35}$$

This finally leads to

$$\boldsymbol{r}_i^\mathsf{T} \boldsymbol{M} \boldsymbol{r}_k = 0 \qquad \text{for} \qquad \lambda_i^2 \neq \lambda_k^2. \tag{36a}$$

The eigenvectors associated to different eigenvalues ($\lambda_i^2 \neq \lambda_k^2$) are orthogonal with respect to the mass matrix. Similary one also has

$$\boldsymbol{r}_i^\mathsf{T} \boldsymbol{K} \boldsymbol{r}_k = 0 \qquad \text{for} \qquad \lambda_i^2 \neq \lambda_k^2, \tag{36b}$$

i.e. the eigenvectors are also orthogonal with respect to the stiffness matrix. In general the eigenvectors will however not be orthogonal in the usual sense, i.e. with respect to the identity matrix!

For multiple eigenvalues, i.e. for $\lambda_i^2 = \lambda_k^2$, (35) does not imply the orthogonality of the corresponding eigenvectors. However, if the matrices $\boldsymbol{M}$ and $\boldsymbol{K}$ are symmetric, as we always assume according to our definition (5), it can be shown that to each pair of eigenvalues of multiplicity m, exactly m linearly independent eigenvectors exist, which are all orthogonal to the remaining n–m eigenvectors. Each linear combination of these eigenvectors is again an eigenvector λ_i^2. Using for example the GRAM-SCHMIDT orthogonalisation procedure, one can now construct an orthogonal basis (orthogonal with respect to $\boldsymbol{M}$) forming a basis of the subspace of $\mathbb{R}^n$ spanned by the m eigenvectors. Doing that for all eigenvectors pertaining to multiple eigenvalues, again results in n eigenvectors mutually orthogonal with respect to $\boldsymbol{M}$ and $\boldsymbol{K}$.

The orthogonality of the eigenvectors has the consequence that $\boldsymbol{r}_1, \boldsymbol{r}_2, \ldots$ $, \boldsymbol{r}_n$ is a basis of $\mathbb{R}^n$. In fact, consider

$$e_1 \boldsymbol{r}_1 + e_2 \boldsymbol{r}_2 + \ldots + e_n \boldsymbol{r}_n = \boldsymbol{0}, \tag{37}$$

i.e. a linear combination of the eigenvectors giving the zero vector, than, multiplication with $\boldsymbol{r}_i^\mathsf{T} \boldsymbol{M}$ from the left leads to

$$e_i \, \boldsymbol{r}_i^\mathsf{T} \boldsymbol{M} \boldsymbol{r}_i = 0, \tag{38}$$

i.e.

$$e_i = 0, \qquad i = 1, \ldots, n \tag{39}$$

and this is the condition for linear independence. Since the eigenvectors form a basis of $\mathbb{R}^n$, any vector $\boldsymbol{a}$ of $\mathbb{R}^n$ has a unique representation in form of

$$\boldsymbol{a} = a_1 \boldsymbol{r}_1 + a_2 \boldsymbol{r}_2 + \ldots + a_n \boldsymbol{r}_n. \tag{40}$$

Multiplying from the left with $\boldsymbol{r}_i^\mathsf{T} \boldsymbol{M}$ and using orthogonality of the eigenvectors leads to

$$\boldsymbol{r}_i^\mathsf{T} \boldsymbol{M} \boldsymbol{a} = a_i \, \boldsymbol{r}_i^\mathsf{T} \boldsymbol{M} \boldsymbol{r}_i, \tag{41}$$

so that the coefficients a_i are obviously given by

$$a_i = \frac{\boldsymbol{r}_i^{\mathsf{T}} \boldsymbol{M} \boldsymbol{a}}{\boldsymbol{r}_i^{\mathsf{T}} \boldsymbol{M} \boldsymbol{r}_i} = \frac{\boldsymbol{r}_i^{\mathsf{T}} \boldsymbol{K} \boldsymbol{a}}{\boldsymbol{r}_i^{\mathsf{T}} \boldsymbol{K} \boldsymbol{r}_i}, \qquad i = 1, \ldots, n. \tag{42}$$

Modal Coordinates The eigenvectors can be assembled column wise in the so called *modal matrix*

$$\boldsymbol{R} = \begin{bmatrix} \boldsymbol{r}_1 & \boldsymbol{r}_2 & \cdots & \boldsymbol{r}_n \end{bmatrix} \tag{43}$$

of dimension $n \times n$. With the coordinate transformation

$$\boldsymbol{q} = \boldsymbol{R}\boldsymbol{p} \tag{44}$$

the equations of motion of a $\boldsymbol{M}$-$\boldsymbol{K}$-system (26) can be written in the form

$$\boldsymbol{M}\boldsymbol{R}\ddot{\boldsymbol{p}} + \boldsymbol{K}\boldsymbol{R}\boldsymbol{p} = \boldsymbol{0} \tag{45}$$

in the modal coordinates $\boldsymbol{p}$. Multiplication with $\boldsymbol{R}^{\mathsf{T}}$ from the left gives

$$\boldsymbol{R}^{\mathsf{T}}\boldsymbol{M}\boldsymbol{R}\ddot{\boldsymbol{p}} + \boldsymbol{R}^{\mathsf{T}}\boldsymbol{K}\boldsymbol{R}\boldsymbol{p} = \boldsymbol{0}. \tag{46}$$

Due to the orthogonality of the eigenvectors with respect to $\boldsymbol{M}$ and $\boldsymbol{K}$ the matrices

$$\tilde{\boldsymbol{M}} = \boldsymbol{R}^{\mathsf{T}}\boldsymbol{M}\boldsymbol{R} = \begin{bmatrix} \tilde{m}_1 & & & \boldsymbol{0} \\ & \tilde{m}_2 & & \\ & & \ddots & \\ \boldsymbol{0} & & & \tilde{m}_n \end{bmatrix}, \quad \tilde{\boldsymbol{K}} = \boldsymbol{R}^{\mathsf{T}}\boldsymbol{K}\boldsymbol{R} = \begin{bmatrix} \tilde{k}_1 & & & \boldsymbol{0} \\ & \tilde{k}_2 & & \\ & & \ddots & \\ \boldsymbol{0} & & & \tilde{k}_n \end{bmatrix}$$

are diagonal. The diagonal elements

$$\tilde{m}_i = \boldsymbol{r}_i^{\mathsf{T}}\boldsymbol{M}\boldsymbol{r}_i, \tag{47a}$$

$$\tilde{k}_i = \boldsymbol{r}_i^{\mathsf{T}}\boldsymbol{K}\boldsymbol{r}_i \tag{47b}$$

are called *modal mass* and *modal stiffness*. Both quantities of course depend on the normalization of the eigenvectors and moreover depend on an arbitrary common factor (since (46) can be multiplied by an arbitrary factors). The quotient

$$\omega_i = \sqrt{\frac{\tilde{k}_i}{\tilde{m}_i}} = \sqrt{\frac{\boldsymbol{r}_i^{\mathsf{T}}\boldsymbol{K}\boldsymbol{r}_i}{\boldsymbol{r}_i^{\mathsf{T}}\boldsymbol{M}\boldsymbol{r}_i}} = \operatorname{Im}\lambda_i, \qquad i = 1, \ldots, n \tag{48}$$

is however independent of the normalization, so that the equations of motion (46) can be written as

$$\ddot{p}_i + \omega_i^2 p_i = 0, \qquad i = 1, \ldots, n. \tag{49}$$

The term ω_i is called the i-th *circular eigenfrequency* of the system, and in the case of semidefinte stiffness matrix there will also be zero circular eigenfrequenciss. In general the eigenfrequencies are ordered as

$$0 \leq \omega_1 \leq \omega_2 \leq \ldots \leq \omega_n \tag{50}$$

according to their magnitude.

The zero eigenfrequencies can be understood intuitively with the aid of the equations of motion of the $\boldsymbol{M}$-$\boldsymbol{K}$-system. The corresponding eigenvectors fulfill the condition $\boldsymbol{K}\boldsymbol{r}_i = \boldsymbol{0}$. They therefore correspond to motions in which there are no restoring forces (or restoring torques). It follows from the equations of motion that also no 'inertia forces' are present. Such motions are called rigid body displacements and the corresponding systems are called unconstrained or flying. The property of a system to be unconstrained or flying is therefore equivalent to a positive semidefinite stiffness matrix.

General Solution If the system has m zero circular eigenfrequencies ($\omega_i = 0$ for $i = 1, \ldots, m$) then the solution of (49) is

$$p_i(t) = A_i + B_i t\,, \qquad i = 1, \ldots, m \tag{51a}$$

$$p_i(t) = C_i \cos\omega_i t + S_i \sin\omega_i t \qquad i = m+1, \ldots, n, \tag{51b}$$

where (51b) can of course also be written as

$$p_i(t) = \hat{p}_i \cos(\omega_i t + \gamma_i) \qquad i = m+1, \ldots, n. \tag{51c}$$

The general solution of the equations of motion of an $\boldsymbol{M}$-$\boldsymbol{K}$-system can then of course be written in the form

$$\boldsymbol{q}(t) = \boldsymbol{R}\boldsymbol{p}(t) = \sum_{i=1}^{n} p_i(t)\, \boldsymbol{r}_i. \tag{52}$$

The integration constants A_i, B_i and C_i, S_i respectively $\hat{p}_i$, γ_i are determined from the initial conditions and the transformations (44) as

$$\begin{bmatrix} A_1 & \cdots & A_m & C_{m+1} & \cdots & C_n \end{bmatrix}^\mathsf{T} = \boldsymbol{R}^{-1}\boldsymbol{q}(0), \tag{53a}$$

$$\begin{bmatrix} B_1 & \cdots & B_m & \omega_1 S_1 & \cdots & \omega_n S_n \end{bmatrix}^\mathsf{T} = \boldsymbol{R}^{-1}\dot{\boldsymbol{q}}(0). \tag{53b}$$

The different integration constants are of course related via

$$\hat{p}_i = \sqrt{C_i^2 + S_i^2}, \qquad \tan\gamma_i = -\frac{S_i}{C_i}, \tag{53c}$$

$$C_i = \hat{p}_i \cos\gamma_i, \qquad S_i = \hat{p}_i \sin\gamma_i. \tag{53d}$$

Energy Integral Multiplying the equations of motion (26) from the left by $\dot{\boldsymbol{q}}^{\mathsf{T}}$ gives

$$\dot{\boldsymbol{q}}^{\mathsf{T}}\boldsymbol{M}\ddot{\boldsymbol{q}} + \dot{\boldsymbol{q}}^{\mathsf{T}}\boldsymbol{K}\boldsymbol{q} = 0, \tag{54}$$

which can also be written as

$$\frac{\mathrm{d}}{\mathrm{d}t}\left(\frac{1}{2}\dot{\boldsymbol{q}}^{\mathsf{T}}\boldsymbol{M}\dot{\boldsymbol{q}} + \frac{1}{2}\boldsymbol{q}^{\mathsf{T}}\boldsymbol{K}\boldsymbol{q}\right) = 0. \tag{55}$$

The expression

$$\frac{1}{2}\dot{\boldsymbol{q}}^{\mathsf{T}}\boldsymbol{M}\dot{\boldsymbol{q}} + \frac{1}{2}\boldsymbol{q}^{\mathsf{T}}\boldsymbol{K}\boldsymbol{q} = h \tag{56}$$

with $h = \text{const.}$ is therefore a first integral of the equations of motion (26). The system is conservative, since the sum of the kinetic and the potential energy is conserved ($T + U = \text{const.}$).

RAYLEIGH's Quotient and RAYLEIGH's Theorem For an $\boldsymbol{M}$-$\boldsymbol{K}$-system with n degrees of freedom we define RAYLEIGH*'s quotient*[1]

$$R(\boldsymbol{u}) = \frac{\boldsymbol{u}^{\mathsf{T}}\boldsymbol{K}\boldsymbol{u}}{\boldsymbol{u}^{\mathsf{T}}\boldsymbol{M}\boldsymbol{u}}, \tag{57}$$

where $\boldsymbol{u} \in \mathbb{R}^n$ is an arbitrary non-zero vector. Obviously the value of RAYLEIGH's quotient does not depend on the normalization of the vector $\boldsymbol{u}$. It is therefore sufficient to consider vectors $\boldsymbol{u}$ on the 'sphere' $\boldsymbol{u}^{\mathsf{T}}\boldsymbol{u} = 1$ or on $\boldsymbol{u}^{\mathsf{T}}\boldsymbol{M}\boldsymbol{u} = 1$.

RAYLEIGH*'s principle* says that for the smallest eigenfrequency of a $\boldsymbol{M}$-$\boldsymbol{K}$-system

$$\omega_1^2 = \min_{\boldsymbol{u}^{\mathsf{T}}\boldsymbol{u}=1} R(\boldsymbol{u}) = \min_{\boldsymbol{u}^{\mathsf{T}}\boldsymbol{M}\boldsymbol{u}=1} R(\boldsymbol{u}) \tag{58}$$

[1]Named after the physicist John William Strutt, 3rd Baron RAYLEIGH, *1842 in Langford (Essex), †1919 in Witham (Essex).

holds. This is very easy to prove. Since the eigenvectors $\boldsymbol{r}_i$ $(i=1,\ldots,n)$ of an $\boldsymbol{M}$-$\boldsymbol{K}$-system form a basis of $\mathbb{R}^n$, any vector $\boldsymbol{u}$ can be written as

$$\boldsymbol{u} = u_1\boldsymbol{r}_1 + u_2\boldsymbol{r}_2 + \ldots + u_n\boldsymbol{r}_n, \tag{59}$$

where we assume that the eigenvectors are normalized via

$$\boldsymbol{r}_i^{\mathsf{T}}\boldsymbol{M}\boldsymbol{r}_i = 1. \tag{60a}$$

Then also

$$\boldsymbol{r}_i^{\mathsf{T}}\boldsymbol{K}\boldsymbol{r}_i = \omega_i^2 \tag{60b}$$

holds. Substituting (59) in (57) gives

$$R(\boldsymbol{u}) = \frac{u_1^2\,\boldsymbol{r}_1^{\mathsf{T}}\boldsymbol{K}\boldsymbol{r}_1 + u_2^2\,\boldsymbol{r}_2^{\mathsf{T}}\boldsymbol{K}\boldsymbol{r}_2 + \ldots + u_n^2\,\boldsymbol{r}_n^{\mathsf{T}}\boldsymbol{K}\boldsymbol{r}_n}{u_1^2\,\boldsymbol{r}_1^{\mathsf{T}}\boldsymbol{M}\boldsymbol{r}_1 + u_2^2\,\boldsymbol{r}_2^{\mathsf{T}}\boldsymbol{M}\boldsymbol{r}_2 + \ldots + u_n^2\,\boldsymbol{r}_n^{\mathsf{T}}\boldsymbol{M}\boldsymbol{r}_n} \tag{61}$$

and with (60) we get

$$\begin{aligned} R(\boldsymbol{u}) &= \frac{u_1^2\,\omega_1^2 + u_2^2\,\omega_2^2 + \ldots + u_n^2\,\omega_n^2}{u_1^2 + u_2^2 + \ldots + u_n^2} \\ &= \omega_1^2\,\frac{u_1^2 + u_2^2\,(\omega_2^2/\omega_1^2) + \ldots + u_n^2\,(\omega_n^2/\omega_1^2)}{u_1^2 + u_2^2 + \ldots + u_n^2}. \end{aligned} \tag{62}$$

Since by definition $\omega_i^2/\omega_1^2 \geq 1$ for $i = 2,\ldots,n$, the numerator in (62) will always be larger or equal than the denominator. For $u_1 = 1$ numerator and denominator assume the same value and $R(\boldsymbol{u})$ assumes its minimum. This proves (58).

As a consequence, inserting an arbitrary vector in the right hand side of (57) always results in an upper bound for the square of the first circular eigenfrequency of the $\boldsymbol{M}$-$\boldsymbol{K}$-system. In many cases the form of the first eigenmode, i.e. the first eigenvector can be intuitively estimated, and then RAYLEIGH's quotient gives an upper bound for ω_1^2, which often is very close to the exact value.

Finding the first eigenvector and the first eigenfrequency of a $\boldsymbol{M}$-$\boldsymbol{K}$-system is transformed into a minimization problem via RAYLEIGH's principle (58)[2]. Sometimes this is referred to as a variational problem. Similarly,

[2]The existence of at least one minimum of $R(\mathbf{u})$ and at least one vector $\mathbf{u}$ leading to this minimum is guaranteed by the WEIERSTRASS theorem on extrema of continous functions in closed domains. Named after the mathematician Karl Theodor Wilhelm WEIERSTRASS, *1985 in Ostenfelde/Münsterland, †1897 in Berlin

the second eigenfrequency can be characterized by minimal properties: Obviously

$$\omega_2^2 = \min_{\boldsymbol{u}^{\mathsf{T}}\boldsymbol{M}\boldsymbol{r}_1=0} R(\boldsymbol{u}) \tag{63}$$

holds, i.e. ω_2^2 is the minimum of $R(\boldsymbol{u})$ over all the vectors orthogonal to the first eigenvector $\boldsymbol{r}_1$ (with respect to $\boldsymbol{M}$), which of course have the representation

$$\boldsymbol{u} = u_2\boldsymbol{r}_2 + u_3\boldsymbol{r}_3 + \ldots + u_n\boldsymbol{r}_n. \tag{64}$$

Generalizing, this leads to a *recursive characterization of the eigenfrequencies and eigenvectors* of a $\boldsymbol{M}$-$\boldsymbol{K}$-system. The k-th eigenfrequency and the k-th eigenvector of a $\boldsymbol{M}$-$\boldsymbol{K}$-system are given by the minimum of RAYLEIGH's quotient

$$\omega_k^2 = \min_{\substack{\boldsymbol{u}^{\mathsf{T}}\boldsymbol{M}\boldsymbol{r}_1=0\\ \vdots \\ \boldsymbol{u}^{\mathsf{T}}\boldsymbol{M}\boldsymbol{r}_{k-1}=0}} R(\boldsymbol{u}), \tag{65}$$

where the minimization is carried out with respect to all vectors orthogonal to the first $k-1$ eigenvectors with respect to $\boldsymbol{M}$. The validity of the recursive characterization immediately follows from (62), where now $u_1=u_2=\ldots=u_{k-1}=0$ and the factor ω_1^2 is substituted by ω_k^2.

The extremal properties of eigenvalues are particularly useful in the study of systems with additional constraints. If we introduce the additional (holonomic) constraint

$$b_1q_1 + b_2q_2 + \ldots + b_nq_n = 0 \tag{66}$$

with $b_1, b_2, \ldots, b_n$ for the $\boldsymbol{M}$-$\boldsymbol{K}$-system under consideration, then the equilibrium at $\boldsymbol{q}=\boldsymbol{0}$ remains unchanged, but the number of degrees of freedom is reduced from n to $n-1$. It is possible to use (66) to express one of the n generalized coordinates through the remaining $n-1$ coordinates. The original $\boldsymbol{M}$-$\boldsymbol{K}$-system is now reduced to a $n-1$ degree of freedom system with eigenfrequencies $\bar{\omega}_i$ and eigenvectors $\bar{\boldsymbol{r}}_i$. This leads to a RAYLEIGH quotient depending only on $n-1$ variables. It is however also possible to use the original RAYLEIGH quotient (57) and only change the range of definition of $\boldsymbol{u}\in\mathbb{R}^n$. It this manner it can be shown that the lowest eigenfrequency $\bar{\omega}_i$ of the system with the additional constraint (66) lies between the first and the second eigenvalue of the original system, i.e. $\omega_1\leq\bar{\omega}_1\leq\omega_2$. This statement is known as RAYLEIGH*'s theorem.*

These ideas can be generalized and the original system can be subjected to not only one, but h constraints of the type (66). Also in this more general case it can be shown that

$$\bar{\omega}_k \leq \omega_{k+h} \tag{67}$$

holds. The $(k+h)$-th circular eigenfrequency ω_{k+h} of the original system is therefore an upper bound for the k-th eigenfrequency $\bar{\omega}_k$ of the system subjected to h additional constraints.

A simple example will be given during the lecture.

1.3 $\boldsymbol{M}$-$\boldsymbol{D}$-$\boldsymbol{K}$-Systems

An important special case of (6) are the $\boldsymbol{M}$-$\boldsymbol{D}$-$\boldsymbol{K}$-systems corresponding to equations of motion of the form

$$\boldsymbol{M}\ddot{\boldsymbol{q}} + \boldsymbol{D}\dot{\boldsymbol{q}} + \boldsymbol{K}\boldsymbol{q} = \boldsymbol{0}. \tag{68}$$

They describe free damped linear vibrations and we assume the positive definiteness of $\boldsymbol{M}$ and the positive semidefiniteness of $\boldsymbol{D}$ and $\boldsymbol{K}$.

Properties of the Eigenvalues The characteristic equation of (10) for the $\boldsymbol{M}$-$\boldsymbol{D}$-$\boldsymbol{K}$-systems simplifies to

$$\det(\lambda^2\boldsymbol{M} + \lambda\boldsymbol{D} + \boldsymbol{K}) = 0. \tag{69}$$

Contrary to what was the case for $\boldsymbol{M}$-$\boldsymbol{K}$-systems in (27), the characteristic equation now also contains odd powers of λ. We therefore have to deal with a polynomial of degree $2n$ in λ, whose $2n$ roots, the eigenvalues of the $\boldsymbol{M}$-$\boldsymbol{D}$-$\boldsymbol{K}$-system λ_i $(i = 1, \ldots, 2n)$ are real and/or occur in complex conjugate pairs. Let us first again assume that the eigenvalues are simple, so that for each eigenvalue λ_i there is exactly one nontrivial eigenvector $\boldsymbol{r}_i$, fulfilling

$$(\lambda_i^2\boldsymbol{M} + \lambda_i\boldsymbol{D} + \boldsymbol{K})\boldsymbol{r}_i = \boldsymbol{0}. \tag{70}$$

Multiplying from the left by the vector $\boldsymbol{r}_i^*$, the complex conjugate transposed to $\boldsymbol{r}$, leads to

$$\lambda_i^2\,\boldsymbol{r}_i^*\boldsymbol{M}\boldsymbol{r}_i + \lambda_i\,\boldsymbol{r}_i^*\boldsymbol{D}\boldsymbol{r}_i + \boldsymbol{r}_i^*\boldsymbol{K}\boldsymbol{r}_i = 0. \tag{71}$$

From $\boldsymbol{M}=\boldsymbol{M}^\mathsf{T}>0$ follows that $\boldsymbol{r}_i^*\boldsymbol{M}\boldsymbol{r}_i$ is real and for $\boldsymbol{r}_i \neq 0$ positive, and from $\boldsymbol{D}=\boldsymbol{D}^\mathsf{T}\geq 0$, $\boldsymbol{K}=\boldsymbol{K}^\mathsf{T}\geq 0$ follows that also $\boldsymbol{r}_i^*\boldsymbol{D}\boldsymbol{r}_i$, $\boldsymbol{r}_i^*\boldsymbol{K}\boldsymbol{r}_i$ are real and non-negative. Solving (71) for λ_i gives

$$\lambda_i = -\frac{1}{2}\frac{\boldsymbol{r}_i^*\boldsymbol{D}\boldsymbol{r}_i}{\boldsymbol{r}_i^*\boldsymbol{M}\boldsymbol{r}_i} \pm \sqrt{\left(\frac{1}{2}\frac{\boldsymbol{r}_i^*\boldsymbol{D}\boldsymbol{r}_i}{\boldsymbol{r}_i^*\boldsymbol{M}\boldsymbol{r}_i}\right)^2 - \frac{\boldsymbol{r}_i^*\boldsymbol{K}\boldsymbol{r}_i}{\boldsymbol{r}_i^*\boldsymbol{M}\boldsymbol{r}_i}}\,. \tag{72}$$

Here again we have to observe that this information is obtained from the projected systems of equations and that in general we cannot be sure that both signs in front of the root will actually represent solutions of (70). Certainly however all the eigenvalues have a non-positive real part. In many cases in mechanical systems the damping will be 'small', which can be represented by

$$\boldsymbol{r}_i^* \boldsymbol{D} \boldsymbol{r}_i < 2\sqrt{(\boldsymbol{r}_i^* \boldsymbol{M} \boldsymbol{r}_i)(\boldsymbol{r}_i^* \boldsymbol{K} \boldsymbol{r}_i)}\,, \qquad i = 1, \ldots, n. \tag{73}$$

This condition is comparable to the case of undercritical damping in one degree of freedom systems. We therefore write (72) as

$$\lambda_{i,n+i} = -\frac{1}{2}\frac{\boldsymbol{r}_i^* \boldsymbol{D} \boldsymbol{r}_i}{\boldsymbol{r}_i^* \boldsymbol{M} \boldsymbol{r}_i} \pm j\sqrt{\frac{\boldsymbol{r}_i^* \boldsymbol{K} \boldsymbol{r}_i}{\boldsymbol{r}_i^* \boldsymbol{M} \boldsymbol{r}_i} - \left(\frac{1}{2}\frac{\boldsymbol{r}_i^* \boldsymbol{D} \boldsymbol{r}_i}{\boldsymbol{r}_i^* \boldsymbol{M} \boldsymbol{r}_i}\right)^2}\,, \qquad i = 1, \ldots, n \tag{74}$$

with a positive radicand and where under this condition

$$\lambda_i = \lambda_{n+i}^* \tag{75}$$

form a complex conjugate pair. Therefore both λ_i and λ_{n+i} are eigenvalues of the $\boldsymbol{M}$-$\boldsymbol{D}$-$\boldsymbol{K}$-system. The corresponding eigenvectors $\boldsymbol{r}_i = \boldsymbol{r}_{n+i}^*$ are complex conjugate and due to the symmetry of $\boldsymbol{M}$, $\boldsymbol{D}$ and $\boldsymbol{K}$ also

$$\boldsymbol{r}_i^* \boldsymbol{M} \boldsymbol{r}_i = \boldsymbol{r}_{n+i}^* \boldsymbol{M} \boldsymbol{r}_{n+i}, \quad \boldsymbol{r}_i^* \boldsymbol{D} \boldsymbol{r}_i = \boldsymbol{r}_{n+i}^* \boldsymbol{D} \boldsymbol{r}_{n+i}, \quad \boldsymbol{r}_i^* \boldsymbol{K} \boldsymbol{r}_i = \boldsymbol{r}_{n+i}^* \boldsymbol{K} \boldsymbol{r}_{n+i} \tag{76}$$

holds.

General Solution, Eigenforms and Eigensolutions It clearly follows from (70) that an eigenvector associated to a complex eigenvalue in general can not be chosen as real. With the already known abbreviations

$$\delta_i = -\operatorname{Re}\lambda_i = \frac{1}{2}\frac{\boldsymbol{r}_i^* \boldsymbol{D} \boldsymbol{r}_i}{\boldsymbol{r}_i^* \boldsymbol{M} \boldsymbol{r}_i} \tag{77a}$$

$$\omega_{\mathrm{d}i} = \operatorname{Im}\lambda_i = \sqrt{\frac{\boldsymbol{r}_i^* \boldsymbol{K} \boldsymbol{r}_i}{\boldsymbol{r}_i^* \boldsymbol{M} \boldsymbol{r}_i} - \left(\frac{1}{2}\frac{\boldsymbol{r}_i^* \boldsymbol{D} \boldsymbol{r}_i}{\boldsymbol{r}_i^* \boldsymbol{M} \boldsymbol{r}_i}\right)^2} \tag{77b}$$

and

$$\boldsymbol{r}_i = \boldsymbol{a}_i + j\boldsymbol{b}_i\,, \tag{77c}$$

the general solution in real form according to (25) can be written as

$$\boldsymbol{q}(t) = \sum_{i=1}^{n} K_i\, e^{-\delta_i t}\Big[\boldsymbol{a}_i \cos(\omega_{\mathrm{d}i} t + \gamma_i) - \boldsymbol{b}_i \sin(\omega_{\mathrm{d}i} t + \gamma_i)\Big]\,, \tag{78}$$

where the integration constants K_i, γ_i are determined by the initial conditions. The $\boldsymbol{M}$-$\boldsymbol{D}$-$\boldsymbol{K}$-system therefore has n eigensolutions of the type

$$\boldsymbol{q}_i(t) = K_i e^{-\delta_i t}\Big[\boldsymbol{a}_i \cos(\omega_{\mathrm{d}i} t + \gamma_i) - \boldsymbol{b}_i \sin(\omega_{\mathrm{d}i} t + \gamma_i)\Big]\ , \ i = 1, \dots, n. \quad (79)$$

With the argument

$$\epsilon_{ik} = \arg r_{ik} = \arg(a_{ik} + j b_{ik}) \quad (80)$$

of the k-th component of the i-th eigenvector $\boldsymbol{r}_i$, the i-th eigensolution can also be written as

$$\boldsymbol{q}_i(t) = K_i e^{-\delta_i t} \begin{bmatrix} |r_{i1}| \cos(\omega_{\mathrm{d}i} t + \gamma_i - \epsilon_{i1}) \\ |r_{i2}| \cos(\omega_{\mathrm{d}i} t + \gamma_i - \epsilon_{i2}) \\ \vdots \\ |r_{in}| \cos(\omega_{\mathrm{d}i} t + \gamma_i - \epsilon_{in}) \end{bmatrix} . \quad (81)$$

Each component of such a solution obviously is a 'damped harmonic oscillation' with 'circular eigenfrequency' $\omega_{\mathrm{d}i}$. The essential difference to the case of the undamped $\boldsymbol{M}$-$\boldsymbol{K}$-systems is that the 'amplitude' decreases exponentially with $e^{-\delta_i t}$ and that each component has a different phase angle ϵ_{ik}. Therefore the elements of $\boldsymbol{q}_i(t)$ do not simultaneously reach their zero position of their local maxima. If we observe an oscillation corresponding to an eigensolution of a $\boldsymbol{M}$-$\boldsymbol{D}$ -$\boldsymbol{K}$-system, in general we see a periodic change of the form of the oscillations and a 'periodic' change of the coordinates with exponential decay.

Complete and Pervasive Damping In engineering applications it is in general important to know to which extent oscillations are damped. From (77) we recognize that a positive semidefinite damping matrix implies $\delta_i \geq 0$, $i = 1, \dots, n$. Similarly, it is clear that with *complete damping*, i.e. a positive definite damping matrix, the stronger inequality $\delta_i > 0$, $i = 1, \dots, n$ holds, so that all the eigensolutions and therefore all the motions of the system are damped. The condition of a positive definite damping matrix is however not necessary for $\delta_i > 0$, $i = 1, \dots, n$, but only sufficient. This is the case for example in the $\boldsymbol{M}$-$\boldsymbol{D}$-$\boldsymbol{K}$-system of Figure 1 and the equations of motion

$$\begin{bmatrix} m & 0 & 0 \\ 0 & m & 0 \\ 0 & 0 & m \end{bmatrix} \begin{bmatrix} \ddot{q}_1 \\ \ddot{q}_2 \\ \ddot{q}_3 \end{bmatrix} + \begin{bmatrix} d & 0 & 0 \\ 0 & 0 & 0 \\ 0 & 0 & 0 \end{bmatrix} \begin{bmatrix} \dot{q}_1 \\ \dot{q}_2 \\ \dot{q}_3 \end{bmatrix} + \begin{bmatrix} 2k & -k & 0 \\ -k & 2k & -k \\ 0 & -k & 2k \end{bmatrix} \begin{bmatrix} q_1 \\ q_2 \\ q_3 \end{bmatrix} = \begin{bmatrix} 0 \\ 0 \\ 0 \end{bmatrix} . \quad (82)$$

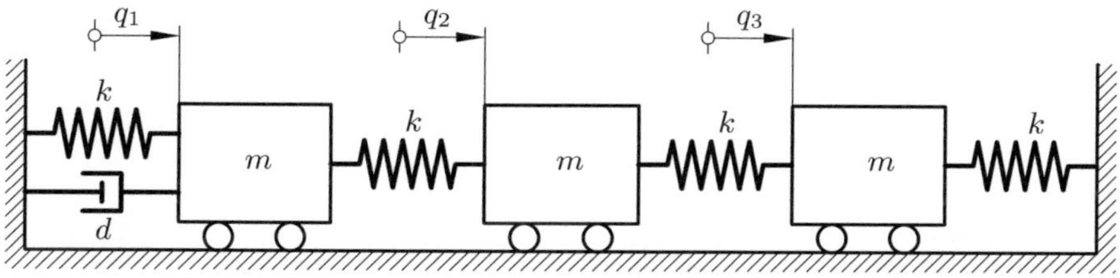

Figure 1: Pervasively but not completely damped $\boldsymbol{M}$-$\boldsymbol{D}$-$\boldsymbol{K}$-system

Although $\boldsymbol{D}$ here is only positive semidefinite, there certainly exists no undamped solution, for which obviously $q_1(t) \equiv 0$ would have to hold. From the first equation of motion

$$m\ddot{q}_1 + d\dot{q}_1 + 2kq_1 - kq_2 = 0 \tag{83}$$

we conclude that $q_1(t) \equiv 0$ implies $q_2(t) \equiv 0$ and therefore the second equation of motion

$$m\ddot{q}_2 - kq_1 + 2kq_2 - kq_3 = 0 \tag{84}$$

would also imply $q_3(t) \equiv 0$, so that the only undamped 'motion' is the trivial solution $q_1 = q_2 = q_3 = 0$. Therefore, in this example we have $\delta_{1,2,3} > 0$, although the damping matrix is only semidefinite.

If the damping matrix is such that all $\delta_i > 0$, then we say that the damping is 'pervasive'. The term indicates that the damping pervades all the eigensolutions. If this is not the case, there is at least one eigenvector $\boldsymbol{r}_k$ satisfying

$$\boldsymbol{D}\boldsymbol{r}_k = \boldsymbol{0}. \tag{85a}$$

Since the corresponding eigensolutions describes an undamped motion, $\boldsymbol{r}_k$ can be chosen real and the eigenpair $(\lambda_k, \boldsymbol{r}_k)$ of the $\boldsymbol{M}$-$\boldsymbol{D}$-$\boldsymbol{K}$-system fulfils

$$(\lambda_k^2 \boldsymbol{M} + \boldsymbol{K})\boldsymbol{r}_k = \boldsymbol{0}, \tag{85b}$$

i.e. it is an eigenpair of the undamped system, which is recognized immediately by inserting (85a) into (70). With known eigenvalue λ_k, (85) forms a homogeneous system of $2n$ linear equations for the determination of the n unknown components of the eigenvector $\boldsymbol{r}_k$. Nontrivial solutions exist if and only if the condition

$$\operatorname{rank}\left[\lambda_k^2 \boldsymbol{M} + \boldsymbol{K},\ \boldsymbol{D}\right] < n \tag{86}$$

is fulfilled. Obviously one has $\operatorname{rank}(\lambda_k^2 \boldsymbol{M} + \boldsymbol{K}) < n$, since λ_k is an eigenvalue of the undamped system, since decay in rank should however not be compensated by 'adding' the matrix $\boldsymbol{D}$. If on the other hand

$$\operatorname{rank}\left[\lambda_i^2 \boldsymbol{M} + \boldsymbol{K},\ \boldsymbol{D}\right] = n, \qquad i = 1, \ldots, n \tag{87}$$

holds for each eigenvalue of the undamped problem, then (85) admits no nontrivial solutions, i.e. there are no undamped motions and the damping is pervasive. For positive definite damping matrices already $\operatorname{rank} \boldsymbol{D} = n$ holds, so that (87) is automatically fulfilled, independently of $\boldsymbol{M}$ and $\boldsymbol{K}$. If however $\boldsymbol{D}$ is only semidefinite, it will depend on the matrices $\boldsymbol{M}$ and $\boldsymbol{K}$ if the damping is pervasive or not.

The condition (87) is analogous to the criteria on controllability in control theory. The question of pervasiveness of damping can be considered in the following way: Is the undamped system controllable via forces acting in the way of damping forces in the damped system? If the answer is positive, then the damped system is pervasively damped. The rank condition (87) is however not really very practical and in some practical problems, as in our example above, the pervasiveness of damping can be examined directly at the equations of motion, or with other controllability criteria known from control theory.

Modal Damping While in the undamped case, that is for $\boldsymbol{M}$-$\boldsymbol{K}$-systems, the (real) eigenvectors define a coordinate transformation (44) uncoupling the system into n uncouples differential equations, there is no such (real) transformation for all other systems, in particularly also not for $\boldsymbol{M}$-$\boldsymbol{D}$-$\boldsymbol{K}$-systems.

Following the steps (33) to (36) in an analogous way for $\boldsymbol{M}$-$\boldsymbol{D}$-$\boldsymbol{K}$-systems, it becomes clear that the eigenvectors in general, i.e. for an arbitrary damping matrix $\boldsymbol{D}$, will no longer be orthogonal with respect to $\boldsymbol{M}$ and $\boldsymbol{K}$. Obviously however we can define a linear coordinate transformation

$$\boldsymbol{q} = \boldsymbol{T}\boldsymbol{z} \tag{88}$$

reducing the equations of motion (68) of an $\boldsymbol{M}$-$\boldsymbol{D}$-$\boldsymbol{K}$-System to

$$\boldsymbol{M}\boldsymbol{T}\,\ddot{\boldsymbol{z}} + \boldsymbol{D}\boldsymbol{T}\,\dot{\boldsymbol{z}} + \boldsymbol{K}\boldsymbol{T}\,\boldsymbol{z} = \boldsymbol{0}. \tag{89}$$

Multiplication from the left with $\boldsymbol{T}^{\mathsf{T}}$ leads to

$$\boldsymbol{T}^{\mathsf{T}}\boldsymbol{M}\boldsymbol{T}\,\ddot{\boldsymbol{z}} + \boldsymbol{T}^{\mathsf{T}}\boldsymbol{D}\boldsymbol{T}\,\dot{\boldsymbol{z}} + \boldsymbol{T}^{\mathsf{T}}\boldsymbol{K}\boldsymbol{T}\,\boldsymbol{z} = \boldsymbol{0}. \tag{90}$$

In (90) the new mass matrix $\boldsymbol{T}^{\mathsf{T}}\boldsymbol{M}\boldsymbol{T}$, the new damping matrix $\boldsymbol{T}^{\mathsf{T}}\boldsymbol{D}\boldsymbol{T}$ and the new stiffness matrix $\boldsymbol{T}^{\mathsf{T}}\boldsymbol{K}\boldsymbol{T}$ are again symmetric.

We now examine for which matrices $\boldsymbol{D}$ there is a transformation matrix leading to the simultaneous diagonalization of the new mass, damping and stiffness matrices in (90). To this end we again examine the eigenvalue problem

$$\left(\lambda^2 \boldsymbol{M} + \lambda \boldsymbol{D} + \boldsymbol{K}\right)\boldsymbol{r} = \boldsymbol{0} \tag{91}$$

of a $\boldsymbol{M}$-$\boldsymbol{D}$-$\boldsymbol{K}$-system, which can also be written as

$$\left(\lambda^2 + \lambda \boldsymbol{M}^{-1}\boldsymbol{D} + \boldsymbol{M}^{-1}\boldsymbol{K}\right)\boldsymbol{r} = \boldsymbol{0} \tag{92}$$

and after multiplication by $\boldsymbol{M}^{-1}\boldsymbol{K}$ from the left as

$$\left(\lambda^2(\boldsymbol{M}^{-1}\boldsymbol{K}) + \lambda(\boldsymbol{M}^{-1}\boldsymbol{K})(\boldsymbol{M}^{-1}\boldsymbol{D}) + (\boldsymbol{M}^{-1}\boldsymbol{K})^2\right)\boldsymbol{r} = \boldsymbol{0}. \tag{93}$$

If $\boldsymbol{M}^{-1}\boldsymbol{K}$ and $\boldsymbol{M}^{-1}\boldsymbol{D}$ commute, i.e. if

$$(\boldsymbol{M}^{-1}\boldsymbol{K})(\boldsymbol{M}^{-1}\boldsymbol{D}) = (\boldsymbol{M}^{-1}\boldsymbol{D})(\boldsymbol{M}^{-1}\boldsymbol{K}) \tag{94}$$

holds,

$$\left(\lambda^2 \boldsymbol{E} + \lambda \boldsymbol{M}^{-1}\boldsymbol{D} + \boldsymbol{M}^{-1}\boldsymbol{K}\right)\boldsymbol{M}^{-1}\boldsymbol{K}\,\boldsymbol{r} = \boldsymbol{0} \tag{95}$$

follows, and therefore also

$$\left(\lambda^2 \boldsymbol{M} + \lambda \boldsymbol{D} + \boldsymbol{K}\right)\boldsymbol{M}^{-1}\boldsymbol{K}\,\boldsymbol{r} = \boldsymbol{0}. \tag{96}$$

Therefore, if $(\lambda, \boldsymbol{r})$ is an eigenpair of the $\boldsymbol{M}$-$\boldsymbol{D}$-$\boldsymbol{K}$-system, the commutivity of $\boldsymbol{M}^{-1}\boldsymbol{K}$ and $\boldsymbol{M}^{-1}\boldsymbol{D}$ implies that also $(\lambda, \boldsymbol{M}^{-1}\boldsymbol{K}\boldsymbol{r})$ is an eigenpair. If the corresponding eigenvector has multiplicity one, both vectors differ only be a scalar factor

$$\boldsymbol{M}^{-1}\boldsymbol{K}\,\boldsymbol{r} = \mu\,\boldsymbol{r}. \tag{97}$$

But this also implies

$$(-\mu \boldsymbol{M} + \boldsymbol{K})\boldsymbol{r} = \boldsymbol{0}. \tag{98}$$

If $\boldsymbol{r}$ is an eigenvector of the $\boldsymbol{M}$-$\boldsymbol{D}$-$\boldsymbol{K}$-system, then it follows from the commutativity condition (94) that $\boldsymbol{r}$ also is an eigenvector of the $\boldsymbol{M}$-$\boldsymbol{K}$-system and can therefore be chosen real. This holds for all eigenvectors. If we multiply (91) from the left with a different eigenvector, the orthogonality of the eigenvectors with respect to $\boldsymbol{M}$ and $\boldsymbol{K}$ (36), we see that the commutativity also implies orthogonality with respect to $\boldsymbol{D}$.

We thus have proved that the commutativity according to (94) is a sufficient condition for the diagonalization and uncoupling of the system. It can be shown without much difficulty that it also is a necessary condition. Also the case of multiple eigenvalues, which has so far been excluded, can be dealt with in similar manner.

If the commutativity condition (94) is fulfilled, a $\boldsymbol{M}$-$\boldsymbol{D}$-$\boldsymbol{K}$-system has the same real eigenvectors as the $\boldsymbol{M}$-$\boldsymbol{K}$-system and the equations of motion can be uncoupled using the modal matrix $\boldsymbol{R}$ and the coordinat transformation $\boldsymbol{q}=\boldsymbol{R}\boldsymbol{p}$. This is the case of *modal damping*. With the modal masses, dampings and stiffnesses

$$\tilde{m}_i = \boldsymbol{r}_i^\mathsf{T}\boldsymbol{M}\boldsymbol{r}_i, \qquad \tilde{d}_i = \boldsymbol{r}_i^\mathsf{T}\boldsymbol{D}\boldsymbol{r}_i, \qquad \tilde{k}_i = \boldsymbol{r}_i^\mathsf{T}\boldsymbol{K}\boldsymbol{r}_i, \tag{99}$$

the uncoupled equations of motion assume the form

$$\tilde{m}_i\ddot{p}_i + \tilde{d}_i\dot{p}_i + \tilde{k}_i p_i = 0, \qquad i = 1, \dots, n, \tag{100}$$

which, using (48) and the *modal damping ratio*

$$\tilde{\vartheta}_i = \frac{\tilde{d}_i}{2\sqrt{\tilde{k}_i\tilde{m}_i}} \tag{101}$$

can also be written as

$$\ddot{p}_i + 2\tilde{\vartheta}_i\omega_i\,\dot{p}_i + \omega_i^2\,p_i = 0, \qquad i = 1, \dots, n. \tag{102}$$

Assuming modal damping, there are exactly n linearly independent parameters $\tilde{d}_i$, or $\tilde{\vartheta}_i$, $i=1,\dots,n$ for the description of the damping of a system with n degrees of freedom with given mass and stiffness matrix. The question arises, how $\boldsymbol{D}$ can be expressed as a function of n parameters, so that on one side the commutativity condition (94) is fulfilled, and on the other hand n arbitrary modal damping rations $\tilde{\vartheta}_i$ can be represented. It can be shown that both requirements are fulfilled if $\boldsymbol{D}$ is represented as a CAUGHEY sum[3]

$$\boldsymbol{D} = \sum_{s=1}^{n} \alpha_s\,\boldsymbol{M}\left(\boldsymbol{M}^{-1}\boldsymbol{K}\right)^{s-1}. \tag{103}$$

[3]Named after the Scottish engineer and physicist Thomas K. CAUGHEY, *1927 in Rutherglen, Scottland,†2004 in Pasadena, CA, USA.

It is clear that (103) satisfies the commutativity condition, since each term in the sum satisfies (94). Therefore

$$\tilde{\boldsymbol{D}} = \boldsymbol{R}^{\mathsf{T}}\boldsymbol{D}\boldsymbol{R} = \sum_{s=1}^{n} \alpha_s\, \boldsymbol{R}^{\mathsf{T}}\boldsymbol{M}\left(\boldsymbol{M}^{-1}\boldsymbol{K}\right)^{s-1}\boldsymbol{R} \tag{104}$$

is a diagonal matrix. If the eigenvectors are normalized via $\boldsymbol{r}_i^{\mathsf{T}}\boldsymbol{r}_i = 1$, $i = 1, \ldots, n$, then $\boldsymbol{R}^{\mathsf{T}}\boldsymbol{R} = \boldsymbol{E}$ and (104) implies

$$\begin{aligned}\tilde{\boldsymbol{D}} &= \sum_{s=1}^{n} \alpha_s\, \boldsymbol{R}^{\mathsf{T}}\boldsymbol{M}\boldsymbol{R}\,\boldsymbol{R}^{\mathsf{T}}\left(\boldsymbol{M}^{-1}\boldsymbol{R}\,\boldsymbol{R}^{\mathsf{T}}\boldsymbol{K}\right)^{s-1}\boldsymbol{R} \\ &= \sum_{s=1}^{n} \alpha_s\, \boldsymbol{R}^{\mathsf{T}}\boldsymbol{M}\boldsymbol{R}\,\left(\boldsymbol{R}^{\mathsf{T}}\boldsymbol{M}^{-1}\boldsymbol{R}\,\boldsymbol{R}^{\mathsf{T}}\boldsymbol{K}\boldsymbol{R}\right)^{s-1} \\ &= \sum_{s=1}^{n} \alpha_s\, \tilde{\boldsymbol{M}}\left(\tilde{\boldsymbol{M}}^{-1}\tilde{\boldsymbol{K}}\right)^{s-1}\end{aligned} \tag{105}$$

and finally

$$\tilde{\boldsymbol{M}}^{-1}\tilde{\boldsymbol{D}} = \sum_{s=1}^{n} \alpha_s \left(\tilde{\boldsymbol{M}}^{-1}\tilde{\boldsymbol{K}}\right)^{s-1}, \tag{106}$$

where the diagonal elements can also be written in the form of a linear system of algebraic equations

$$\begin{bmatrix} \tilde{d}_1/\tilde{m}_1 \\ \tilde{d}_2/\tilde{m}_2 \\ \vdots \\ \tilde{d}_n/\tilde{m}_n \end{bmatrix} = \begin{bmatrix} \omega_1^0 & \omega_1^2 & \cdots & \omega_1^{2n-2} \\ \omega_2^0 & \omega_2^2 & \cdots & \omega_2^{2n-2} \\ \vdots & \vdots & \ddots & \vdots \\ \omega_n^0 & \omega_n^2 & \cdots & \omega_n^{2n-2} \end{bmatrix} \begin{bmatrix} \alpha_1 \\ \alpha_2 \\ \vdots \\ \alpha_n \end{bmatrix}. \tag{107}$$

It can be shown that this algebraic system can always be solved for the coefficients α_s provided $\omega_i^2 \neq \omega_k^2$, $i \neq k$ [4]. For arbitrarily given $\tilde{d}_i$ and therefore also modal damping ratios ϑ_i, a representation by the α_s is always possible. The CAUGHEY sum (103) is therefore the most general form of modal damping.

[4]The coefficient matrix is a VANDERMONDE-Matrix, named after the French musician, mathematician and chemist Alexandre-Théophile VANDERMONDE, *1735 in Paris, †1796 in Paris.

► A Special Case

In many engineering applications not only is the damping small, but also the damping terms are not well known and not easy to measure. Therefore in systems with many degrees of freedom rather arbitrary assumptions are often made on the matrix $\boldsymbol{D}$. Since the uncoupling in the real domain is rather convenient, frequently it is assumed that the damping is modal and that

$$\boldsymbol{D} = \alpha_1 \boldsymbol{M} + \alpha_2 \boldsymbol{K} \tag{108}$$

is valid, i.e. the CAUGHEY sum is truncated after the second term. This is a special case which sometimes is referred to as *convenience hypothesis* (in German: *Bequemlichkeitshypothese*). The part proportional to the mass matrix $\alpha_1 \boldsymbol{M}$ is often called *external damping* and the part $\alpha_2 \boldsymbol{K}$ is called *inner* or *material damping.*

If the damping matrix is unknown, in general one will first solve the eigenvalue problem of the undamped system, which leads to the principal coordinates and the ω_i, $i = 1, \ldots, n$, and next in (102) direct assumptions on the values of the individual damping rations $\tilde{\vartheta}_i$ are made. For example, in vibrations studies of civil engineering structures, such as the vibrations excited by earthquakes in nuclear power stations, one sometimes assumes $\tilde{\vartheta}_i = 0.05$, for all i. In steel constructions on the other hand, values of $\tilde{\vartheta}_i = 0.002$ may be reasonable. This corresponds to an amplification of the amplitudes in the forced oscillations by the factor 300 approximately, in resonance. One should note that these damping ratios contain the total structural damping, which in reality is nonlinear, i.e. also the energy dissipation in the hinges and connections, as well as the energy losses through the foundations, and not only the losses due to nonelastic material behavior (material damping). This material damping is often taken into account by considering a complex modulus of elasticity or a loss factor. These are material parameters; the resulting modal damping ratios will also depend on other system parameters. In most structures the effect of the material damping *strictu senso* is in general much smaller than those caused by other loss mechanisms. In artificially damped systems, for example in a vehicle suspension, the damping matrix may be relatively well known. The convenience hypothesis as a rule is not fulfilled in these cases. The following procedure is then common: First, the eigenvectors $\boldsymbol{r}_i$, $i = 1, \ldots, n$ and the modal matrix $\boldsymbol{R}$ are determined for the undamped system. The damped system is then transformed to the modal coordinates of the undamped system. If now the nondiagonal terms in the matrix $\boldsymbol{R}^\mathsf{T} \boldsymbol{D} \boldsymbol{R}$ are neglected, then again an uncoupled system of the type (102) is obtained. These equations in many cases lead to a very good approximation to the exact solution of the original undamped system, in particular if the terms on the main diagonal of $\boldsymbol{R}^{\mathsf{T}}\boldsymbol{D}\boldsymbol{R}$ are substantially larger than the off diagonal terms. This should at least give good starting values for a possible iterative determination of the complex eigenvectors.

1.4 $\boldsymbol{M}$-$\boldsymbol{G}$-$\boldsymbol{K}$-Systems

Gyroscopic terms represented by the skew-symmetric matrix $\boldsymbol{G}$ occur for example in conservative systems if the equations of motion are formulated in a rotating coordinate system, or as a consequence of the elimination of cyclic coordinates. In gyroscopic systems the kinetic energy contains terms linear in the generalized velocities. The equations of motion of the free vibrations of such a system are represented by

$$\boldsymbol{M}\ddot{\boldsymbol{q}} + \boldsymbol{G}\dot{\boldsymbol{q}} + \boldsymbol{K}\boldsymbol{q} = \boldsymbol{0}, \tag{109}$$

where the gyroscopic matrix is skew-symmetric, that is, $\boldsymbol{G}=-\boldsymbol{G}^{\mathsf{T}}$, and, in particular all elements on the main diagonal of this matrix vanish.

Properties of the Eigenvalues The characteristic equation (10) now is simplified to

$$\det(\lambda^2\boldsymbol{M} + \lambda\boldsymbol{G} + \boldsymbol{K}) = 0. \tag{110}$$

The determinant of a matrix is equal to the one of its transposed, so that for all values of λ

$$\det(\lambda^2\boldsymbol{M} + \lambda\boldsymbol{G} + \boldsymbol{K}) = \det(\lambda^2\boldsymbol{M} - \lambda\boldsymbol{G} + \boldsymbol{K}). \tag{111}$$

This means one has

$$\det(\lambda^2\boldsymbol{M} + \lambda\boldsymbol{G} + \boldsymbol{K}) = \det((-\lambda)^2\boldsymbol{M} + (-\lambda)\boldsymbol{G} + \boldsymbol{K}), \tag{112}$$

which is only possible if exclusively even powers of λ occur, so that, if λ is an eigenvalue, so is $-\lambda$. This is different from the damped systems, already examined, in which we had the symmetric matrix $\boldsymbol{D}$ instead of $\boldsymbol{G}$, and in which the eigenvalues (λ, λ^*) occurred in complex conjugate pairs; now we have quadruples $(\pm\lambda, \pm\lambda^*)$ of eigenvalues! We now project the eigenvalue problem

$$(\lambda_i^2\boldsymbol{M} + \lambda_i\boldsymbol{G} + \boldsymbol{K})\boldsymbol{r}_i = \boldsymbol{0} \tag{113}$$

on $\boldsymbol{r}_i^*$ and get

$$\lambda_i^2 \, \boldsymbol{r}_i^* \boldsymbol{M} \boldsymbol{r}_i + \lambda_i \, \boldsymbol{r}_i^* \boldsymbol{G} \boldsymbol{r}_i + \boldsymbol{r}_i^* \boldsymbol{K} \boldsymbol{r}_i = 0. \tag{114}$$

The positive symmetric mass mass matrix and the semidefinite stiffness matrix imply $\boldsymbol{r}_i^* \boldsymbol{M} \boldsymbol{r}_i > 0$ and $\boldsymbol{r}_i^* \boldsymbol{K} \boldsymbol{r}_i \geq 0$. Moreover, $\boldsymbol{r}_i^* \boldsymbol{G} \boldsymbol{r}_i$ is purely imaginary, since $\boldsymbol{G}$ is skew-symmetric. Therefore

$$\begin{aligned} \lambda_i &= -\frac{1}{2} \frac{\boldsymbol{r}_i^* \boldsymbol{G} \boldsymbol{r}_i}{\boldsymbol{r}_i^* \boldsymbol{M} \boldsymbol{r}_i} \pm \sqrt{\left(\frac{1}{2} \frac{\boldsymbol{r}_i^* \boldsymbol{G} \boldsymbol{r}_i}{\boldsymbol{r}_i^* \boldsymbol{M} \boldsymbol{r}_i}\right)^2 - \frac{\boldsymbol{r}_i^* \boldsymbol{K} \boldsymbol{r}_i}{\boldsymbol{r}_i^* \boldsymbol{M} \boldsymbol{r}_i}} \\ &= j \left(\frac{1}{2} \frac{j \boldsymbol{r}_i^* \boldsymbol{G} \boldsymbol{r}_i}{\boldsymbol{r}_i^* \boldsymbol{M} \boldsymbol{r}_i} \pm \sqrt{\left(\frac{1}{2} \frac{j \boldsymbol{r}_i^* \boldsymbol{G} \boldsymbol{r}_i}{\boldsymbol{r}_i^* \boldsymbol{M} \boldsymbol{r}_i}\right)^2 + \frac{\boldsymbol{r}_i^* \boldsymbol{K} \boldsymbol{r}_i}{\boldsymbol{r}_i^* \boldsymbol{M} \boldsymbol{r}_i}} \right) \end{aligned} \tag{115}$$

is purely imaginary, since $j \boldsymbol{r}_i^* \boldsymbol{G} \boldsymbol{r}_i$ is real and the radicand is non-negative. Note that in general only one of the signs in (115) will give a solution of (113), since (115) does not represent a complex conjugate pair, contrary to what happened in (32) for the $\boldsymbol{M}$-$\boldsymbol{K}$-system and in (74) in the $\boldsymbol{M}$-$\boldsymbol{D}$-$\boldsymbol{K}$-system.

Since the eigenvalues are purely imaginary, of course $-\lambda$ and λ^* are identical, so that the eigenvalues and the eigenvectors

$$\lambda_i = \lambda_{n+i}^*, \qquad \boldsymbol{r}_i^{\mathsf{T}} = \boldsymbol{r}_{n+i}^* \tag{116}$$

again occur in complex conjugate pairs.

General Solution, Eigenforms and Eigensolutions From (113) it is obvious that the eigenvector associated to a complex eigenvalue can in general not be made real. We order the eigenvalues so that

$$\omega_i = \operatorname{Im} \lambda_i, \qquad i = 1, \ldots, n \tag{117}$$

form an positively increasing sequence and set

$$\boldsymbol{r}_i = \boldsymbol{a}_i + j \boldsymbol{b}_i, \qquad i = 1, \ldots, n. \tag{118}$$

According to (25) this leads to the general solution in real form

$$\boldsymbol{q}(t) = \sum_{i=1}^{n} K_i \Big[\boldsymbol{a}_i \cos(\omega_i t + \gamma_i) - \boldsymbol{b}_i \sin(\omega_i t + \gamma_i) \Big], \tag{119}$$

where the integration constants K_i, γ_i are determined by the initial conditions. The $\boldsymbol{M}$-$\boldsymbol{G}$-$\boldsymbol{K}$-system therefore has eigensolutions of the form

$$\boldsymbol{q}_i(t) = K_i \Big[\boldsymbol{a}_i \cos(\omega_i t + \gamma_i) - \boldsymbol{b}_i \sin(\omega_i t + \gamma_i) \Big], \qquad i = 1, \ldots, n. \tag{120}$$

With the argument

$$\epsilon_{ik} = \arg r_{ik} = \arg(a_{ik} + jb_{ik}) \tag{121}$$

of the k-th component of the i-th eigenvector $\boldsymbol{r}_i$, the i-th eigensolution can also be written as

$$\boldsymbol{q}_i(t) = K_i \begin{bmatrix} |r_{i1}|\cos(\omega_i t + \gamma_i - \epsilon_{i1}) \\ |r_{i2}|\cos(\omega_i t + \gamma_i - \epsilon_{i2}) \\ \vdots \\ |r_{in}|\cos(\omega_i t + \gamma_i - \epsilon_{in}) \end{bmatrix}. \tag{122}$$

Therefore, each component of this eigensolution is a harmonic oscillation with circular eigenfrequency ω_i. However, exactly as in the $\boldsymbol{M}$-$\boldsymbol{D}$-$\boldsymbol{K}$-systems, each component in general has a different phase angle ϵ_{ik}. The elements of $\boldsymbol{q}_i(t)$ therefore do not reach their maxima or their zero position simultaneously. Observing the eigensolutions of a $\boldsymbol{M}$-$\boldsymbol{G}$-$\boldsymbol{K}$-system, in general one sees a periodic change of the vibration form and a periodic change of the coordinate values.

Gyroscopic Stabilization Up to now we have always assumed that the stiffness matrix is at least positive semidefinite. For the time being we relax this assumption in discussing the eigenvalues of $\boldsymbol{M}$-$\boldsymbol{G}$-$\boldsymbol{K}$-systems in comparison to those of underlying $\boldsymbol{M}$-$\boldsymbol{K}$-system (32). In the undamped $\boldsymbol{M}$-$\boldsymbol{K}$-systems, an indefinite stiffness matrix $\boldsymbol{K}$ with at least one negative eigenvalue (no minimum of the potential energy!) leads to at least two eigenvalues of the $\boldsymbol{M}$-$\boldsymbol{K}$-system with non-vanishing real part (i.e. one with positive and one with negative real part). In $\boldsymbol{M}$-$\boldsymbol{G}$-$\boldsymbol{K}$-systems this can be different, as can be seen from (115). Also for $\boldsymbol{r}_i^*\boldsymbol{K}\boldsymbol{r}_i < 0$ with $(j\boldsymbol{r}_i^*\boldsymbol{G}\boldsymbol{r}_i)^2 > -4\boldsymbol{r}_i^*\boldsymbol{M}\boldsymbol{r}_i\,\boldsymbol{r}_i^*\boldsymbol{K}\boldsymbol{r}_i$ the radicand is positive and the eigenvalues may possibly be purely imaginary. This is the case of gyroscopic stabilization. Note that even very small damping, as it is present in almost all engineering systems, may destroy the gyroscopic stabilization. This led to problems with one of the first artificial satellites, which was stabilized gyroscopically, since the damping was neglected in the calculations (the satellite lost its stability in attitude and started tumbling).

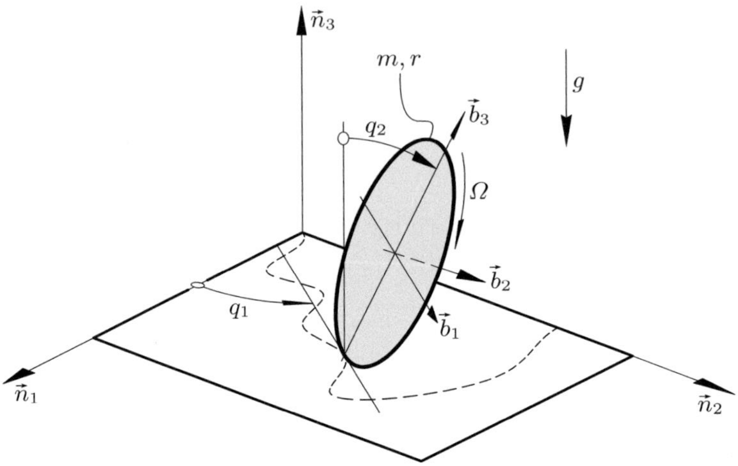

Figure 2: Rolling Penny

► **The Rolling Penny**

As an example for gyroscopic stabilization consider the rolling penny represented in Figure 2. The linearized quations of motion are

$$mr^2\begin{bmatrix}\frac{3}{4} & 0\\ 0 & \frac{5}{4}\end{bmatrix}\begin{bmatrix}\ddot{q}_1\\ \ddot{q}_2\end{bmatrix}+mr^2\begin{bmatrix}0 & \frac{3}{2}\Omega\\ -\frac{3}{2}\Omega & 0\end{bmatrix}\begin{bmatrix}\dot{q}_1\\ \dot{q}_2\end{bmatrix}+\begin{bmatrix}0 & 0\\ 0 & -mgr\end{bmatrix}\begin{bmatrix}q_1\\ q_2\end{bmatrix}=\begin{bmatrix}0\\ 0\end{bmatrix}. \quad (123)$$

Obviously the system has a positive definite mass matrix, a *negative* semidefinite stiffness matrix and is gyroscopic.

We first treat the case $\Omega = 0$. The two equations of motion are then uncoupled and the system has the eigenvalues

$$\lambda_1 = 0, \qquad \lambda_2 = 0, \qquad \lambda_3 = \sqrt{\frac{4}{5}\frac{g}{r}}, \qquad \lambda_4 = -\sqrt{\frac{4}{5}\frac{g}{r}} \quad (124)$$

and therefore an unstable trivial solution. This result is also intuitively clear. The two zero eigenvalues $\lambda_{1,2}$ are associated to the (unconstrained) rotation about the vertical axis, and $\lambda_{3,4}$ are related to the falling penny.

In the case $\Omega \neq 0$, i.e. for small motions about the vertical rolling penny, the behavior can be different. The *ansatz* $\boldsymbol{q}(t) = \boldsymbol{r}e^{\lambda t}$ leads to the characteristic equation

$$\lambda^4 + \left(\frac{12}{5}\Omega^2 - \frac{4}{5}\frac{g}{r}\right)\lambda^2 = 0 \quad (125)$$

with the eigenvalues

$$\lambda_1 = 0, \qquad \lambda_2 = 0, \qquad \lambda_{3,4} = \pm j\sqrt{\frac{12}{5}\,\Omega^2 - \frac{4}{5}\frac{g}{r}}. \tag{126}$$

For $\Omega^2 > g/3r$ all eigenvalues have vanishing real parts. The penny no longer tips over, but performs oscillations about the rolling motion. A typical example is the coin rolling on a table. Due to dissipative effects in the contact between coin and table, omega decreases, and finally the coin ends up tipping over. But also with constant rotational speed, dissipative effects may destroy the gyroscopic stabilization, as we shall see in the next paragraph.

Gyroscopic Systems with Damping In the previous paragraph we mentioned that a system with an indefinite stiffness matrix may be stabilized by gyroscopic forces, and we noted that damping may hinder such a stabilization. For a better understanding the THOMSON-TAIT theorem[5] is helpful:

> The eigenvalues of a $\boldsymbol{M}$-$\boldsymbol{D}$-$\boldsymbol{G}$-$\boldsymbol{K}$-system with positive definite damping matrix have negative real parts iff the stiffness matrix is positive definite.

The THOMSON-TAIT theorem can be generalized for positive semidefinite damping matrices, provided the damping is pervasive [6] in the sense of (87). With (115) it can then be said that with a positive definite stiffness matrix in the undamped or pervasively damped case the sign of the real parts of the eigenvalues is not affected by the gyroscopic matrix $\boldsymbol{G}$.

The THOMSON-TAIT theorem being a necessary (and sufficient) condition, the problem of gyroscopic stabilization in real systems becomes obvious. If a system with indefinite stiffness matrix is gyroscopically stabilized, pervasive damping as defined by (87) will lead to instability, i.e. to eigenvalues with positive real part. This phenomenon is sometimes known as destabilization due to damping or also as a damping paradoxon.

► The Rolling Penny with Damping

The damping paradoxon can be illustrated on the gyroscopically stabilized wheel dealt with in the previous paragraph. Without examining the details

[5]Named after the Irish physicist William THOMSON, 1. Baron Kelvin, *1824 in Belfast,†1907 in Netherhall and the Scottish physicist Peter Guthrie TAIT, *1831 in Dalkeith, †1901 in Edinburgh.

[6]In this context, pervasive damping means that the damping matrix fulfills the rank condition (87) and not necessarily all the eigenvalues of the $\boldsymbol{M}$-$\boldsymbol{D}$-$\boldsymbol{K}$-Systems have negative real parts.

of the physics, we add the damping matrix

$$\boldsymbol{D} = \begin{bmatrix} d & 0 \\ 0 & 0 \end{bmatrix} \tag{127}$$

to the system (123) and obtain the equations of motion

$$mr^2 \begin{bmatrix} \frac{3}{4} & 0 \\ 0 & \frac{5}{4} \end{bmatrix} \begin{bmatrix} \ddot{q}_1 \\ \ddot{q}_2 \end{bmatrix} + \begin{bmatrix} d & \frac{3}{2}mr^2\Omega \\ -\frac{3}{2}mr^2\Omega & 0 \end{bmatrix} \begin{bmatrix} \dot{q}_1 \\ \dot{q}_2 \end{bmatrix} + \begin{bmatrix} 0 & 0 \\ 0 & -mgr \end{bmatrix} \begin{bmatrix} q_1 \\ q_2 \end{bmatrix} = \begin{bmatrix} 0 \\ 0 \end{bmatrix}. \tag{128}$$

The characteristic equation now assumes the form

$$\lambda^4 + \frac{4}{3}\frac{d}{mr^2}\lambda^3 + \left(\frac{12}{5}\Omega^2 - \frac{4}{5}\frac{g}{r}\right)\lambda^2 - \frac{16}{15}\frac{dg}{mr^3}\lambda = 0. \tag{129}$$

According to the HURWITZ criterion [7], there is at least one root with positive real part, since the coefficient of λ is negative and the coefficient of λ^3 does not vanish. The trivial solution of (128) is therefore unstable. This example shows that damping may in some cases have a negative effect on stability.

The Energy Integral As in the $\boldsymbol{M}$-$\boldsymbol{K}$-systems, also for $\boldsymbol{M}$-$\boldsymbol{G}$-$\boldsymbol{K}$-systems a first integral can easily be derived. Multiplying the equations of motion (109) from the left with $\dot{\boldsymbol{q}}^{\mathsf{T}}$ gives

$$\dot{\boldsymbol{q}}^{\mathsf{T}}\boldsymbol{M}\ddot{\boldsymbol{q}} + \dot{\boldsymbol{q}}^{\mathsf{T}}\boldsymbol{G}\dot{\boldsymbol{q}} + \dot{\boldsymbol{q}}^{\mathsf{T}}\boldsymbol{K}\boldsymbol{q} = 0. \tag{130}$$

Since $\dot{\boldsymbol{q}}^{\mathsf{T}}\boldsymbol{G}\dot{\boldsymbol{q}}=0$, it follows that the power of the gyroscopic forces vanishes. As in (56) we therefore have here

$$\frac{1}{2}\dot{\boldsymbol{q}}^{\mathsf{T}}\boldsymbol{M}\dot{\boldsymbol{q}} + \frac{1}{2}\boldsymbol{q}^{\mathsf{T}}\boldsymbol{K}\boldsymbol{q} = h \tag{131}$$

as a first integral of (109). Due to this first integral, also the eigenvalues of $\boldsymbol{M}$-$\boldsymbol{G}$-$\boldsymbol{K}$-systems have certain extremality properties, which are however more complicated than in $\boldsymbol{M}$-$\boldsymbol{K}$-systems and we therefore will not dwell on them. Also the orthogonality conditions of the eigenvectors of $\boldsymbol{M}$-$\boldsymbol{K}$-systems are no longer valid for $\boldsymbol{M}$-$\boldsymbol{G}$-$\boldsymbol{K}$-systems, at least not in their originally very simple form.

[7] Named after the German mathematician Adolf HURWITZ, *1859 in Hildesheim, †1919 in Zürich.

1.5 $\boldsymbol{M}$-$\boldsymbol{K}$-$\boldsymbol{N}$-Systems

In the previous paragraphs it was shown that the eigenvalues of $\boldsymbol{M}$-$\boldsymbol{D}$-$\boldsymbol{G}$-$\boldsymbol{K}$-systems with a positive definite stiffness matrix have a non-positive real part. For the undamped case the energy integrals (56) and (131) were derived. We now examine the case of generalized forces given by the product of a skew-symmetric matrix $\boldsymbol{N}=-\boldsymbol{N}^{\mathsf{T}}$ with the vector of generalized coordinates $\boldsymbol{q}$. Forces of this type are called *circulatory* forces and are non-conservative, i.e. their work depends not only on the end points but also on the path in the configuration space. Systems leading to equations of motion with a *circulatory matrix* $\boldsymbol{N}$ are for example systems with frictional contact with a rotor rotating with fixed angular speed, as in disk brakes, etc. In this case the associated energy source is given by the rotor. In other systems it may be given by fluid flow and fluid-structure interaction.

In the present paragraph we restrict our attention to the free vibrations of undamped and non-gyroscopic systems with equations of motion of the type

$$\boldsymbol{M}\ddot{\boldsymbol{q}} + (\boldsymbol{K} + \boldsymbol{N})\boldsymbol{q} = \boldsymbol{0}. \tag{132}$$

Properties of the Eigenvalues The characteristic equation (10) for $\boldsymbol{M}$-$\boldsymbol{K}$-$\boldsymbol{N}$-systems assume the form

$$\det(\lambda^2\boldsymbol{M} + \boldsymbol{K} + \boldsymbol{N}) = 0, \tag{133}$$

i.e. a polynomial of degree n in λ^2. Since only even powers of λ are present, the eigenvalues occur in pairs, λ and $-\lambda$. Also, the coefficients of the polynomial are all real, so that, as in $\boldsymbol{M}$-$\boldsymbol{G}$-$\boldsymbol{K}$-systems, the eigenvalues will form quadruples $(\pm\lambda, \pm\lambda^*)$. If the eigenvalues are simple, than for each nontrivial λ_i^2 there is exactly one eigenvector $\boldsymbol{r}_i$ fulfilling

$$(\lambda_i^2\boldsymbol{M} + \boldsymbol{K} + \boldsymbol{N})\boldsymbol{r}_i = \boldsymbol{0}. \tag{134}$$

Multiplication with $\boldsymbol{r}_i^*$ from the left and projection of the equations on $\boldsymbol{r}_i^*$ gives

$$\lambda_i^2\boldsymbol{r}_i^*\boldsymbol{M}\boldsymbol{r}_i + \boldsymbol{r}_i^*\boldsymbol{K}\boldsymbol{r}_i + \boldsymbol{r}_i^*\boldsymbol{N}\boldsymbol{r}_i = 0 \tag{135}$$

and with a positive mass matrix leads to

$$\lambda_i^2 = -\frac{\boldsymbol{r}_i^*\boldsymbol{K}\boldsymbol{r}_i + \boldsymbol{r}_i^*\boldsymbol{N}\boldsymbol{r}_i}{\boldsymbol{r}_i^*\boldsymbol{M}\boldsymbol{r}_i}. \tag{136}$$

For positive semidefinite $\boldsymbol{K}$ one has $\boldsymbol{r}_i^* \boldsymbol{K} \boldsymbol{r}_i \geq 0$ and this expression is real. The product $\boldsymbol{r}_i^* \boldsymbol{N} \boldsymbol{r}_i$ is purely imaginary since $\boldsymbol{N} = -\boldsymbol{N}^\mathsf{T}$. We therefore write (136) in the form

$$\lambda_i^2 = -\frac{\boldsymbol{r}_i^* \boldsymbol{K} \boldsymbol{r}_i}{\boldsymbol{r}_i^* \boldsymbol{M} \boldsymbol{r}_i} + j \left(\frac{j \boldsymbol{r}_i^* \boldsymbol{N} \boldsymbol{r}_i}{\boldsymbol{r}_i^* \boldsymbol{M} \boldsymbol{r}_i} \right), \tag{137}$$

where the first summand as well as the expression in brackets is real. Here of course we have to examine which of the solutions of the projected system also fulfill (134). For $\boldsymbol{r}_i^* \boldsymbol{N} \boldsymbol{r}_i \neq 0$, λ_i^2 is complex, so that the quadruple $(\pm\lambda, \pm\lambda^*)$ contains solutions which are point symmetric about the origin in the complex plane. Therefore all the solutions of (137) are also solutions of (134) and (133). For $\boldsymbol{r}_i^* \boldsymbol{N} \boldsymbol{r}_i = 0$, (137) leads to a pair of purely imaginary solutions. Two of the elements of the quadruple $(\pm\lambda, \pm\lambda^*)$ are therefore identical but each one fulfills both (134) and (133). For the further discussion of the eigenvalues and the resulting eigenmotions we distinguish between the following cases:

$\boldsymbol{r}_i^* \boldsymbol{N} \boldsymbol{r}_i = 0$: The corresponding eigenvalues are purely imaginary. The corresponding eigenmotions are harmonic oscillations with the imaginary part of the eigenvalue being the circular frequency. Zero eigenfrequencies can only occur with non-positive definite stiffness matrix.

$\boldsymbol{r}_i^* \boldsymbol{N} \boldsymbol{r}_i \neq 0$: The eigenvalues occur in quadruples $(\pm\lambda, \pm\lambda^*)$ with not vanishing real and imaginary (for $\boldsymbol{K} > 0$) parts. The quadruple therefore contains a complex conjugate eigenpair with positive real part, so that for $t \to \infty$ the corresponding eigensolutions grows beyond all bounds. The trivial solution of the equations of motion is therefore unstable and self-excited vibrations are present.

From (134) it is clear that eigenvectors associated to $\boldsymbol{r}_i^* \boldsymbol{N} \boldsymbol{r}_i = 0$, i.e. to purely imaginary eigenvalues, can be chosen real, while real eigenvectors are not possible for $\boldsymbol{r}_i^* \boldsymbol{N} \boldsymbol{r}_i \neq 0$ in general. In any case the eigenvectors associated to the eigenvalue pair $\pm\lambda$ can be chosen to be identical, if we continue to assume that all eigenvalues are simple. The eigenmotions can be written in real form (25b).

► A M-K-N-System with Two Degrees of Freedom

In order to illustrate the basic ideas, we consider the double pendulum with follower force as shown in Figure 3 on the right. The equations of motion

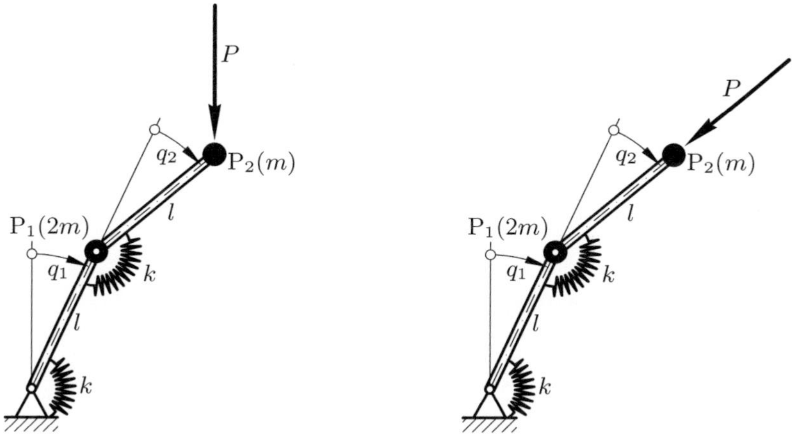

Figure 3: Double pendulum with conservative loads (left) and with follower force (right)

linearized about the trivial equilibrium position read

$$\begin{bmatrix} 6ml^2 & 2ml^2 \\ 2ml^2 & ml^2 \end{bmatrix} \begin{bmatrix} \ddot{q}_1 \\ \ddot{q}_2 \end{bmatrix} + \begin{bmatrix} k & Pl \\ 0 & k \end{bmatrix} \begin{bmatrix} q_1 \\ q_2 \end{bmatrix} = \begin{bmatrix} 0 \\ 0 \end{bmatrix} . \tag{138}$$

For the following discussion we introduce the abbreviations

$$p = \frac{Pl}{k}, \qquad \tau = t\sqrt{\frac{k}{ml^2}} \tag{139}$$

and make the equations non-dimensional. We decompose the matrix of the coordinate proportional forces into a symmetric and a skew-symmetric part (stiffness matrix and circulatory matrix), so that the equations of motion assume the form

$$\begin{bmatrix} 6 & 2 \\ 2 & 1 \end{bmatrix} \begin{bmatrix} q_1'' \\ q_2'' \end{bmatrix} + \left(\begin{bmatrix} 1 & p/2 \\ p/2 & 1 \end{bmatrix} + \begin{bmatrix} 0 & p/2 \\ -p/2 & 0 \end{bmatrix} \right) \begin{bmatrix} q_1 \\ q_2 \end{bmatrix} = \begin{bmatrix} 0 \\ 0 \end{bmatrix}, \tag{140}$$

where the apostrophe indicates derivatives with respect to the nondimensional time. The follower force here appears both in the stiffness matrix as well as in the circulatory matrix. The eigenvalue problem

$$\left(\lambda^2 \begin{bmatrix} 6 & 2 \\ 3 & 1 \end{bmatrix} + \begin{bmatrix} 1 & p \\ 0 & 1 \end{bmatrix} \right) \begin{bmatrix} r_1 \\ r_2 \end{bmatrix} = \begin{bmatrix} 0 \\ 0 \end{bmatrix} \tag{141}$$

has the characteristic equation

$$\det\begin{bmatrix} 6\lambda^2+1 & 2\lambda^2+p \\ 2\lambda^2 & \lambda^2+1 \end{bmatrix} = 2\lambda^4 + (7-2p)\lambda^2 + 1 = 0. \tag{142}$$

The solution of this second order equation in λ^2 gives

$$\lambda_1^2 = \frac{-7+4p+\sqrt{(7-2p)^2-8}}{4}, \quad \lambda_2^2 = \frac{-7+4p-\sqrt{(7-2p)^2-8}}{4}. \tag{143}$$

Figure 4(left) shows the eigenvalues depending on the parameter: The eigenvalues are imaginary for $0<p<7/2-\sqrt{2}$, as can be seen from a short analysis of (143). With increasing p, they move pairwise towards each other, reaching a critical value at $p=p_{CF}=7/2-\sqrt{2}$ (the index F denotes the follower force and the C stands for 'critical'), where there are double eigenvalues, since the radicand in (143) vanishes. For $p>p_{CF}$ the radicand in negative, $\lambda_{1,2}^2$ becomes complex and the eigenvalues appear in quadruples with non-vanishing real parts: one pair with positive and one pair with negative real part. This means that for $p>p_{CF}$ the trivial solution of (140) is unstable. There will be exponentially growing solutions. If p is sufficiently close to p_{CF}, the system will oscillate with slowly increasing amplitude and with a 'circular frequency' approximately given by

$$\omega = |\operatorname{Im}\lambda_{1,2,3,4}| = \sqrt{\frac{7-2p_{CF}}{4}} = \frac{1}{\sqrt[4]{2}}. \tag{144}$$

For $p=p_{CF}$ the system begins to flutter, and we also say that it becomes unstable through flutter[8].

We now compare these results with those obtained if a vertical load P acts at the upper particle, such as the weight, instead of the follower force (see left part of Figure 4). Such a force is conservative and can be derived from a potential. The equations of motion are now of the form

$$\begin{bmatrix} 6ml^2 & 2ml^2 \\ 2ml^2 & ml^2 \end{bmatrix}\begin{bmatrix} \ddot q_1 \\ \ddot q_2 \end{bmatrix} + \begin{bmatrix} k-2Pl & -Pl \\ -Pl & k-Pl \end{bmatrix}\begin{bmatrix} q_1 \\ q_2 \end{bmatrix} = \begin{bmatrix} 0 \\ 0 \end{bmatrix}. \tag{145}$$

The non-dimensionalization with (139) leads to

$$\begin{bmatrix} 6 & 2 \\ 2 & 1 \end{bmatrix}\begin{bmatrix} q_1'' \\ q_2'' \end{bmatrix} + \begin{bmatrix} 1-2p & -p \\ -p & 1-p \end{bmatrix}\begin{bmatrix} q_1 \\ q_2 \end{bmatrix} = \begin{bmatrix} 0 \\ 0 \end{bmatrix} \tag{146}$$

[8]The stability of the trivial solution with $p<p_{CF}$ should really be examined with a nonlinear theory, since here we have weak stability and a critical case in the sence of Liapounov stability theory. Named after the Russian mathematician and physicist Alexander Michailowitsch Liapounov, *1875 in Jaroslawl, †1918 in Odessa.

and therefore to the characteristic equation

$$\det\begin{bmatrix}6\lambda^2+1-2p & 2\lambda^2-p\\ 2\lambda^2-p & \lambda^2+1-p\end{bmatrix}=2\lambda^4+(7-4p)\lambda^2+p^2-3p+1=0. \quad (147)$$

The solution of this quadratic equation in λ^2 gives

$$\lambda_1^2=\frac{-7+2p+\sqrt{8p^2-32p+41}}{4},\quad \lambda_2^2=\frac{-7+2p-\sqrt{8p^2-32p+41}}{4}. \quad (148)$$

Figure 4(right) shows the eigenvalues depending on the parameter p. For $p=0$ the eigenvalues are imaginary and of course coincide with the one's of the system subject to the follower force. It turns out that for $p=p_{\mathrm{CK}}$ (the index K indicates conservative loading) the first of the two solutions (148) vanishes, i.e. at the critical load one pair of eigenvalues is equal to zero. Therefore also the corresponding circular frequency is equal to zero. The eigenvalue with the smaller magnitude therefore reaches the origin of the complex plane for $p=p_{\mathrm{CK}}$ and for $p>p_{\mathrm{CK}}$ assumes a real part different from zero. The stability boundaries therefore are essentially different from the one's with follower forces: Flutter, which marked the transition to instability, is now no longer present and in the present case we refer to instability through *divergence*. For a larger value of p, which can easily be calculated for the second solution of (148), also the second pair of eigenvalues reaches the real axis and also the second eigenmode becomes unstable.

This concludes the example, but the following is of general importance: While with conservative loading a structure always looses stability through divergence, and the stability boundary can always be computed from the stiffness matrix alone, without even looking at the equations of motion, in the case of circulatory forces flutter is possible, and the stability in general has to be decided on the basis of the equations of motion[9].

It is true that the follower force as assumed in this example does not really occur in engineering problems. It was chosen for a very clear and simple clarification of the basic problems which may occur with circulatory forces. Circulatory forces however appear in many engineering problems and can be responsible e.g. for the flutter of an aircraft wing or the self-excited vibrations leading to brake squeal.

1.6 *M*-*G*-*K*-*N*-Systems

If in addition to the circulatory forces also gyroscopic forces are present in a system, we have the case of a ***M*-*G*-*K*-*N*-system**. These systems are

[9]It can easily be checked in the example that for $p=p_{\mathrm{CK}}$ the stiffness matrix has a zero eigenvalue and is positive semi-definite, for $p>p_{\mathrm{CK}}$ it is indefinite.

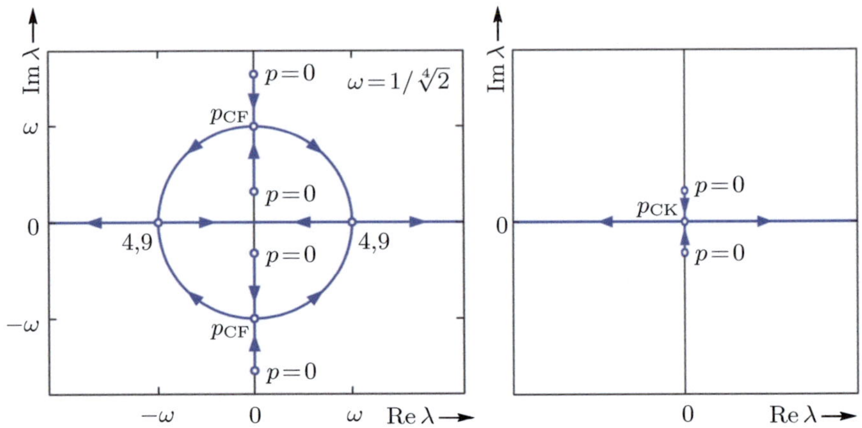

Figure 4: Eigenvalues of the double pendulum with follower force (left, solution of (143)) and with conservative load (right, first solution of (148))

rather common in engineering applications and therefore merit our attention. Here we examine the free vibrations of a system with equations of motion of the type

$$\boldsymbol{M}\ddot{\boldsymbol{q}} + \boldsymbol{G}\dot{\boldsymbol{q}} + (\boldsymbol{K} + \boldsymbol{N})\boldsymbol{q} = \boldsymbol{0}, \tag{149}$$

where we make the usual assumption on the matrices.

Properties of the Eigenvalues The characteristic equation now assumes the form

$$\det(\lambda^2\boldsymbol{M} + \lambda\boldsymbol{G} + \boldsymbol{K} + \boldsymbol{N}) = 0. \tag{150}$$

This polynomial of degree $2n$ in λ can be written as

$$a_{2n}\lambda^{2n} + a_{2n-1}\lambda^{2n-1} + \ldots + a_2\lambda^2 + a_1\lambda + a_0 = 0. \tag{151}$$

From the properties of determinants we know that $a_{2n}\lambda^{2n}$ is equal to the product of the principal diagonal's elements of the matrix $(\lambda^2\boldsymbol{M}+\lambda\boldsymbol{G}+\boldsymbol{K}+\boldsymbol{N})$. The term a_{2n} is therefore different from zero for a positive mass matrix. Since the principal diagonal of $\boldsymbol{G}$ is formed exclusively by zeros ($\boldsymbol{G}=-\boldsymbol{G}^\mathsf{T}$), the coefficient a_{2n-1} vanishes, so that the first eventually non-vanishing coefficient is a_{2n-2}. Therefore, according the HURWITZ criterion not all the eigenvalues of a $\boldsymbol{M}$-$\boldsymbol{G}$-$\boldsymbol{K}$-$\boldsymbol{N}$-system have negative real parts, i.e. there are purely imaginary eigenvalues and/or eigenvalues with positive real parts. Furthermore it can be stated that at least one eigenvalue with positive real part will occur, if not all the coefficients of the odd powers in λ disappear in (151). These conditions $a_{2i-1}=0$ $(i=1,\ldots,n)$ have the consequence that in general $\boldsymbol{M}$-$\boldsymbol{G}$-$\boldsymbol{K}$-$\boldsymbol{N}$-systems have at least one eigenvalue with positive real part and that the trivial solution is unstable.

We again clarify this with an example of a two-degree of freedom $\boldsymbol{M}$-$\boldsymbol{G}$-$\boldsymbol{K}$-$\boldsymbol{N}$-system. Without loss of generality a positive definite mass matrix can be assumed as diagonal. The equations of motion therefore read

$$\begin{bmatrix} m_1 & 0 \\ 0 & m_2 \end{bmatrix} \begin{bmatrix} \ddot{q}_1 \\ \ddot{q}_2 \end{bmatrix} + \begin{bmatrix} 0 & g \\ -g & 0 \end{bmatrix} \begin{bmatrix} \dot{q}_1 \\ \dot{q}_2 \end{bmatrix} + \left(\begin{bmatrix} k_1 & k_3 \\ k_3 & k_2 \end{bmatrix} + \begin{bmatrix} 0 & n \\ -n & 0 \end{bmatrix} \right) \begin{bmatrix} q_1 \\ q_2 \end{bmatrix} = \begin{bmatrix} 0 \\ 0 \end{bmatrix}, \quad (152)$$

where n is not to be confused with the number of degrees of freedom. In can be seen that the characteristic equation

$$m_1 m_2 \lambda^4 + (g^2 + m_1 k_2 + m_2 k_1)\lambda^2 + 2gn\lambda + (n^2 + k_1 k_2 - k_3^2) = 0 \quad (153)$$

has in fact no term in λ^3, so that according to the HURWITZ criterion not all eigenvalues can have negative real parts. Eigenvalues with zero real parts are possible only for $ng=0$, namely only if exclusively even powers of λ are present. This however means that the gyroscopic matrix $\boldsymbol{G}$ and/or the circulatory matrix $\boldsymbol{N}$ have to vanish, so that we would no longer have a $\boldsymbol{M}$-$\boldsymbol{G}$-$\boldsymbol{K}$-$\boldsymbol{N}$-system.

Systems with $\boldsymbol{G}$ and $\boldsymbol{N}$ matrices are very common in applications. Well known examples are the vibrations of a rotor in hydrodynamic bearings and the squealing disk brake, already mentioned before. The interaction of circulatory forces with the gyroscopic effects in the undamped case always leads to self-excited vibrations, except in some special cases. The trivial solution can then only be stabilized by damping.

1.7 $\boldsymbol{M}$-$\boldsymbol{D}$-$\boldsymbol{G}$-$\boldsymbol{K}$-$\boldsymbol{N}$-Systems

In the general case, the free vibrations of a linear mechanical system are described by equations of motion of the type

$$\boldsymbol{M}\ddot{\boldsymbol{q}} + (\boldsymbol{D} + \boldsymbol{G})\dot{\boldsymbol{q}} + (\boldsymbol{K} + \boldsymbol{N})\boldsymbol{q} = \boldsymbol{0}, \quad (154)$$

where $\boldsymbol{D}$ and $\boldsymbol{K}$ are not necessarily positive (semi-)definite. Simple statements on the eigenvalues and eigenvectors can then no longer be made on the basis of the matrices, as was the case in particular situations examined above. It is then necessary to solve the eigenvalue problem numerically in order to get information on the system's stability.

2 Equations of Motion of Continuous Mechanical Systems

2.1 Equation of Motion of Beams

In this section we discuss the equations of elastic beams, which are a simple paradigm of more complex elastic structures. We study a beam theory which considers only the effect of bending moment on the dynamics of the beam. Thus, it is required that the shear forces be small so that the shear deformation of the beam is negligible. Another way of stating this assumption is to say that the beam is almost infinitely stiff in shear. The effect of shear deformation on the transverse dynamics of a beam will be considered separately later in this chapter.

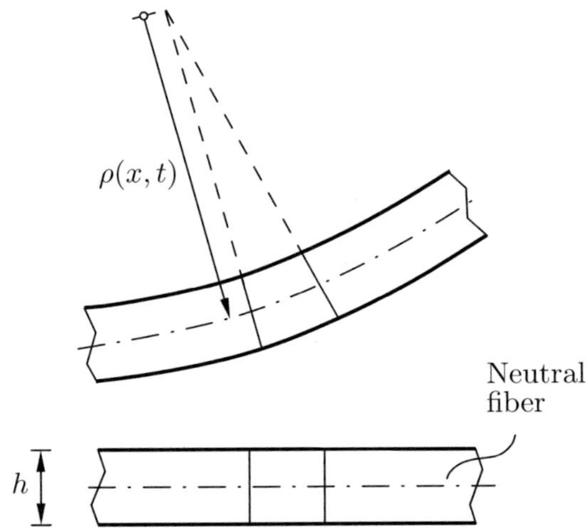

Figure 5: Schematic representation of a beam under planar deflection

The NEWTONian Formulation Consider a straight beam undergoing a planar deflection in uni-axial bending as represented schematically in Fig. 5. The simplest of all beam theories starts with the assumption that planar cross-sections of the undeformed beam remain planar even after the beam undergoes a deformation, as illustrated in the figure. From elementary theory of elasticity (see, for example, Timoshenko and Goodier (1970)), it is known that when the beam is deflected, certain hypothetical longitudinal lines or *fibers* are elongated, while others are compressed. There, however, exist fibers which are neither elongated nor compressed, but are merely deflected. Such a fiber is called the *neutral fiber*, and is shown in Fig. 5. It is assumed that cross-sections orthogonal to the neutral fiber before deformation are are also orthogonal to the neutral fiber in the deformed beam. These assumptions are referred to as the EULER-BERNOULLI *hypotheses*[10]. The EULER-BERNOULLI hypotheses hold good as long as the ratio of the height of the beam h to the radius of curvature $\rho(x,t)$ of the neutral fiber after deformation is much smaller than unity. In terms of the forces, the assumptions remain meaningful for small bending moment gradient (*i.e.*, for small shear) along the length of the beam. With this assumption, and

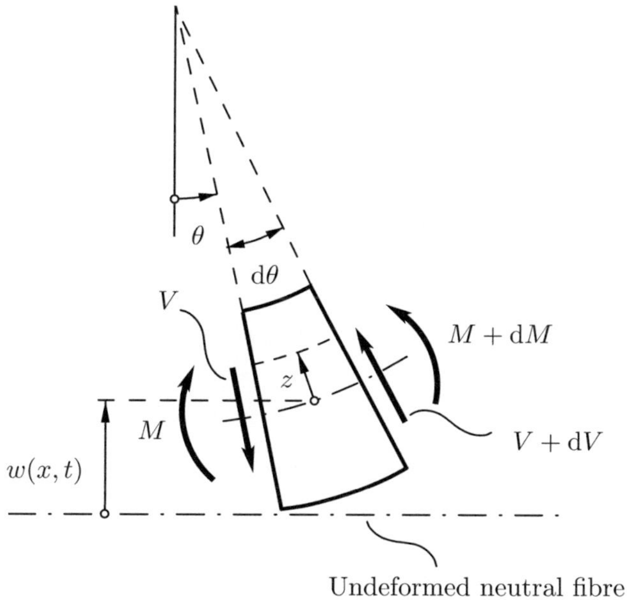

Figure 6: Infinitesimal element of a deflected beam

referring to Fig. 6, the strain-displacement relation at any height z measured from the plane of the neutral fibers can be written from the theory of

[10]Named after Jacob BERNOULLI, a Swiss mathematician, *1654/55 in Basel, Switzerland, †1705 in Basel, Switzerland; Leonhard EULER, a Swiss mathematician and physicist, *1707 in Basel, Switzerland, †1783 in Saint Petersburg, Russia, and Daniel BERNOULLI, again a Swiss mathematician and physicist, *1700 in Groningen, Netherlands, †1782 in Basel, Switzerland.

elasticity as

$$
\begin{aligned}
\epsilon_x(x,z,t) &= \frac{(\rho(x,t)-z)\,\mathrm{d}\theta - \rho(x,t)\,\mathrm{d}\theta}{\rho(x,t)\,\mathrm{d}\theta} = -\frac{z}{\rho(x,t)} \\
&= -\frac{z w_{,xx}(x,t)}{[1+w_{,x}^2(x,t)]^{3/2}} \\
&\approx -z w_{,xx}(x,t) \qquad (\text{assuming } w_{,x} \ll 1),
\end{aligned} \tag{155}
$$

where $w(x,t)$ is the transverse deflection field. Next, the constitutive relation for a linearly elastic material can be written from Hooke's[11] law as

$$
\sigma_x(x,z,t) = E\epsilon_x(x,z,t) = -Ezw_{,xx}(x,t), \tag{156}
$$

where E is Young's modulus[12]. The bending moment at any section can then be written as

$$
\begin{aligned}
M(x,t) &= -\int_{-h/2}^{h/2} z\sigma_x(x,z,t)\,\mathrm{d}A \\
&= \int_{-h/2}^{h/2} Ew_{,xx}(x,t)z^2\,\mathrm{d}A \\
&= EI(x)w_{,xx}(x,t),
\end{aligned} \tag{157}
$$

where $I(x)$ is the second moment of area of cross-section of the beam about the *neutral axis*. The neutral axis is the line of intersection of the plane of the neutral fibers, and the plane of the cross-section of the beam. Now, the equation of translational dynamics of an infinitesimal element can be written as

$$
(\rho A(x)\,\mathrm{d}x)w_{,tt} = p(x,t)\mathrm{d}x + (V+\mathrm{d}V)\cos(\theta+\mathrm{d}\theta) - V\cos\theta,
$$

$$
\text{or} \quad \rho A w_{,tt} = p(x,t) + V_{,x}, \tag{158}
$$

where $p(x,t)$ is the external transverse force density, V is the shear force at any cross-section, and it is assumed that $\cos\theta \approx 1$. The rotational dynamics

[11] Named after Robert HOOKE, an English natural philosopher, physicist and chemist, *1635 on Isle of Wight, England, †1705 in London, England

[12] Named after Thomas YOUNG, a British scientist, *1773 in Somerset, England, †1829 in London, England; developed by Leonhard EULER in 1727

of the infinitesimal element is represented by

$$(\rho I(x)\,\mathrm{d}x)\theta_{,tt} = (M+\mathrm{d}M) - M + (V+\mathrm{d}V)\frac{\mathrm{d}x}{2} + V\frac{\mathrm{d}x}{2},$$

$$\text{or} \quad \rho I(x)\theta_{,tt} = M_{,x} + V. \tag{159}$$

Using the relation $\tan\theta = w_{,x}$, one can write

$$\theta_{,t} = \frac{w_{,xt}}{(1+w_{,x}^2)} \approx w_{,xt}, \tag{160}$$

$$\text{and} \quad \theta_{,tt} = \frac{w_{,xtt}}{(1+w_{,x}^2)} - \frac{2w_{,x}w_{,xt}^2}{(1+w_{,x}^2)^2} \approx w_{,xtt}, \tag{161}$$

where all non-linear terms have been dropped. Using (157) and (161) in (159), and subsequently eliminating V between (159) and (158) yields on simplification

$$\rho A w_{,tt} + [EIw_{,xx}]_{,xx} - [\rho I w_{,xtt}]_{,x} = p(x,t). \tag{162}$$

This equation of motion is known as the RAYLEIGH *beam equation.* The term $(EIw_{,xx})_{,xx}$ is usually referred to as the *flexure term*, where EI is called the *flexural stiffness*, and $(\rho I w_{,xtt})_{,x}$ is known as the rotary inertia term. When the rotary inertia term is neglected, we obtain

$$\rho A w_{,tt} + [EIw_{,xx}]_{,xx} = p(x,t), \tag{163}$$

which is referred to as the EULER-BERNOULLI *beam model.* It is observed that the equations of motion (162) or (163) are fourth order partial differential equations in space, and second order in time. Thus, we require four boundary conditions, and two initial conditions. The boundary conditions are discussed in a later section.

The Variational Formulation The variational method provides an alternate convenient approach for obtaining the equation of motion and the boundary conditions for beams. The total kinetic energy $\mathcal{T}$ due to translation and rotation of an infinitesimal beam element can be written as

$$\begin{aligned}\mathcal{T} &= \frac{1}{2}\int_0^l \left[\rho A w_{,t}^2 + \rho I \theta_{,t}^2\right] \mathrm{d}x \\ &= \frac{1}{2}\int_0^l \left[\rho A w_{,t}^2 + \rho I w_{,xt}^2\right] \mathrm{d}x \qquad \text{(using (160))}.\end{aligned} \tag{164}$$

The potential energy $\mathcal{V}$ can be written from the theory of elasticity as

$$\begin{aligned}\mathcal{V} &= \frac{1}{2}\int_0^l \int_A \sigma_x \epsilon_x \,\mathrm{d}A\,\mathrm{d}x \\ &= \frac{1}{2}\int_0^l \int_A E w_{,xx}^2 z^2 \,\mathrm{d}A\,\mathrm{d}x \qquad \text{(using (155) and (156))} \\ &= \frac{1}{2}\int_0^l EI w_{,xx}^2 \,\mathrm{d}x. \end{aligned} \tag{165}$$

The Lagrangian is given by $\mathcal{L} = \mathcal{T} - \mathcal{V}$, and the variational formulation yields

$$\delta \int_{t_1}^{t_2} \mathcal{L}\,\mathrm{d}t = 0,$$

$$\text{or} \quad \delta \int_{t_1}^{t_2} \frac{1}{2}\int_0^l \left[\rho A w_{,t}^2 + \rho I w_{,xt}^2 - EI w_{,xx}^2\right] \mathrm{d}x\mathrm{d}t = 0. \tag{166}$$

Following the procedure discussed in the appendix of Hagedorn and DasGupta (2007), we have from (166)

$$\int_{t_1}^{t_2}\int_0^l [\rho A w_{,t}\,\delta w_{,t} + \rho I w_{,xt}\,\delta w_{,xt} - EI w_{,xx}\,\delta w_{,xx}]\,\mathrm{d}x\mathrm{d}t = 0,$$

$$\begin{aligned}\text{or} \quad & -\int_{t_1}^{t_2} EI w_{,xx}\,\delta w_{,x}\Big|_0^l \mathrm{d}t - \int_{t_1}^{t_2} [(EI w_{,xx})_{,x} - \rho I w_{,xtt}]\,\delta w\Big|_0^l \mathrm{d}t \\ & + \int_{t_1}^{t_2}\int_0^l [-\rho A w_{,tt} + (\rho I w_{,xtt})_{,x} - (EI w_{,xx})_{,xx}]\,\delta w\,\mathrm{d}x\mathrm{d}t = 0,\end{aligned} \tag{167}$$

where, we have used the fact that the variation of the field variable and its spatial derivatives at the initial and final times is zero, *i.e.*, $\delta w|_{t_i} = \delta w_{,x}|_{t_i} \equiv 0$ for $i = 0, 1$. The condition (167) must holds for arbitrary variations δw. This yields, from the last integral in (167), the equation of motion

$$\rho A w_{,tt} + [EI w_{,xx}]_{,xx} - [\rho I w_{,xtt}]_{,x} = 0. \tag{168}$$

Similarly, boundary conditions are obtained from the first and second integrals in (167). For example, one possible set of boundary conditions is given

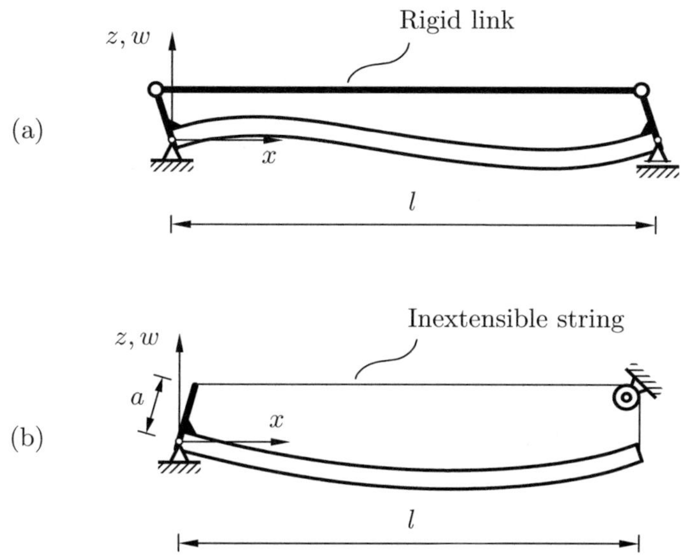

Figure 7: Special boundary conditions for beams

by

$$EIw_{,xx}(0,t) \equiv 0 \qquad \text{or} \qquad w_{,x}(0,t) \equiv 0, \tag{169}$$

$$\text{and} \quad EIw_{,xx}(l,t) \equiv 0 \qquad \text{or} \qquad w_{,x}(l,t) \equiv 0, \tag{170}$$

$$\text{and} \quad [(EIw_{,xx})_{,x} - \rho I w_{,xtt}](0,t) \equiv 0 \qquad \text{or} \qquad w(0,t) \equiv 0, \tag{171}$$

$$\text{and} \quad [(EIw_{,xx})_{,x} - \rho I w_{,xtt}](l,t) \equiv 0 \qquad \text{or} \qquad w(l,t) \equiv 0. \tag{172}$$

The first condition in (169) and (170) implies zero moment at the ends, while the first condition in (171) and (170) implies zero shear force. The second equation in each of the above conditions is a geometric boundary condition which implies either a zero displacement, or a zero slope.

Other kinds of boundary conditions are also possible, involving a linear combination of the boundary terms in (167). For example, for the beam shown in Fig. 7(a), the geometric boundary conditions are given by $w(0,t) \equiv 0$, $w(l,t) \equiv 0$, and $w_{,x}(0,t) = w_{,x}(l,t)$. In this case, the difference of the boundary term evaluated at $x = l$ and $x = 0$ under the first integral in (167) must vanish. This leads to the fourth (natural) boundary conditions $EIw_{,xx}(0,t) = EIw_{,xx}(l,t)$ (since $\delta w_{,x}(0,t) = \delta w_{,x}(l,t)$). In Fig. 7(b), the boundary condition are $w(0,t) \equiv 0$, $EIw_{,xx}(l,t) \equiv 0$, and $aw_{,x}(0,t) = w(l,t)$. The natural boundary condition in this case also can be obtained easily from the boundary terms in (167).

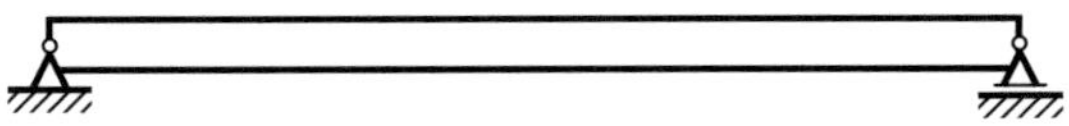

(a) Simply supported beam

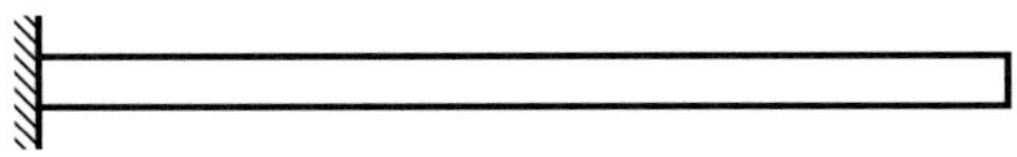

(b) Cantilever beam

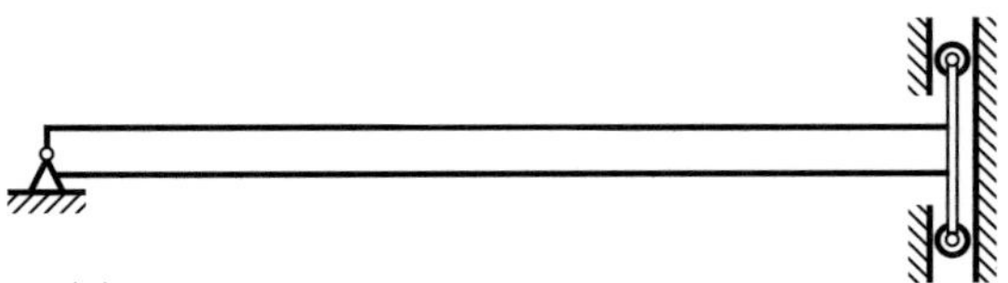

(c) Beam with a sliding boundary

Figure 8: Various boundary conditions for a beam

Various Boundary Conditions for a Beam Some of the above boundary conditions are realized in various combinations in beams depending on the support, as illustrated in Fig. 8. The simplest support conditions can be either pinned, clamped, free, or sliding, as illustrated in Fig. 8. When the end is pinned without friction, there is zero transverse displacement (geometric boundary condition), and zero moment (dynamic boundary condition) at that end. Thus, we have in Fig. 8(a)

$$w(0,t) \equiv 0, \quad w(l,t) \equiv 0, \quad EIw_{,xx}(0,t) \equiv 0, \quad \text{and} \quad EIw_{,xx}(l,t) \equiv 0. \tag{173}$$

At a clamped end, as shown in Fig. 8(b), the displacement and slope of the beam are zero (both are geometric boundary conditions). Therefore, we have

$$w(0,t) \equiv 0, \qquad \text{and} \qquad w_{,x}(0,t) \equiv 0. \tag{174}$$

At a free boundary, it is evident that the moment and the shear force vanish (both dynamic boundary conditions). Hence, one can write

$$EIw_{,xx}(l,t) \equiv 0, \qquad \text{and} \qquad \rho I w_{,xtt}(l,t) - [EIw_{,xx}(l,t)]_{,x} \equiv 0. \tag{175}$$

A sliding boundary is characterized by zero slope and zero shear. Thus, the mathematical conditions for the right boundary of the beam shown in Fig. 8(c) are

$$w_{,x}(l,t) \equiv 0, \qquad \text{and} \qquad EIw_{,xxx}(l,t) \equiv 0. \tag{176}$$

When there are external forces over the beam, or at the boundaries, appropriate forcing terms can be added to, respectively, (168), or in the moment and shear boundary conditions.

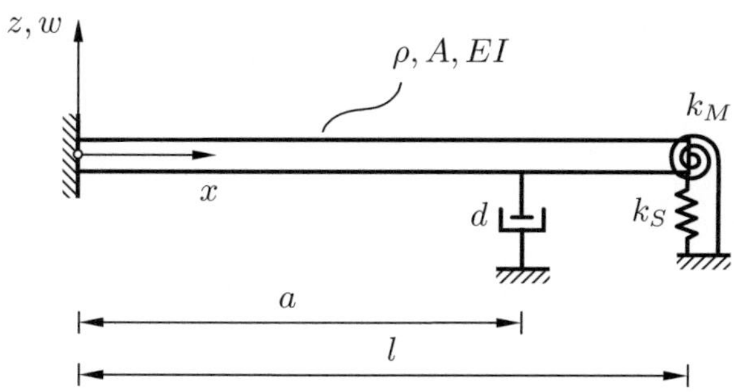

Figure 9: Cantilever beam with discrete damping and stiffened free-end

For example, consider a uniform cantilever beam with discrete damping and a stiffened free-end, as shown in Fig. 9. The equation of motion, and the boundary conditions can be written as

$$\rho A w_{,tt} + [EIw_{,xx}]_{,xx} - [\rho I w_{,xtt}]_{,x} + d\,\delta(x-a)w_{,t} = 0,$$

$$w(0,t) = 0, \qquad w_{,x}(0,t) = 0, \qquad EIw_{,xx}(l,t) = -k_M w_{,x}(l,t),$$

$$\text{and} \quad EIw_{,xxx}(l,t) - \rho I w_{,xtt}(l,t) = -k_S w(l,t).$$

Taut String and Tensioned Beam Before proceeding further, let us compare a taut string and a tensioned beam. Consider a uniform beam under axial tension. One can easily show that the equation of motion of this beam is obtained as

$$\rho A w_{,tt} - T w_{,xx} + EI w_{,xxxx} - \rho I w_{,ttxx} = 0, \tag{177}$$

where T is the tension in the beam. Consider a non-dimensionalization scheme

$$\bar{w} = \frac{w}{r_g}, \qquad \bar{x} = \frac{x}{l}, \qquad \text{and} \qquad \bar{t} = \frac{tc}{l} = \frac{t}{l}\sqrt{\frac{T}{\rho A}},$$

where $r_g := \sqrt{I/A}$ is the radius of gyration of the cross-section about the neutral axis of the beam. Using these non-dimensionalized variables, (177) can be written as

$$\bar{w}_{,\bar{t}\bar{t}} - \bar{w}_{,\bar{x}\bar{x}} + \frac{EI}{Tl^2}\bar{w}_{,\bar{x}\bar{x}\bar{x}\bar{x}} - \frac{I}{Al^2}\bar{w}_{,\bar{t}\bar{t}\bar{x}\bar{x}} = 0. \tag{178}$$

Thus, the non-dimensional quantity EI/Tl^2 decides the relative importance of the flexure term $w_{,xxxx}$, while I/Al^2 reflects the relative importance of the rotary inertia term $w_{,ttxx}$. One can also write these non-dimensional numbers as

$$\frac{EI}{Tl^2} = \frac{EI/\rho A}{Tl^2/\rho A} = \frac{1}{T/EA}\,\frac{1}{l^2 A/I} = \frac{1}{\epsilon_x}\frac{1}{s_r^2}, \qquad \text{and} \qquad \frac{I}{Al^2} = \frac{1}{s_r^2}, \tag{179}$$

where ϵ_x is the longitudinal strain in the x-axis direction due to pre-tension, and $s_r := l/r_g$ is defined as the slenderness ratio. Then, one can rewrite (178) as

$$\bar{w}_{,\bar{t}\bar{t}} - \bar{w}_{,\bar{x}\bar{x}} + \frac{1}{\epsilon_x}\frac{1}{s_r^2}\bar{w}_{,\bar{x}\bar{x}\bar{x}\bar{x}} - \frac{1}{s_r^2}\bar{w}_{,\bar{t}\bar{t}\bar{x}\bar{x}} = 0. \tag{180}$$

It is clear from (179) that when the beam is very slender (*i.e.*, $s_r \gg 1$), the third and fourth terms in (180) become insignificant. In that case, the beam can be treated as a string with no flexural stiffness, and no rotary inertia. Further, since $\epsilon_x \ll 1$, it follows that the rotary inertia term is relatively small significant compared to the flexure term. However, as we shall see later, the rotary inertia term gains in importance with increasing curvature of the beam. In the case of a moderate slenderness ratio, due to the fact that $\epsilon_x \ll 1$ (*i.e.*, $T \ll EA$), we have $1/\epsilon_x s_r^2 \gg 1$, and the flexure term becomes the most important term in the dynamics. In that case, we may drop the second term in (180), and consider the simple beam equation (162), or (163). It may be further noticed that the third term in (180) becomes important for large curvatures of the continuum, which typically occurs at the support points. In such cases, one may use a mixed string-beam model in which the bulk of the continuum is modeled as a string, while near the support points, a beam model is used. Such analysis can be found in Anderson and Hagedorn (1995).

2.2 Free Vibration Problem

The free vibration problem is essentially the determination of the eigenfrequencies and the corresponding eigenfunctions of the system. Hence, we begin here with the modal analysis of the beam models derived above. The solution of the initial value problem can be formulated using the eigenfunctions obtained from the modal analysis.

Modal Analysis

► The Eigenvalue Problem

Consider a RAYLEIGH beam described by the equation of motion

$$\rho A w_{,tt} + (EI w_{,xx})_{,xx} - (\rho I w_{,xtt})_{,x} = 0. \tag{181}$$

Assume a modal solution of (181) in the form

$$w(x,t) = W(x)\mathrm{e}^{i\omega t}, \tag{182}$$

where ω is the circular eigenfrequency, and $W(x)$ is the eigenfunction. The actual real solution is obtained by taking the real part, or the imaginary part of the complex expression in (182). Substituting the modal solution in the field equation (181) yields on rearrangement the eigenvalue problem for a RAYLEIGH beam as

$$-\omega^2[\rho A W - (\rho I W')'] + (EIW'')'' = 0. \tag{183}$$

One may consider (183) as a general eigenvalue problem of the form

$$-\omega^2 \mathcal{M}[W] + \mathcal{K}[W] = 0, \tag{184}$$

where

$$\mathcal{M}[\cdot] = \left[\rho A - \frac{\mathrm{d}}{\mathrm{d}x}\left(\rho I \frac{\mathrm{d}}{\mathrm{d}x}\right)\right][\cdot], \quad \text{and} \quad \mathcal{K}[\cdot] = \frac{\mathrm{d}^2}{\mathrm{d}x^2}\left(EI\frac{\mathrm{d}^2}{\mathrm{d}x^2}\right)[\cdot]. \tag{185}$$

In the case of an EULER-BERNOULLI beam described by

$$\rho A w_{,tt} + (EI w_{,xx})_{,xx} = 0, \tag{186}$$

substituting the solution form (182) leads to the eigenvalue problem

$$-\omega^2 \rho A W + (EIW'')'' = 0. \tag{187}$$

It is evident that (187) is a special case of (184) with

$$\mathcal{M}[\cdot] = \rho A[\cdot], \qquad \text{and} \qquad \mathcal{K}[\cdot] = \frac{\mathrm{d}^2}{\mathrm{d}x}\left(EI\frac{\mathrm{d}^2}{\mathrm{d}x^2}\right)[\cdot].$$

The general solution of (184) (and hence (187)) cannot be obtained in closed form for arbitrary $EI(x)$ and/or $\rho A(x)$. Therefore, we will solve the eigenvalue problem for uniform beams only. Before proceeding further to solve the eigenvalue problem, let us first discuss the orthogonality property of eigenfunctions of (184).

▸ Orthogonality Relations

Consider the eigenvalue problem of the RAYLEIGH beam described by (184). For two different modes j and k, one can write (184) as

$$-\omega_j^2 \mathcal{M}[W_j] + \mathcal{K}[W_j] = 0, \tag{188}$$

$$\text{and} \quad -\omega_k^2 \mathcal{M}[W_k] + \mathcal{K}[W_k] = 0, \tag{189}$$

where $\mathcal{M}[\cdot]$ and $\mathcal{K}[\cdot]$ are given by (185). Multiplying (188) by W_k, (188) by W_j, subtracting one equation from the other, and integrating the result over the length of the beam gives

$$\begin{aligned}&[((EIW_j'')' - \omega_j^2 \rho I W_j')W_k - ((EIW_k'')' - \omega_j^2 \rho I W_k')W_j]\big|_0^l + \\ &[EIW_k''W_j' - EIW_j''W_k']\big|_0^l + (\omega_j^2 - \omega_k^2)\int_0^l [\rho A W_k - (\rho I W_k')']W_j \,\mathrm{d}x = 0.\end{aligned} \tag{190}$$

Using the boundary conditions defined by (169)-(172), it can be easily checked that the boundary terms in (190) disappear. Hence, we immediately obtain the orthogonality relation from (190) as

$$\int_0^l [\rho A W_k - (\rho I W_k')']W_j \,\mathrm{d}x = 0, \qquad j \neq k, \tag{191}$$

$$\text{or} \quad \int_0^l \mathcal{M}[W_k]W_j \,\mathrm{d}x = 0, \qquad j \neq k. \tag{192}$$

In the case of an EULER-BERNOULLI beam, (191) simplifies further to

$$\int_0^l \rho A W_k W_j \, \mathrm{d}x = 0, \qquad j \neq k. \tag{193}$$

One may normalize the eigenfunctions with respect to an inner product such that

$$\int_0^l \mathcal{M}[W_k] W_j \, \mathrm{d}x = \delta_{jk}, \tag{194}$$

where δ_{jk} represents the Kronecker delta function. The eigenfunctions so normalized form an orthonormal basis. As a consequence of this orthonormality, from (184) and (192), one can easily write

$$\int_0^l \mathcal{K}[W_k] W_j \mathrm{d}x = \omega_k^2 \delta_{jk}.$$

► Modal Analysis of Uniform Beams

Consider the eigenvalue problem of a uniform RAYLEIGH beam described by

$$-\omega^2[\rho A W - \rho I W''] + EIW'''' = 0. \tag{195}$$

Substituting in (195) a solution of the form

$$W(x) = B\mathrm{e}^{\tilde{\beta} x}, \tag{196}$$

where B and $\tilde{\beta}$ are constants, one can write

$$EI\tilde{\beta}^4 - \omega^2 \rho I \tilde{\beta}^2 - \omega^2 \rho A = 0$$

$$\Rightarrow \quad \tilde{\beta}^2 = \frac{1}{2EI}\left[\omega^2 \rho I \pm \sqrt{\omega^4 \rho^2 I^2 + 4\omega^2 EI\rho A}\right]. \tag{197}$$

It is easily observed that the bracketed term in (197) will take both, a positive, and a negative value. Therefore, $\tilde{\beta}$ has four solutions give as $\tilde{\beta} = \pm\beta_1, \pm i\beta_2$, where

$$\beta_1 = \frac{1}{\sqrt{2EI}}\left[\omega^2 \rho I + \sqrt{\omega^4 \rho^2 I^2 + 4\omega^2 EI\rho A}\right]^{1/2}, \tag{198}$$

$$\text{and} \quad \beta_2 = \frac{1}{\sqrt{2EI}}\left[-\omega^2 \rho I + \sqrt{\omega^4 \rho^2 I^2 + 4\omega^2 EI\rho A}\right]^{1/2}. \tag{199}$$

Thus, the general (complex) solution of (183) is obtained as

$$W(x) = A_1 \mathrm{e}^{\beta_1 x} + A_2 \mathrm{e}^{-\beta_1 x} + A_3 \mathrm{e}^{i\beta_2 x} + A_4 \mathrm{e}^{-i\beta_2 x}, \tag{200}$$

where A_i, $i = 1, \ldots, 4$ are (complex) constants. Alternatively, the solution may also be expressed in the real form as

$$W(x) = B_1 \cosh \beta_1 x + B_2 \sinh \beta_1 x + B_3 \cos \beta_2 x + B_4 \sin \beta_2 x, \tag{201}$$

where B_i, $i = 1, \ldots, 4$ are real constants to be obtained from the boundary conditions.

Next, we consider the case of a uniform EULER-BERNOULLI beam. Substituting the solution (196) in the eigenvalue problem

$$-\omega^2 \rho A W + EIW'''' = 0, \tag{202}$$

where ρA and EI are constants, we obtain

$$-\rho A \omega^2 + EI\tilde{\beta}^4 = 0$$

$$\Rightarrow \quad \tilde{\beta}^2 = \sqrt{\frac{\omega^2 \rho A}{EI}}. \tag{203}$$

Therefore, we have the four solutions $\tilde{\beta} = \pm\beta, \pm i\beta$, where

$$\beta = (\omega^2 \rho A / EI)^{1/4}. \tag{204}$$

Now, one can write the general (complex) solution (for $\omega \neq 0$) of the eigenvalue problem (187) as

$$W(x) = A_1 \mathrm{e}^{\beta x} + A_2 \mathrm{e}^{-\beta x} + A_3 \mathrm{e}^{i\beta x} + A_4 \mathrm{e}^{-i\beta x}, \tag{205}$$

where A_i, $i = 1, \ldots, 4$ are (complex) constants, or in the real form as

$$W(x) = B_1 \cosh \beta x + B_2 \sinh \beta x + B_3 \cos \beta x + B_4 \sin \beta x, \tag{206}$$

where B_i, $i = 1, \ldots, 4$ are real constants of integration which are determined by the boundary conditions of the problem.

In the following, we consider beams with some typical support conditions, and determine their eigenfrequencies and eigenfunctions.

(a) Uniform Simply-Supported Beam

Consider a simply-supported (pinned-pinned) uniform RAYLEIGH beam. The boundary conditions for the corresponding eigenvalue problem (184) are given by

$$W(0) = 0, \quad W''(0) = 0, \quad W(l) = 0, \quad \text{and} \quad W''(l) = 0. \tag{207}$$

Using the first two conditions from (207) in (201) yield $B_1 = B_3 = 0$. The last two boundary conditions in (207) yield

$$B_2 \sinh \beta_1 l + B_4 \sin \beta_2 l = 0, \tag{208}$$

$$\text{and} \quad B_2 \sinh \beta_1 l - B_4 \sin \beta_2 l = 0. \tag{209}$$

For non-trivial solutions of (B_2, B_4) from (208)-(209), one must have

$$\sinh \beta_1 l \, \sin \beta_2 l = 0$$

$$\Rightarrow \quad \sin \beta_2 l = 0 \qquad (\text{since } \sinh \beta_1 l \neq 0 \text{ for any } \beta_1 l \neq 0), \tag{210}$$

which is the characteristic equation for the problem. The solutions of the characteristic equation are obtained as

$$\beta_2 = \frac{n\pi}{l}, \qquad n = 1, 2, \ldots, \infty. \tag{211}$$

Substituting this expression of β_2 in (199), and solving for ω yield the circular natural frequencies of a simply-supported uniform RAYLEIGH beam as

$$\omega_n^R = \frac{n^2\pi^2}{l^2} \frac{1}{\left[1 + n^2\pi^2 \dfrac{I}{l^2 A}\right]^{1/2}} \sqrt{\frac{EI}{\rho A}}, \qquad n = 1, 2, \ldots, \infty. \tag{212}$$

Taking $n \gg 1$ such that $1 + n^2\pi^2 I/l^2 A \approx n^2\pi^2 I/l^2 A$, one obtains from (212) the approximation $\omega_n^R \approx (n\pi/l)\sqrt{E/\rho}$. As can be easily checked, these are the circular eigenfrequencies of longitudinal vibrations of a fixed-fixed bar.

The final step of modal analysis is to determine the eigenfunctions. From (208) and (209), and the characteristic equation (210) one can easily conclude that $B_2 = 0$. Substituting this in (201), along with $B_1 = B_3 = 0$ and (211), the eigenfunctions of a simply-supported uniform RAYLEIGH beam can be written as

$$W_n(x) = B \sin \frac{n\pi x}{l}, \qquad n = 1, 2, \ldots, \infty, \tag{213}$$

where B is an arbitrary constant. These eigenfunctions are clearly orthogonal, and can be normalized to make them orthonormal.

In the case of a simply-supported uniform EULER-BERNOULLI beam, we have the same expression for β_n given by (211), as one can easily check. Therefore, the circular natural frequencies of an EULER-BERNOULLI beam are obtained by substituting the expression of β from (211) in (204), and solving for ω_n. This yields

$$\omega_n^{EB} = \frac{n^2\pi^2}{l^2}\sqrt{\frac{EI}{\rho A}}, \qquad n = 1, 2, \ldots, \infty. \tag{214}$$

It may be observed by comparing (212) and (214) that, in the case of a very slender beam (*i.e.*, $s_r = l^2 A/I \gg 1$), the natural frequencies of the lower modes given by the RAYLEIGH beam model tend to be the same as those of those obtained from the EULER-BERNOULLI beam model (since $1+n^2\pi^2 I/l^2 A \approx 1$). Thus, for lower modes of very slender beams, the effect of rotary inertia is insignificant. It can be checked that the eigenfunctions for the simply-supported uniform RAYLEIGH and EULER-BERNOULLI beams are the same.

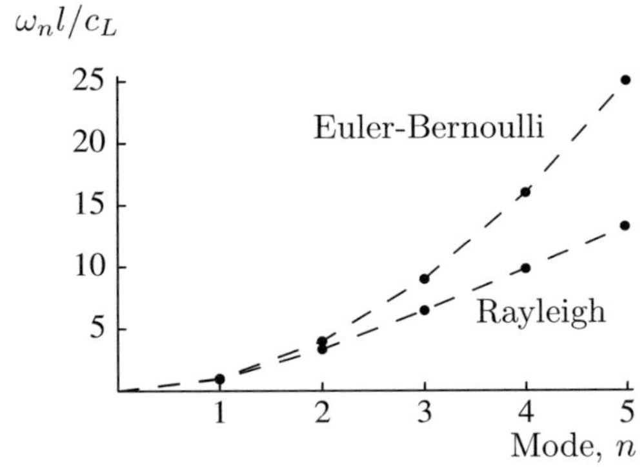

Figure 10: Comparison of natural frequencies of a simply-supported RAYLEIGH beam and an EULER-BERNOULLI beam for a fixed slenderness ratio $s_r = 10$

The two non-dimensional natural frequencies $\omega_n^R l/c_L$, and $\omega_n^{EB} l/c_L$, where $c_L = \sqrt{E/\rho}$, are compared for first few modes in Fig. 10. It is observed that for lower modes, the two frequencies tend to match. However, divergence is observed at higher modes. This is primarily due to the effect of rotary inertia in the RAYLEIGH beam. It is easy to conclude from the eigenfunctions

(213) that, for higher modes, the curvature of the beam increases, thereby increasing the influence of rotary inertia on the dynamics of the beam.

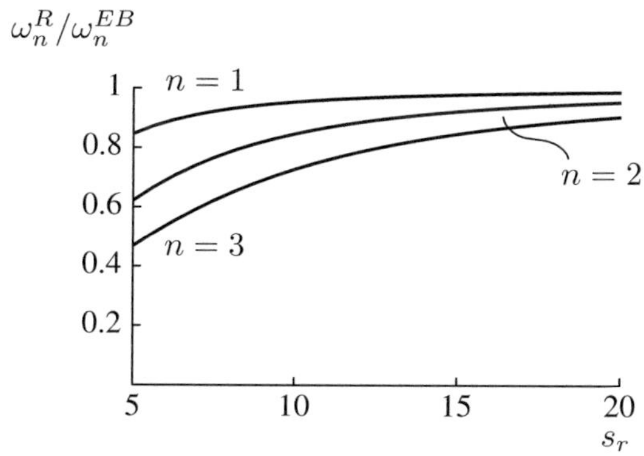

Figure 11: Variation of ω_n^R/ω_n^{EB} with slenderness ratio s_r for the first three natural frequencies

The ratio ω_n^R/ω_n^{EB} as a function of the slenderness ratio s_r is plotted in Fig. 11 for the first three modes. At low slenderness ratios, the frequency ratio is widely different for different modes. However, as the beam gets slender, the two frequencies tend to agree as can be observed from the figure. Further, for the lower modes, the effect of rotary inertia becomes more pronounced at low slenderness ratios.

(b) Uniform Cantilever Beam

Here we consider a uniform EULER-BERNOULLI cantilever beam for which the boundary conditions are given by

$$W(0) = 0, \quad W'(0) = 0, \quad W''(l) = 0, \quad \text{and} \quad W'''(l) = 0. \tag{215}$$

Substituting the solution form (206) in these boundary conditions yields

$$B_1 + B_3 = 0, \tag{216}$$

$$B_2 + B_4 = 0, \tag{217}$$

$$B_1 \cosh \beta l + B_2 \sinh \beta l - B_3 \cos \beta l - B_4 \sin \beta l = 0, \tag{218}$$

$$\text{and} \quad B_1 \sinh \beta l + B_2 \cosh \beta l + B_3 \sin \beta l - B_4 \cos \beta l = 0. \tag{219}$$

For a non-trivial solution of the $(B_1, \ldots, B_4)$, we must have

$$\begin{vmatrix} 1 & 0 & 1 & 0 \\ 0 & 1 & 0 & 1 \\ \cosh\beta l & \sinh\beta l & -\cos\beta l & -\sin\beta l \\ \sinh\beta l & \cosh\beta l & \sin\beta l & -\cos\beta l \end{vmatrix} = 0$$

$$\Rightarrow \quad \cos\beta l \cosh\beta l + 1 = 0, \tag{220}$$

which is the characteristic equation of a cantilever EULER-BERNOULLI beam.

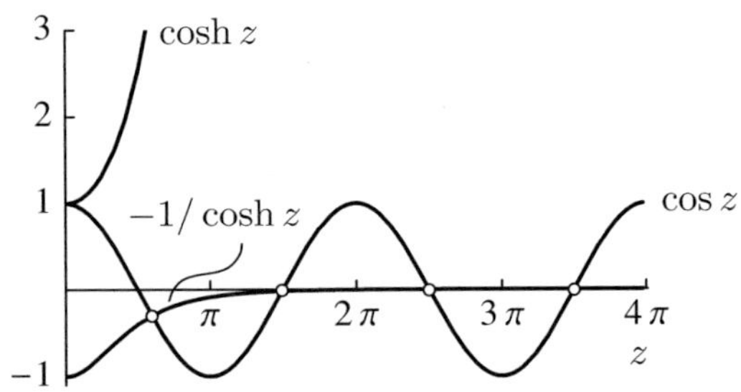

Figure 12: Graphical representation of the solutions of the characteristic equation of a cantilever beam

The solutions of the characteristic equation (220) are visualized graphically by circles in Fig. 12. It can be observed that the function $1/\cosh z$ converges to zero rapidly, and the characteristic equation (220) essentially reduces to $\cos\beta l = 0$ for higher modes. The analytical solution can be expressed in the form

$$\beta_n = \omega_n \sqrt{\frac{\rho A}{EI}} = \left(\frac{2n-1}{2}\pi + e_n\right)\frac{1}{l} \tag{221}$$

$$\Rightarrow \quad \omega_n = \left(\frac{2n-1}{2}\pi + e_n\right)^2 \frac{1}{l^2}\sqrt{\frac{EI}{\rho A}}, \qquad n = 1, 2, \ldots, \infty, \tag{222}$$

where e_n are small correction terms, and obtained as $e_1 = 0.3042$, $e_2 = -0.018$, $e_3 = 0.001, \ldots$. The corrections in the higer modes tend to zero rapidly, and can be neglected.

For values of β given by (221), a non-trivial solution of $(B_1, \ldots, B_4)$ can be determined from (216)-(219) by considering any three of the four equations. For example, using (216) and (217) we can eliminate B_3 and B_4 from (218) to obtain

$$B_1 = -\frac{\sinh\beta_n l + \sin\beta_n l}{\cosh\beta_n l + \cos\beta_n l} B_2 := \alpha_n B_2. \tag{223}$$

Therefore, taking $B_2 = 1$, one possible solution is given by

$$B_1 = \alpha_n, \qquad B_2 = 1, \qquad B_3 = -\alpha_n, \qquad \text{and} \qquad B_4 = -1, \tag{224}$$

which yields the nth eigenfunctions as

$$W_n(x) = \sinh\beta_n x - \sin\beta_n x - \left[\frac{\sinh\beta_n l + \sin\beta_n l}{\cosh\beta_n l + \cos\beta_n l}\right](\cosh\beta_n x - \cos\beta_n x). \tag{225}$$

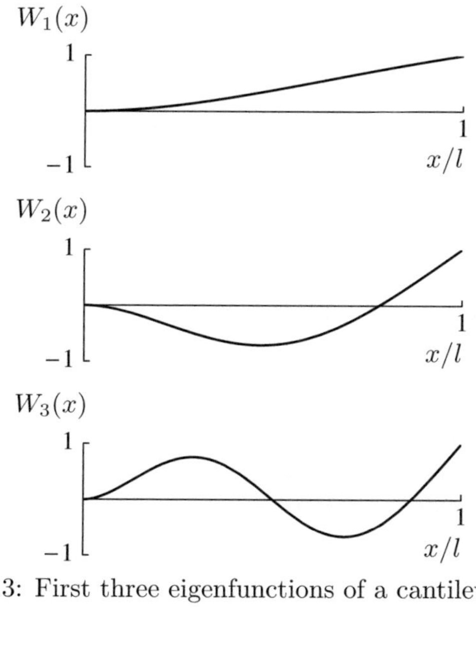

Figure 13: First three eigenfunctions of a cantilever beam

The first three eigenfunctions are shown in Fig. 13. These eigenfunctions satisfy the orthogonality condition (193) as can be checked.

(c) Uniform Free-Free Beam

Consider a free-free EULER-BERNOULLI beam. The boundary conditions in this case are zero moment and shear force at both ends of the beam. This implies

$$W''(0) = 0, \quad W'''(0) = 0, \quad W''(l) = 0, \quad \text{and} \quad W'''(l) = 0. \tag{226}$$

Using these boundary conditions in the shape function (206), we have

$$B_1 - B_3 = 0, \tag{227}$$

$$B_2 - B_4 = 0, \tag{228}$$

$$B_1 \cosh \beta l + B_2 \sinh \beta l - B_3 \cos \beta l - B_4 \sin \beta l = 0, \tag{229}$$

$$\text{and} \quad B_1 \sinh \beta l + B_2 \cosh \beta l + B_3 \sin \beta l - B_4 \cos \beta l = 0. \tag{230}$$

A non-trivial solution of the B_i is obtained if and only if

$$\begin{vmatrix} 1 & 0 & -1 & 0 \\ 0 & 1 & 0 & -1 \\ \cosh \beta l & \sinh \beta l & -\cos \beta l & -\sin \beta l \\ \sinh \beta l & \cosh \beta l & \sin \beta l & -\cos \beta l \end{vmatrix} = 0$$

$$\Rightarrow \quad \cos \beta l \cosh \beta l - 1 = 0. \tag{231}$$

The geometric visualization of the solution of the characteristic equation (231) is done by circles in Fig. 14. It can be observed that $\beta l = 0$ is a solution to (231). However, for $\beta = 0$, (205) is no longer the form of solution to the differential equation. This case, therefore, has to be considered separately. For higer modes, since $\cosh z$ is an exponentially divergent function, the characteristic equation can be approximated by $\cos \beta l = 0$. The solution of (231) can be represented in the form

$$\beta_n = \omega_n \sqrt{\frac{\rho A}{EI}} = \left(\frac{2n+1}{2} \pi + e_n \right) \frac{1}{l} \tag{232}$$

$$\Rightarrow \quad \omega_n = \left(\frac{2n+1}{2} \pi + e_n \right)^2 \frac{1}{l^2} \sqrt{\frac{EI}{\rho A}}, \qquad n = 1, 2, \ldots, \infty, \tag{233}$$

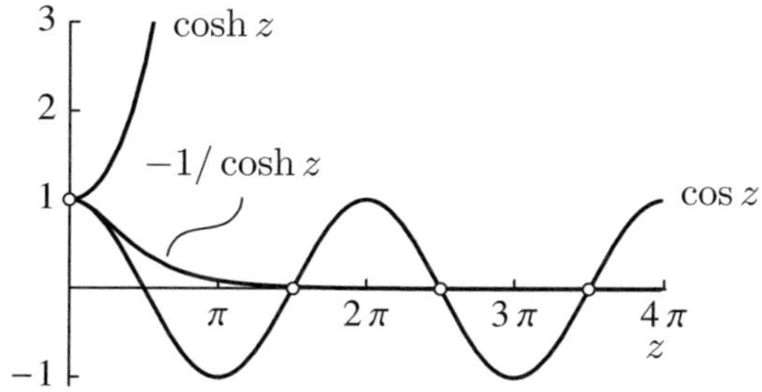

Figure 14: Graphical representation of the solutions of the characteristic equation of a free-free beam

where e_n are small correction terms. For example, $e_1 = 0.01766$, $e_2 = -0.00078, \ldots$. The correction for higher modes are negligibly small, and can be dropped for all practical purposes.

The eigenfunctions for the non-zero eigenfrequencies can be determined from (227)-(230). It can be easily checked that solving for B_1 from (227)-(229) yields

$$B_1 = \frac{-\sinh\beta_n l + \sin\beta_n l}{\cosh\beta_n l - \cos\beta_n l} B_2 := \alpha_n B_2. \tag{234}$$

Therefore, taking $B_2 = 1$, a possible solution is given by

$$B_1 = \alpha_n, \qquad B_2 = 1, \qquad B_3 = \alpha_n, \qquad \text{and} \qquad B_4 = 1, \tag{235}$$

which yields the nth eigenfunctions as

$$W_n(x) = \sinh\beta_n x + \sin\beta_n x + \left[\frac{-\sinh\beta_n l + \sin\beta_n l}{\cosh\beta_n l - \cos\beta_n l}\right](\cosh\beta_n x + \cos\beta_n x). \tag{236}$$

The first three eigenfunctions of the free-free beam are shown in Fig. 15.

Once again it can be checked that these eigenfunctions are orthogonal.

For the case $\beta = 0$ (*i.e.*, $\omega_n = 0$), (187) implies that

$$W'''' = 0$$

$$\Rightarrow \quad W(x) = B_1 + B_2 x + B_3 x^2 + B_4 x^3. \tag{237}$$

This solution corresponds to the rigid-body motion, and consists of translation and rotation of the beam. Using (226), we can easily obtain $B_3 = 0$,

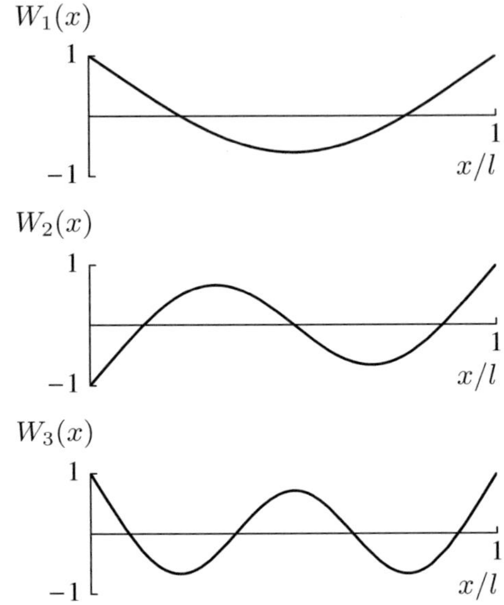

Figure 15: First three eigenfunctions of a free-free beam

and $B_4 = 0$. Then, the solution for the rigid-body motion for small time can be written as

$$w_0(x,t) = (K + Lt) + \left[\frac{x}{l} - \frac{1}{2}\right](G + Ht), \tag{238}$$

where K, L, G, and H are appropriate constants to be determined from the initial conditions. The complete solution of the free-free beam can now be written as

$$w(x,t) = (K+Lt)+\left[\frac{x}{l} - \frac{1}{2}\right](G+Ht)+\sum_{n=1}^{\infty}(C_n \cos\omega_n t + S_n \sin\omega_n t)W_n(x), \tag{239}$$

where $W_n(x)$ are given by (236).

► Approximate Methods

In the case of an arbitrary geometry of the beam, or in the presence of discrete elements, the exact modal analysis becomes at least difficult, and usually even impossible. In such situations, the approximate methods such as the Ritz and the Galerkin methods are useful.

In the both methods, we approximate the solutions of the variational problem (166) in the form

$$w(x,t) = \sum_{j=1}^{N} p_j(t)\psi_j(x) = \boldsymbol{\Psi}^T \mathbf{p}, \tag{240}$$

where $p_j(t)$ are the modal coordinates, and the $\psi_j(x)$ are suitably chosen shape-functions. In the RITZ method, for convergence, the shape-functions must satisfy all the geometric boundary conditions of the problem, and be differentiable at least up to the highest order of space derivative in the Lagrangian (*admissible functions*). Substituting (240) in (166), and following the procedure detailed in Hagedorn and DasGupta (2007), we obtain the discretized equations of motion as

$$\mathbf{M}\ddot{\mathbf{p}} + \mathbf{K}\mathbf{p} = \mathbf{0}, \tag{241}$$

where

$$\mathbf{M} = \int_0^l [\rho A \boldsymbol{\Psi}\boldsymbol{\Psi}^T + \rho I \boldsymbol{\Psi}(\boldsymbol{\Psi}')^T]\,\mathrm{d}x, \quad \text{and} \quad \mathbf{K} = \int_0^l EI\boldsymbol{\Psi}''(\boldsymbol{\Psi}'')^T\,\mathrm{d}x. \tag{242}$$

The approximate eigenfrequencies and eigenfunctions can now be obtained.

In the case of the GALERKIN method, for convergence, $\psi_j(x)$ must satisfy all the boundary conditions of the problem, and must be differentiable at least up to the highest derivative in the equation of motion (*comparison functions*). Substituting the solution form in the equation of motion, we obtain the discretized equations of motion (241). However, the definitions in this case are

$$\mathbf{M} = \int_0^l [\rho A \boldsymbol{\Psi}\boldsymbol{\Psi}^T - \boldsymbol{\Psi}(\rho I(\boldsymbol{\Psi}')^T)']\,\mathrm{d}x, \quad \text{and} \quad \mathbf{K} = \int_0^l \boldsymbol{\Psi}[(EI\boldsymbol{\Psi}'')^T]''\,\mathrm{d}x. \tag{243}$$

The Initial Value Problem The initial value problem for a beam is specified in terms of the initial position and velocity conditions as $w(x,0) = w_0(x)$ and $w_{,t}(x,0) = v_0(x)$. The solution of this problem can be conveniently represented as

$$w(x,t) = \sum_{j=1}^{\infty}(C_j \cos\omega_j t + S_j \sin\omega_j t)W_j(x), \tag{244}$$

where $W_j(x)$ are the eigenfunctions of the beam, and C_j and S_j are unknown constants which are to be determined from the initial conditions.

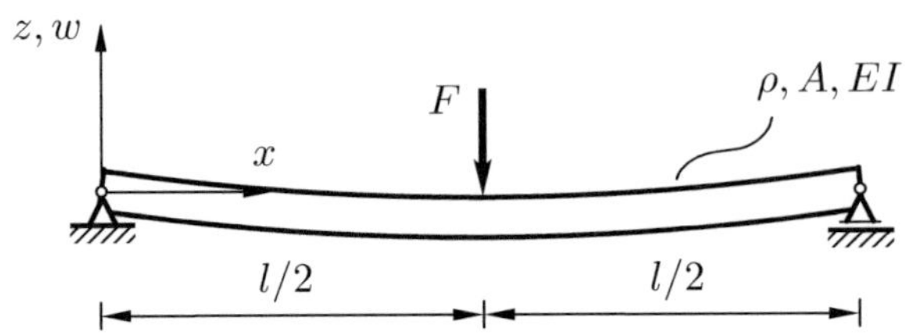

Figure 16: Beam with an initial constant point force

Consider a simply-supported beam of uniform cross-section with a concentrated force at the center, as shown in Fig. 16. If the force is suddenly removed, we have an initial value problem with an initial deflected shape $w_0(x)$, and zero initial velocity. The initial value problem can then be defined as

$$EIw_{,xxxx}(x,t) + \rho A w_{,tt} = 0, \qquad w(0,t) \equiv 0, \text{ and } w(l,t) \equiv 0, \tag{245}$$

with the initial conditions $w(x,0) = w_0(x)$, and $w_{,t}(x,0) = v_0(x) \equiv 0$. The initial deflected shape can be determined from the statics boundary value problem

$$EIw_{,xxxx}(x,0) = -F\delta(x - l/2), \qquad w(0,0) = 0, \text{ and } w(l,0) = 0. \tag{246}$$

From the solution of the eigenvalue problem of a simply-supported beam, it is already known that the eigenfunctions are given by $W_n(x) = \sin n\pi x/l$. Therefore, the solution of the initial value problem (245) can be sought using the expansion

$$w(x,t) = \sum_{n=1}^{\infty} (C_n \cos \omega_n t + S_n \sin \omega_n t) \sin \frac{n\pi x}{l}, \tag{247}$$

where C_n and S_n are unknown constants to be determined from the initial conditions

$$w(x,0) = w_0(x) = \sum_{n=1}^{\infty} C_n \sin \frac{n\pi x}{l}, \qquad \text{and} \qquad w_{,t}(x,0) = v_0(x) \equiv 0. \tag{248}$$

The initial shape $w_0(x)$ should satisfy (246). Therefore, substituting the first condition from (248) in (246), and taking the inner product of both sides with $\sin m\pi x/l$ yields on simplification

$$C_m = \begin{cases} \dfrac{2Fl^3}{m^4\pi^4 EI}(-1)^{(m-1)/2}, & m = 1, 3, 5, \ldots, \infty \\ 0, & m = 2, 4, 6, \ldots, \infty. \end{cases}$$

Using the initial condition on the velocity $w_{,t}(x, 0) \equiv 0$ one can easily obtain

$$S_m = 0, \qquad m = 1, 2, \ldots, \infty.$$

Thus, the solution of the initial value problem is of the form

$$w(x, t) = \sum_{n=1,3,5\ldots}^{\infty} \frac{2Fl^3}{n^4\pi^4 EI}(-1)^{(n-1)/2} \cos\omega_n t \sin\frac{n\pi x}{l}.$$

2.3 Forced Vibration Analysis

The general forced vibration problem for an EULER-BERNOULLI beam can be represented as

$$\rho A w_{,tt} + (EIw_{,xx})_{,xx} = q(x, t), \tag{249}$$

where $q(x, t)$ is a general forcing function. In the following, we will discuss some solution methods for (249).

Eigenfunction Expansion Method The solution of (249) can be written as

$$w(x, t) = w_H(x, t) + w_P(x, t), \tag{250}$$

where $w_H(x, t)$, and $w_P(x, t)$ are, respectively, the general solution to the homogeneous problem (*i.e.*, $q(x, t) = 0$), and a particular solution to the inhomogeneous problem. The homogeneous solution is already discussed in the previous section, and is of the form (244). Consider a particular solution in the form of the eigenfunction expansion

$$w_P(x, t) = \sum_{j=1}^{\infty} p_j(t) W_j(x), \tag{251}$$

where $W_j(x)$ are the eigenfunctions, and $p_j(t)$ are the corresponding modal coordinates. Using (251) in (250), and substituting (250) in (249) yields

$$\sum_{j=1}^{\infty} \rho A \ddot{p}_j W_j + (EIW_j'')'' p_j = q(x,t),$$

$$\text{or} \quad \sum_{j=1}^{\infty} \rho A [\ddot{p}_j + \omega_j^2 p_j] W_j = q(x,t), \qquad \text{(using (202))}. \tag{252}$$

Taking the inner product with $W_k(x)$ on both sides, and using the orthonormality condition (194), we get

$$\ddot{p}_k + \omega_k^2 p_k = f_k(t), \qquad k = 1, 2, \ldots, \infty. \tag{253}$$

where

$$f_k(t) = \int_0^l q(x,t) W_k(x) \, \mathrm{d}x.$$

Thus, (253) represents the modal dynamics of the forced EULER-BERNOULLI beam. These equations can be solved using standard techniques such as GREEN's function method, or the LAPLACE transform method, as illustrated in Hagedorn and DasGupta (2007). The complete solution is then obtained from (250) as

$$w(x,t) = \sum_{j=1}^{\infty} (C_j \cos \omega_j t + S_j \sin \omega_j t) W_j(x) + \sum_{j=1}^{\infty} p_j(t) W_j(x), \tag{254}$$

where C_j and S_j are the constants of integration to be determined from the initial conditions.

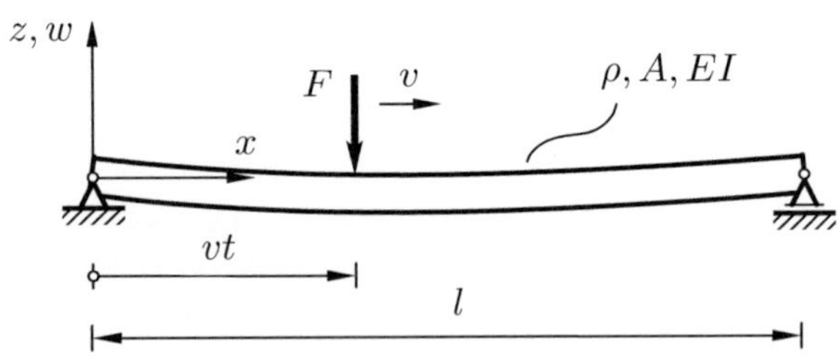

Figure 17: Beam with a traveling constant point force

Consider the case of a traveling force on a uniform EULER-BERNOULLI beam, as shown in Fig. 17. In this case, $q(x,t) = F\delta(x - vt)$, where v is the speed of travel of the constant force of magnitude F. Therefore, from (253), we have the equations of modal dynamics as

$$\ddot{p}_k + \omega_k^2 p_k = \frac{F}{\rho A}\sin\frac{k\pi vt}{l}, \qquad 0 \leq t \leq l/c, \qquad k = 1, 2, \ldots, \infty. \tag{255}$$

The solution of p_k is then obtained as

$$p_k = \frac{2Fl^3}{\pi^4 EI}\frac{1}{k^2(k^2 - \rho A l^2 v^2/\pi^2 EI)}, \qquad k = 1, 2, \ldots, \infty. \tag{256}$$

Using the initial conditions $w(x,0) \equiv 0$ and $w_{,t}(x,0) \equiv 0$ one obtains the complete solution for $0 \leq t \leq l/c$ as

$$w(x,t) = \frac{2Fl^3}{\pi^4 EI}\sum_{j=1}^{\infty}\frac{1}{j^2(j^2 - \rho A l^2 v^2/\pi^2 EI)}(\sin\frac{j\pi vt}{l} - \frac{j\pi v}{l\omega_j}\sin\omega_j t)\sin\frac{j\pi x}{l}. \tag{257}$$

The shape of the beam at certain selected time instants are shown in Fig. 18 for $v/l = \omega_1/4\pi$, and in Fig. 19 for $v/l = \omega_1\pi/4$.

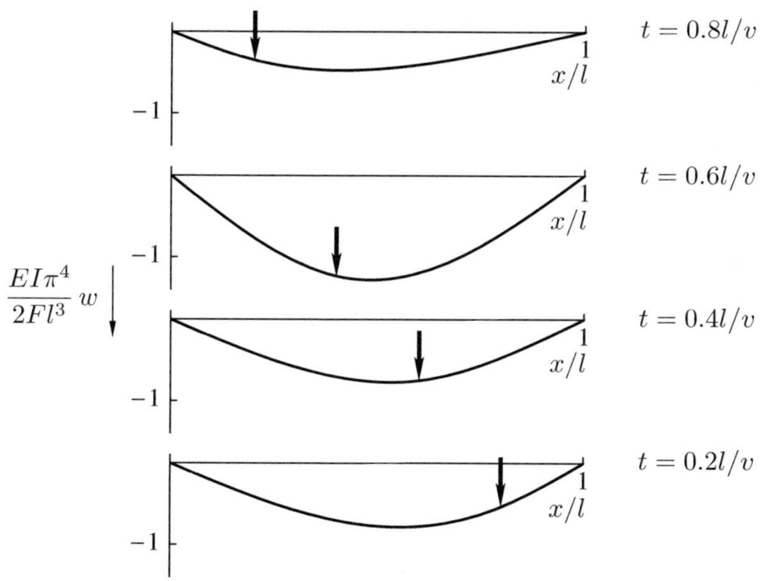

Figure 18: Response of a beam with constant force with $v/l = \omega_1/4\pi$

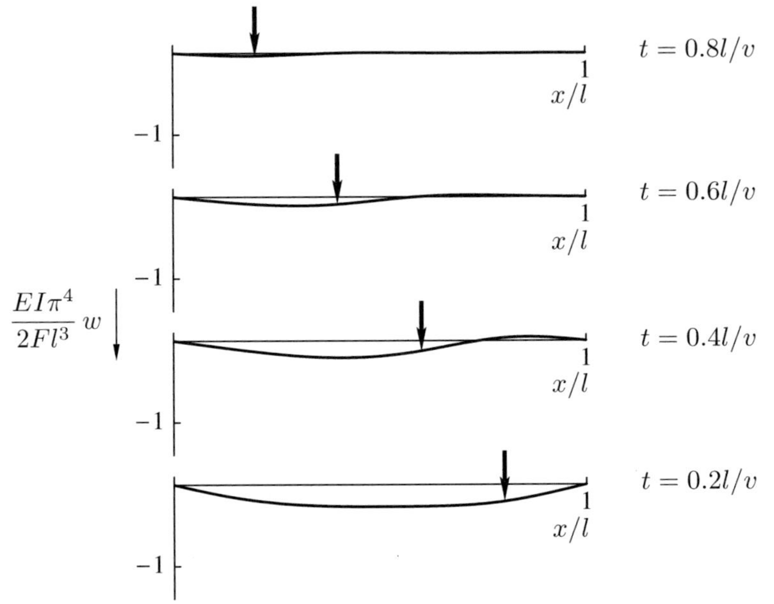

Figure 19: Response of a beam with constant force with $v/l = \omega_1\pi/4$

Approximate Methods The approximate methods of RITZ and GALERKIN can also be used for studying the forced motion of beams. Expressing the solution of (249) as (240), the discretized equations of motion are obtained in the form

$$\mathbf{M}\ddot{\mathbf{p}} + \mathbf{K}\mathbf{p} = \mathbf{f}(t),$$

where $\mathbf{M}$ and $\mathbf{K}$ are as defined in (242) (for the RITZ method), or (243) (for GALERKIN's method), and

$$\mathbf{f}(t) = \int_0^l \boldsymbol{\Psi}(x) q(x,t)\,\mathrm{d}x, \tag{258}$$

where $\boldsymbol{\Psi}(x)$ is the vector of admissible functions (for RITZ method), or comparison functions (for GALERKIN method).

2.4 Non-homogeneous Boundary Conditions

As discussed before, in the presence of non-homogeneous boundary conditions, we cannot use the expansion theorem for studying the dynamics of a continuous system. Moreover, generating the comparison functions for the GALERKIN method also becomes difficult. Here, we use the approach discussed in Section 2.7 to convert a non-homogeneous boundary condition to a homogeneous one, along with an appropriate forcing in the equation of motion.

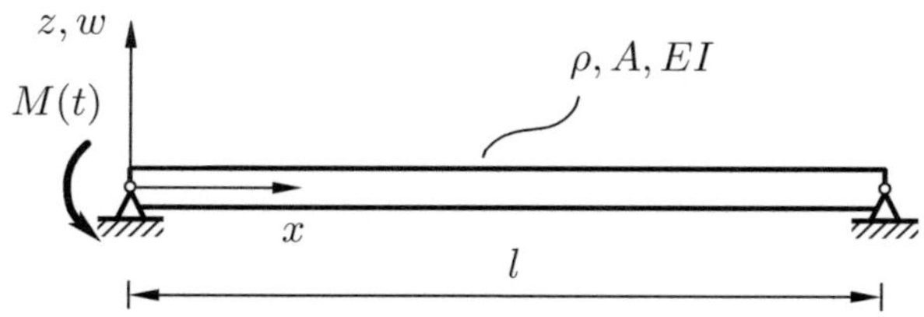

Figure 20: Beam with non-homogeneous boundary condition

Consider a simply-supported EULER-BERNOULLI beam with a specified time-varying moment $M(t)$ at one end, as shown in Fig. 20. The equation of motion and the boundary conditions are given by

$$\rho A w_{,tt} + EI w_{,xxxx} = 0, \tag{259}$$

$$w(0,t) \equiv 0, \quad w_{,xx}(0,t) = \frac{M(t)}{EI}, \quad w(l,t) \equiv 0, \quad \text{and} \quad w_{,xx}(l,t) \equiv 0. \tag{260}$$

Let us rewrite the field variable $w(x,t)$ as

$$w(x,t) = u(x,t) + \eta(x)\frac{M(t)}{EI}, \tag{261}$$

where $u(x,t)$ is a new field variable, and $\eta(x)$ is an unknown function. Substituting (261) in the equation of motion (259) yields

$$\rho A u_{,tt} + EI u_{,xxxx} = -\eta(x)\frac{\ddot{M}(t)}{EI} - \eta''''(x)\frac{M(t)}{EI}. \tag{262}$$

Next, substituting (261) in the boundary conditions (260), one obtains

$$u(0,t) + \eta(0)\frac{M(t)}{EI} = 0, \quad u_{,xx}(0,t) + \eta''(0)\frac{M(t)}{EI} = \frac{M(t)}{EI},$$

$$u(l,t) + \eta(l)\frac{M(t)}{EI} = 0, \quad u_{,xx}(0,t) + \eta''(l)\frac{M(t)}{EI} = 0.$$

We assume homogeneous boundary conditions for (262), *i.e.*, $u(0,t) \equiv 0$, $u_{,xx}(0,t) \equiv 0$, $u(l,t) \equiv 0$, and $u_{,xx}(l,t) \equiv 0$. Then, it is evident from the above that the function $\eta(x)$ must satisfy the conditions

$$\eta(0) = 0, \qquad \eta''(0) = 1, \qquad \eta(l) = 0, \qquad \text{and} \qquad \eta''(l) = 0. \tag{263}$$

Let us assume $\eta(x) = a_0 + a_1 x + a_2 x + a_3 x^3$. Substituting this form of $\eta(x)$ in (263), one can easily obtain

$$\eta(x) = \frac{lx}{6}\left(-\frac{x^2}{l^2} + 3\frac{x}{l} - 2\right).$$

This determines the right hand side of the transformed equation of motion (262) which now represents a simply-supported beam with forcing, and has homogeneous boundary conditions. The transformed problem can be solved easily for $u(x,t)$, and the solution of the original problem (259)-(260) is then obtained from (261). It must be mentioned that $\eta(x)$ is not a unique function. However, the solution of the original problem can be correctly determined by the above procedure.

2.5 An example of a hybrid system

A continuous system may interact with discrete elements as discussed in previous sections. For such hybrid systems, the modal analysis can be performed by analysing the system in parts along with appropriate matching conditions and boundary conditions for each of the parts. Often in these systems, the boundary conditions themselves involve ordinary differential equations, as will be evident in this section.

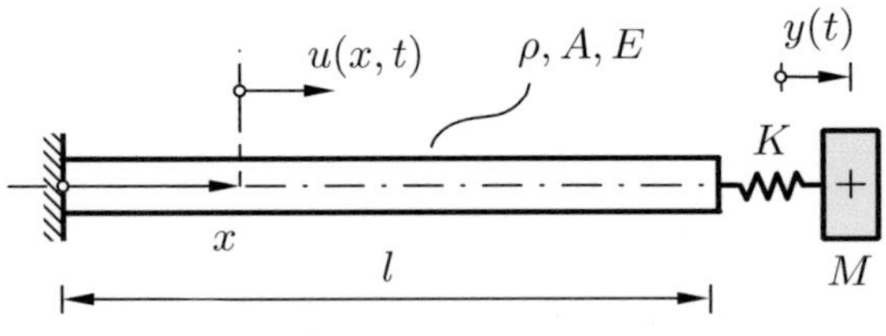

Figure 21: A hybrid system formed by a continuous sub-system and lumped elements

Let us consider the modal analysis of longitudinal vibrations of a bar with a mass-spring system at the right boundary, as shown in Fig. 21. This system can be described by one field variable $u(x,t)$ and one discrete variable $y(t)$. The equations of motion are

$$u_{,tt} - c^2 u_{,xx} = 0, \tag{264}$$

$$\text{and} \quad M\ddot{y} + Ky = Ku(l,t), \tag{265}$$

and the boundary conditions are given by

$$u(0,t) \equiv 0, \qquad \text{and} \qquad EAu_{,x}(l,t) \equiv K(y - u(l,t)). \tag{266}$$

As it is evident, the second boundary condition in (266) involves the ordinary differential equation (265).

Assume a modal solution of the form

$$\left\{ \begin{array}{c} u(x,t) \\ y(t) \end{array} \right\} = \left\{ \begin{array}{c} U(x) \\ Y \end{array} \right\} \mathrm{e}^{i\omega t}. \tag{267}$$

It may be noted that the modal vector for this problem is given by $(U(x), Y)^{\mathsf{T}}$. Substituting this solution in the equations of motion (264)-(265) and simplifying, we obtain the eigenvalue problem

$$U'' + \frac{\omega^2}{c^2} U = 0, \tag{268}$$

$$\text{and} \quad (-M\omega^2 + K)Y = KU(l), \tag{269}$$

with the associated boundary conditions given by (266) as

$$U(0) = 0, \tag{270}$$

$$\text{and} \quad EAU'(l) = K[Y - U(l)] = \frac{KM\omega^2}{K - M\omega^2} U(l) \quad , \tag{271}$$

using (269). Note here that the boundary condition (271) also involves the circular frequency ω. Assuming a solution of (268) in the form

$$U(x) = C \cos \frac{\omega x}{c} + S \sin \frac{\omega x}{c} \tag{272}$$

we have from the boundary conditions (270)-(271)

$$\begin{bmatrix} 1 & 0 \\ \left(\dfrac{KM\omega^2}{K - M\omega^2} \cos \dfrac{\omega l}{c} + \right. & \left(\dfrac{KM\omega^2}{K - M\omega^2} \sin \dfrac{\omega l}{c} - \right. \\ \left. \dfrac{EA\omega}{c} \sin \dfrac{\omega l}{c} \right) & \left. \dfrac{EA\omega}{c} \cos \dfrac{\omega l}{c} \right) \end{bmatrix} \left\{ \begin{array}{c} C \\ S \end{array} \right\} = 0. \tag{273}$$

The non-triviality of the solution of (C, S) implies that the determinant of the matrix in (273) must vanish, which yields the characteristic equation

$$\tan\frac{\omega l}{c} - \frac{EA(K - M\omega^2)}{c\omega MK} = 0.$$

This transcendental equation yields infinitely many circular eigenfrequencies ω_k, $k = 1, 2, \ldots, \infty$. Substituting these eigenfrequencies in (273), one obtains $(C_k, S_k) = (0, 1)$, and correspondingly

$$U_k(x) = \sin\frac{\omega_k x}{c},$$

so that, using (269), the eigenvectors are obtained as

$$\left\{\begin{array}{c} U_k(x) \\ \\ Y_k \end{array}\right\} = \left\{\begin{array}{c} \sin\dfrac{\omega_k x}{c} \\ \\ \dfrac{K\sin(\omega_k l/c)}{-M\omega_k^2 + K} \end{array}\right\}, \qquad k = 1, 2, \ldots, \infty.$$

It is to be noted that these vectors are formed by the displacement field $U_k(x)$ in the rod, and the discrete coordinate Y_k. They are a not vectors in two dimensional Euclidean space, but rather in an $(\infty + 1)$ dimensional space. Since these infinitely many eigenvectors are all linearly independent, one can conveniently express the solution of (264)-(265) using the expansion theorem as

$$\left\{\begin{array}{c} u(x,t) \\ y(t) \end{array}\right\} = \sum_{k=1}^{\infty} p_k(t) \left\{\begin{array}{c} U_k(x) \\ Y_k \end{array}\right\},$$

where $p_k(t)$ is the modal coordinate corresponding to mode k.

The orthogonality relation for the above eigenvectors are obtained from the procedure discussed in Hagedorn and DasGupta (2007) as follows. Consider the modes j and k which satisfy the following equations

$$U_j'' + \frac{\omega_j^2}{c^2} U_j = 0, \qquad Y_j = \frac{KU_j(l)}{-M\omega_j^2 + K}, \tag{274}$$

$$U_k'' + \frac{\omega_k^2}{c^2} U_k = 0, \qquad Y_k = \frac{KU_k(l)}{-M\omega_k^2 + K}, \tag{275}$$

along with appropriate boundary and matching conditions. Multiply the first equation in (274) by U_k and the first equation in (275) by U_j, and subtract the second product from the first and integrate over the length of the beam to obtain

$$\int_0^l \left(U_k U_j'' + \frac{\omega_j^2}{c^2} U_k U_j \right) \mathrm{d}x -$$

$$\int_0^l \left(U_j U_k'' + \frac{\omega_k^2}{c^2} U_j U_k \right) \mathrm{d}x = 0,$$

$$\Rightarrow \quad \int_0^l \left(U_k U_j'' - U_j U_k'' + \frac{\omega_j^2 - \omega_k^2}{c^2} U_j U_k \right) \mathrm{d}x = 0. \qquad (276)$$

Integrating by parts the first term in (276) twice, and using the boundary and matching conditions from (270)-(271) yield on simplification

$$(\omega_j^2 - \omega_k^2) \left[\frac{M}{EA} \left(\frac{KU_j(l)}{K - M\omega_j^2} \right) \left(\frac{KU_k(l)}{K - M\omega_k^2} \right) + \frac{1}{c^2} \int_0^l U_j U_k \, \mathrm{d}x \right] = 0$$

$$\Rightarrow \quad MY_jY_k + \rho A \int_0^l U_j U_k \, \mathrm{d}x = 0, \qquad \text{for } j \neq k,$$

where we have used (274) and (275). These are the orthogonality relations for the system.

2.6 Continuous Systems with Damping

All vibratory systems experience resistance to motion, commonly referred to as damping. Damping forces may arise from external interactions of the system (external damping), or from within the system (internal damping). Damping from aerodynamic drag due to viscosity is the most common example of external damping, while internal damping occurs due to internal friction between the molecular layers as a result of differential straining. In these damping mechanisms, mechanical energy is converted irreversibly into thermal energy which flows out of the system.

Three damping models, namely viscous damping, COULOMB damping (or dry friction), and structural damping (or hysteretic damping) are usually used for engineering purposes. The viscous damping model, which is the most commonly used model, relates the damping forces with the time rate of change of the field variable, or its spatial derivatives. We will use this model only in our discussions below.

Systems with Distributed Damping Consider the longitudinal oscillations of a uniform fixed-free bar. We assume that the internal damping in the material is such that the stresses are a linear function of both, the strain, and the strain rate. Thus, we have

$$\sigma_x(x,t) = E\epsilon_x(x,t) + d_I\epsilon_{x,t}(x,t) = Eu_{,x}(x,t) + d_I u_{,xt}(x,t), \tag{277}$$

where $d_I > 0$ is the coefficient of internal damping in the material. We also assume a distributed external damping force of the usual form $-d_E u_{,t}(x,t)$, where $d_E > 0$ is the coefficient of external damping. Then, proceeding similarly to what was done previously, one obtains the equation of motion of the longitudinal dynamics of a bar with internal and external damping as

$$\rho A u_{,tt} - EAu_{,xx} - d_I A u_{,xxt} + d_E u_{,t} = 0. \tag{278}$$

The boundary conditions are not affected by these damping terms. One can define a damping operator

$$\mathcal{D}[\cdot] = \left(-d_I A \frac{\mathrm{d}^2}{\mathrm{d}x^2} + d_E\right)[\cdot], \tag{279}$$

and represent (278) in a compact form as

$$\rho A u_{,tt} + \mathcal{D}[u_{,t}] + \mathcal{K}[u] = 0, \tag{280}$$

where $\mathcal{K}[\cdot] = -EA[\cdot]_{,xx}$.

Multiplying both sides of (278) by $u_{,t}$ and integrating over the domain of the bar yields

$$\int_0^l (\rho A u_{,t}u_{,tt} - u_{,t}EAu_{,xx} - u_{,t}d_I A u_{,xxt} + d_E u_{,t}^2)\,\mathrm{d}x = 0,$$

$$\Rightarrow \quad \left[u_{,t}EAu_{,x} + u_{,t}d_I A u_{,xt}\right]_0^l +$$

$$\int_0^l \left[\left(\frac{1}{2}\rho A u_{,t}^2\right)_{,t} + u_{,xt}EAu_{,x} + d_I A u_{,xt}^2 + d_E u_{,t}^2\right]\mathrm{d}x = 0. \tag{281}$$

Using the fixed-free boundary conditions, one can rewrite (281) as

$$\frac{\mathrm{d}}{\mathrm{d}t}\int_0^l \left(\frac{1}{2}\rho A u_{,t}^2 + \frac{1}{2}EAu_{,x}^2\right)\mathrm{d}x = -\int_0^l (d_I A u_{,xt}^2 + d_E u_{,t}^2)\,\mathrm{d}x. \tag{282}$$

The integral on the left hand side in (282) can be easily recognized to be the total mechanical energy of the bar. Since the right hand side is always negative, (282) implies that the time rate of change of mechanical energy of the bar is always negative, *i.e.*, mechanical energy monotonically decreases with time.

Consider now a system represented by

$$\mu(x)u_{,tt} + \mathcal{D}[u_{,t}] + \mathcal{K}[u] = 0. \tag{283}$$

We explore the possibility of a solution of (283) in the form

$$u(x,t) = \sum_{k=1}^{\infty} p_k(t) U_k(x), \tag{284}$$

where the eigenfunctions $U_k(x)$ are taken to be the same as for the undamped case, *i.e.*, they are solutions of the self-adjoint eigenvalue problem

$$-\lambda\mu(x)U + \mathcal{K}[U] = 0, \tag{285}$$

with appropriate boundary conditions. We will assume that the eigenfunctions are orthonormal with respect to $\mu(x)$, *i.e.*, $\langle \mu(x)U_j, U_k\rangle = \delta_{jk}$. Substituting (284) in (283) and taking inner product with $U_j(x)$ yields

$$\ddot{p}_j + \sum_{k=1}^{\infty} d_{jk}\dot{p}_k + \lambda_j p_j = 0, \qquad j = 1, 2, \ldots, \infty \tag{286}$$

where

$$d_{jk} = \langle \mu(x)(-d_I A U_{k,xx} + d_E U_k), U_j\rangle. \tag{287}$$

It is evident that, in general, the damping matrix $\mathbf{D} = [d_{jk}]$ will not be diagonal. Therefore, all the modes of the system are coupled through $\mathbf{D}$, and (286) cannot be solved exactly in closed form.

Consider the special situation when

$$\mathcal{D}[U_k(x)] = d_k \mu(x) U_k(x), \tag{288}$$

where d_k are constants. Then, it can be easily checked that the resulting damping matrix $\mathbf{D}$ is diagonal. It can be observed that (288) represents an eigenvalue problem for the damping operator similar to (285). It then follows that if the operators $\mathcal{D}[\cdot]$ and $\mathcal{K}[\cdot]$ have the same eigenfunctions, the resulting damping matrix $\mathbf{D}$ is diagonal. We can determine the condition for the two operators to have the same eigenfunctions as follows. From (288), one can write

$$\begin{aligned}\mathcal{K}[\mu^{-1}(x)\mathcal{D}[U_k(x)]] &= \mathcal{K}[d_k U_k(x)] \\ &= d_k \lambda_k U_k(x) \qquad \text{(using (285))}.\end{aligned} \tag{289}$$

Similarly, from (285), it follows that

$$\begin{aligned}\mathcal{D}[\mu^{-1}(x)\mathcal{K}[U_k(x)]] &= \mathcal{D}[\lambda_k U_k(x)] \\ &= \lambda_k d_k U_k(x) \qquad \text{(using (288)).} \qquad (290)\end{aligned}$$

From (289) and (290), we can conclude that when $\mathcal{K}[\cdot]$ and $\mathcal{D}[\cdot]$ have the same eigenfunctions they satisfy

$$\mathcal{K}[\mu^{-1}(x)\mathcal{D}[U_k]] - \mathcal{D}[\mu^{-1}(x)\mathcal{K}[U_k]] = 0, \qquad k = 1, 2, \ldots, \infty$$

$$\Rightarrow \quad (\mathcal{K}[\mu^{-1}(x)\mathcal{D}] - \mathcal{D}[\mu^{-1}(x)\mathcal{K}])[\cdot] = 0, \qquad (291)$$

i.e., the two operators commute with respect to $\mu^{-1}(x)$. The converse of this result can also be easily established. Let the two operators commute, *i.e.*, (291) is satisfied. From (285), one can easily obtain

$$-\lambda\mathcal{D}[U] + \mathcal{D}[\mu^{-1}(x)\mathcal{K}[U]] = 0$$

$$\Rightarrow \quad -\lambda\mathcal{D}[U] + \mathcal{K}[\mu^{-1}(x)\mathcal{D}[U]] = 0 \qquad \text{(using (291),}$$

$$or \quad -\lambda\mu(x)V + \mathcal{K}[V] = 0, \qquad (292)$$

where

$$V = \mu^{-1}(x)\mathcal{D}[U]. \qquad (293)$$

It is evident that if V satisfies (292), in view of (285) it must be true that $V = \beta U$ for some constant factor β. Hence, from (293) we have

$$\mathcal{D}[U] = \beta\mu(x)U,$$

i.e., U must also be an eigenfunction of the damping operator $\mathcal{D}[\cdot]$. Therefore, (291) is the necessary and sufficient condition for $\mathcal{K}[\cdot]$ and $\mathcal{D}[\cdot]$ to have the same eigenfunctions, and hence for the damping matrix $\mathbf{D}$ to be diagonal. It is not difficult to show that the condition (291) implies that the operator $\mathcal{K}[\mu^{-1}(x)\mathcal{D}[\cdot]]$ is self-adjoint.

One clear advantage obtained if $\mathcal{D}[\cdot]$ satisfies (291) is that the discretized equations of motion are completely decoupled when the solution of the damped system is expanded in terms of the eigenfunctions of the undamped system. This decoupling allows us to solve the discretized equations in an easy manner. One special choice of the damping operator for which the commutation holds is

$$\mathcal{D}[\cdot] = \beta\mu(x) + \gamma\mathcal{K}[\cdot], \qquad (294)$$

where β and γ are arbitrary constants. Such a damping is usually known as *classical damping* or *proportional damping.* The condition (294) is satisfied in the case of the damped bar described by (278). Therefore, the differential equation for the jth modal coordinate of the bar is given by

$$\ddot{p}_j + d_j\dot{p}_j + \lambda_j p_j = 0. \tag{295}$$

which can be easily solved for $p_j(t)$. Finally, the complete solution of the longitudinal vibration of the bar is obtained from (284).

Systems with Discrete Damping In many practical situations, a continuous system may interact with discrete damping elements. For example, certain support points of a structure may provide substantially higher damping to the structure than its internal damping. In that case, the damping can be considered to be due to discrete dampers at such support points. Discrete damper elements are also routinely attached to structures for vibration control. Here we consider two specific cases, and discuss the effects of discrete damping.

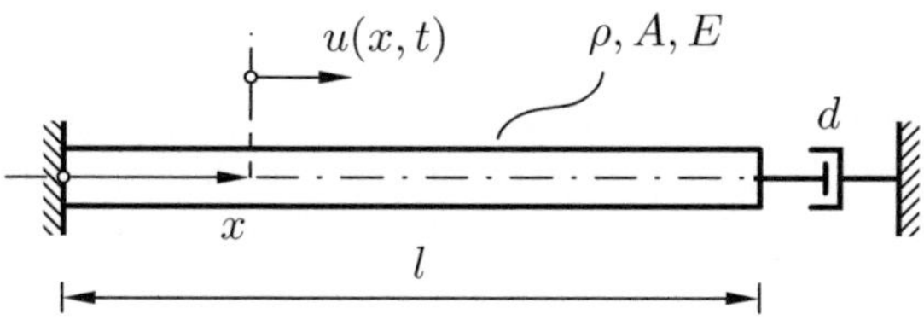

Figure 22: A uniform bar with boundary damping

Consider a uniform bar fixed at one end, and having an external damper at the other end, as shown in Fig. 22. The equation of motion can be written as

$$u_{,tt} - c^2 u_{,xx} = 0, \tag{296}$$

while the boundary conditions are

$$u(0,t) = 0, \qquad \text{and} \qquad EAu_{,x}(l,t) = -du_{,t}(l,t). \tag{297}$$

Assuming a solution form

$$u(x,t) = U(x)\mathrm{e}^{st}, \tag{298}$$

we obtain the eigenvalue problem

$$U'' - \frac{s^2}{c^2}U = 0, \tag{299}$$

$$\text{with} \quad U(0) = 0, \qquad \text{and} \qquad U'(l) = -\frac{sd}{EA}U(l). \tag{300}$$

Consider the general solution of (299) in the form

$$U(x) = B\mathrm{e}^{sx/c} + C\mathrm{e}^{-sx/c}, \tag{301}$$

where B and C are constants of integration. Substituting this solution in the boundary conditions (300) yields on simplification

$$\begin{bmatrix} 1 & 1 \\ \mathrm{e}^{\gamma}(1+a) & -\mathrm{e}^{-\gamma}(1-a) \end{bmatrix} \begin{Bmatrix} B \\ C \end{Bmatrix} = 0, \tag{302}$$

where $\gamma = sl/c$, and $a = cd/EA$. The condition of non-triviality of the solution of (302) yields the characteristic equation as

$$\mathrm{e}^{2\gamma} = \frac{a-1}{a+1}, \tag{303}$$

which can be solved for γ, and hence, the eigenvalues s of the system for $a \neq 1$. When $a = 1$, which occurs for the special value of boundary damping $d = EA/c$, it is observed from (303) that no eigenvalue exists. In this case, there is no solution of the assumed form (298).

When $a \neq 1$, one can rewrite (303) using the definition $\gamma := \alpha + i\beta$ as

$$\begin{aligned} \mathrm{e}^{2(\alpha+i\beta)} &= \frac{a-1}{a+1}, \\ \Rightarrow \ \alpha &= \frac{1}{2}\ln\left|\frac{a-1}{a+1}\right|, \\ \text{and} \ \beta_k &= \begin{cases} (2k-1)\pi/2, & 0 \leq a < 1 \\ k\pi, & a > 1 \end{cases} \qquad k = 1, 2, \ldots, \infty. \end{aligned}$$

It can be easily checked that, when $d = 0$, this gives the eigenvalues of a fixed-free bar, while $d \to \infty$ yields the eigenvalues of a fixed-fixed bar. It is surprising to note that all the modes have the same decay rate since α does not depend on k. Further, the transition in the imaginary part of the eigenvalues is discrete as a crosses unity. The locus of an eigenvalue with a as the parameter is depicted in Fig. 23.

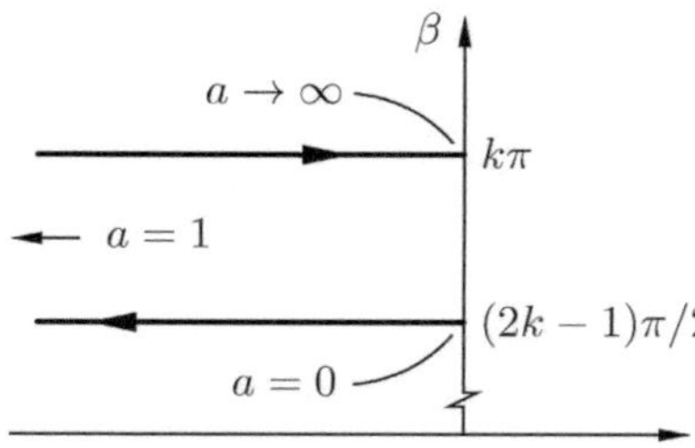

Figure 23: Locus of an eigenvalue with a as a parameter for a bar with boundary damping

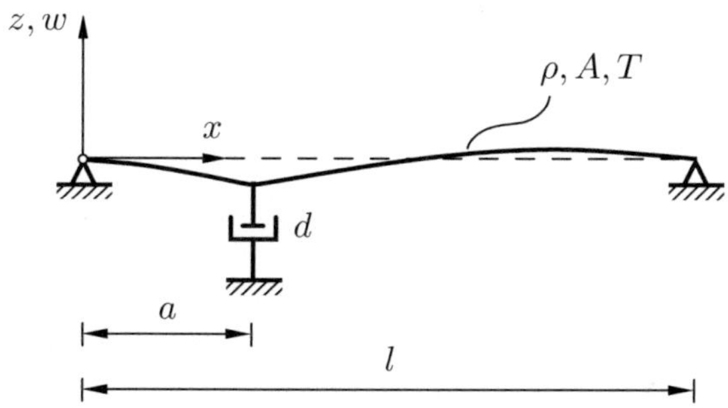

Figure 24: A string with discrete damping

Consider next the case of a taut string with a discrete external damper, as shown in Fig. 24. The equation of motion of the system can be written as

$$\rho A w_{,tt} + d w_{,t} \delta(x - x_d) - T w_{,xx} = 0, \tag{304}$$

where x_d is the location of the damper. Let us expand the solution in terms of the eigenfunctions of an undamped string as

$$w(x,t) = \sum_{k=1}^{\infty} p_k(t) \sin \frac{k\pi x}{l}. \tag{305}$$

Substituting this solution in (304) and taking the inner product with $\sin j\pi x/l$ yields the jth modal coordinate equation as

$$\ddot{p}_j + \sum_{k=1}^{\infty} \left(\frac{d}{\rho A} \sin \frac{k\pi x_d}{l} \sin \frac{j\pi x_d}{l} \right) \dot{p}_k + \frac{T}{\rho A} p_j = 0. \tag{306}$$

It may be observed here that the damping matrix is positive semi-definite with rank one. Further, it couples all the modes of the undamped system. When x_d is chosen such that jx_d/l is never an integer for any j, it can be shown that all the modes are damped. In other words, the total mechanical energy of the string decreases monotonically in time. In this case, the damping is called *pervasive*. Such a damper location is most desirable when we want to damp any arbitrary string motion. In the case, where jx_d/l is an integer for some j, the damping is not pervasive, and certain modes remain undamped since one of the nodes of such modes is at x_d. For example, if $x_d = l/3$, the 3rd, $6^{th}, \ldots$ modes will remain undamped.

2.7 An example of non-homogeneous Boundary Conditions

In all the preceding discussions, the boundary conditions were assumed to be homogeneous. However, there are situations where they are not. Non-homogeneity in boundary conditions occur when either a motion or a force is prescribed at a boundary.

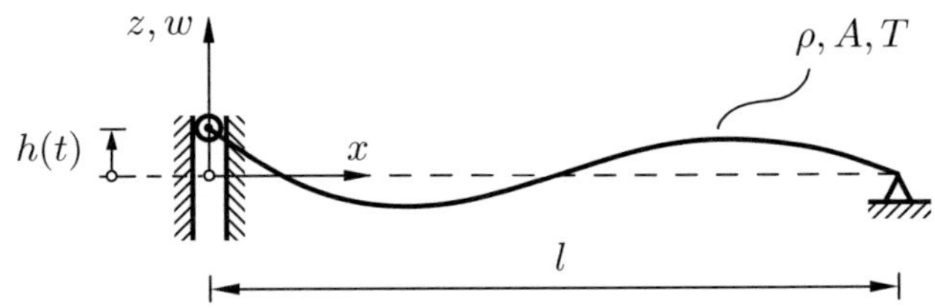

Figure 25: A string with a specified boundary motion

Consider a sliding-fixed string with a specified motion at the left boundary, as shown in Fig. 25. The equation of motion and boundary conditions can be represented as

$$w_{,tt} - c^2 w_{,xx} = 0, \tag{307}$$

$$w(0,t) = h(t), \qquad \text{and} \qquad w(l,t) \equiv 0, \tag{308}$$

where $h(t)$ is an arbitrary function of time. For such non-homogeneous boundary conditions, the solution cannot be directly expanded in a series of eigenfunctions of a problem with homogeneous boundary conditions. However, the methods of integral transforms (such as LAPLACE transforms) may still be applicable. Alternatively, one may also convert a problem with non-homogeneous boundary conditions to an equivalent problem with homogeneous boundary conditions and an appropriate forcing in the equation of motion to take care of the boundary non-homogeneity. Once this is done, the modal expansion method becomes applicable. In the following, we shall pursue this method.

For the problem (307)-(308), let

$$w(x,t) = u(x,t) + h(t)\eta(x), \tag{309}$$

where $u(x,t)$ and $\eta(x)$ are unknown functions. Substituting this form in the boundary conditions (308), we have

$$w(0,t) = u(0,t) + h(t)\eta(0) = h(t), \qquad \text{and} \qquad w(l,t) = u(l,t) + h(t)\eta(l) = 0.$$

If we let

$$u(0,t) \equiv 0 \qquad \text{and} \qquad u(l,t) \equiv 0, \tag{310}$$

then $\eta(x)$ must be chosen such that $\eta(0) = 1$, and $\eta(l) = 0$. The simplest choice is then $\eta(x) = 1 - x/l$. Therefore, from (309),

$$w(x,t) = u(x,t) + h(t)\left(1 - \frac{x}{l}\right).$$

Substituting this in (307), one can write the equation of motion of the string using the field variable $u(x,t)$ as

$$u_{,tt} - c^2 u_{,xx} = -\left(1 - \frac{x}{l}\right)\ddot{h}(t),$$

along with the homogeneous boundary conditions (310). This transformed problem can be easily identified as a fixed-fixed string with distributed forcing, and can be solved using the modal expansion method.

3 Liapounov Stability Theory

3.1 The Concept of Liapounov Stability

The concept of the stability of an equilibrium is somewhat familiar from elementary mechanics. It is known, for example, that in a system whose mechanical energy is conserved - that is, in a 'conservative system' - an equilibrium position corresponding to a minimum of the potential energy is a stable equilibrium position. This is schematically represented in Figure 26, where the 'frictionless' motion of a particle under the influence of gravity on a given curve $y(x)$ in the vertical xy-plane is depicted.

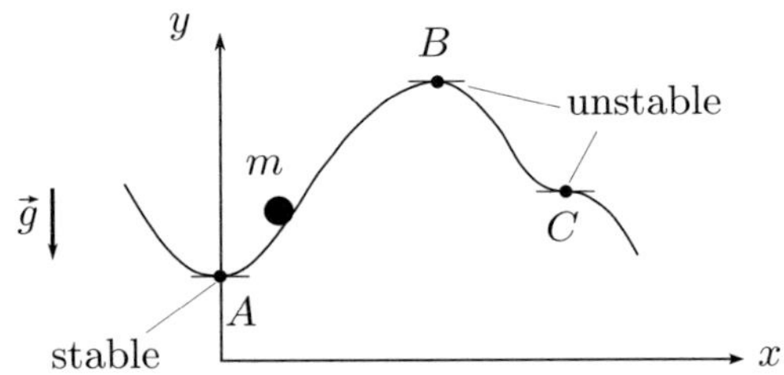

Figure 26: The concept of stability of an equilibrium position

Equilibrium positions are located at all points where the curve has a horizontal tangent, that is where dy/dx is equal to zero. The point A (a relative minimum of the potential energy) corresponds to a stable equilibrium position; the points B (a relative maximum of the potential energy) and C (a point of inflection with horizontal tangent) correspond to unstable equilibrium positions. These considerations intuitively make good sense in spite of the fact that no definition of stability (and of instability) has been given yet.

The theorem implying the stability of an equilibrium for a minimum of the potential energy may already be found in Lagrange's *Mecanique analytique* (1788); a rigorous proof was given later by Dirichlet. The stability definition on which the theorem was based was quite similar to the modern definition as it will shortly be given here. In fact, the Lagrange-Dirichlet stability theorem is often used as the definition of stability in elastomechanics: There, an equilibrium position is called stable when it corresponds to a relative minimum of the potential energy; otherwise, it is called unstable. This approach is unsatisfactory for several reasons which will be discussed in more detail later. Dirichlet's proof of the instability theorem then provided the inspiration for the Russian engineer
A. M. Liapounov to develop his stability theory (especially his 'direct

method') towards the end of the 19th century. Today, the stability concept which he introduced plays an important role in the theory of ordinary differential equations. In the following, this concept will be used unless specifically stated otherwise.

In order to give a clear presentation of general results, it is advisable to use matrix notation and to write all differential equations as first-order systems in the form

$$\dot{\boldsymbol{x}} = \boldsymbol{f}(\boldsymbol{x}, t) \tag{311}$$

where $\dot{\boldsymbol{x}}^{\mathsf{T}} = (\dot{x}_1, \dot{x}_2, \ldots, \dot{x}_n)$, $\boldsymbol{f}^{\mathsf{T}} = (f_1, f_2, \ldots, f_n)$. It is assumed that conditions sufficient to guarantee the existence and uniqueness of the solutions of (311) on their domain of definition are satisfied for $t \in [t_0, \infty)$[13]. A solution of (311) thus is uniquely determined by its initial conditions; one generally denotes this by $\boldsymbol{x} = \boldsymbol{x}(\boldsymbol{x}_0, t_0; t)$ with $\boldsymbol{x}(\boldsymbol{x}_0, t_0; t_0) = \boldsymbol{x}_0$.

Consider now a specific solution $\boldsymbol{x}(\boldsymbol{a}, \boldsymbol{t}_0; t)$. Then $\boldsymbol{x}(\boldsymbol{a}, \boldsymbol{t}_0; t)$ is said to be stable, if the difference $|\boldsymbol{x}(\boldsymbol{x}_0, t_0; t) - \boldsymbol{x}(\boldsymbol{a}, t_0; t)|$ remains smaller than an arbitrarily given small ϵ, for all time, as long as $\boldsymbol{x}_0$ is chosen sufficiently close to $\boldsymbol{a}$. More precisely: $\boldsymbol{x}(\boldsymbol{a}, t_0; t)$ *is stable, if for every (arbitrarily small)* $\epsilon > 0$ *there exists a* $\delta(\epsilon) > 0$ *such that*[14]

$$|\boldsymbol{x}_0 - \boldsymbol{a}| < \delta(\epsilon) \implies |\boldsymbol{x}(\boldsymbol{x}_0, t_0; t) - \boldsymbol{x}(\boldsymbol{a}, t_0; t)| < \epsilon, \ \forall t \geq t_0. \tag{312}$$

The solution $\boldsymbol{x}(\boldsymbol{x}_0, t_0; t)$ thus remains in an arbitrarily thin 'tube' about $\boldsymbol{x}(\boldsymbol{a}, t_0; t)$ in the 'augmented state space' ($\mathbb{R}^{n+1}$ corresponding to $x_1, x_2, \ldots, x_n, t$) for all time, as long as $\boldsymbol{x}_0$ is chosen sufficiently close to a (Figure 27). In this notation the vertical bars, for example, may refer to the Euclidean norm $|\boldsymbol{x}| = \sqrt{(x_1^2 + x_2^2 + \cdots + x_n^2)}$ or to any other norm, such as $|\boldsymbol{x}| = |\boldsymbol{x}_1| + |\boldsymbol{x}_2| + \cdots + |\boldsymbol{x}_n|$ (in *finite*-dimensional spaces all norms are equivalent in the sense that 'smallness' with respect to any of them implies 'smallness' with respect to all others). Within this definition it thus makes no sense to use terms such as 'stable system' or 'stable differential equation', since one and the same differential equation may have stable as well as unstable solutions (linear differential equations are an exception). *A solution* $\boldsymbol{x}(\boldsymbol{a}, t_0; t)$ *of* (311) *is called unstable whenever it is not stable*; in the case of instability, there always exists some $\epsilon > 0$, and some $\boldsymbol{x}_0$ in an arbitrarily small neighborhood of a such that $\boldsymbol{x}(\boldsymbol{x}_0, t_0; t)$ will leave the ϵ-tube for some $t > t_0$ (by now it is apparent that stability is nothing more than *uniformly* continuous dependence on the initial conditions).

[13] $t \in [t_1, t_2)$ corresponds to $t_1 \leq t < t_2$.

[14] $a \implies b$ means that the statement a implies the statement b; $\forall t \geq$ means 'for all values of t greater than or equal to t_0'.

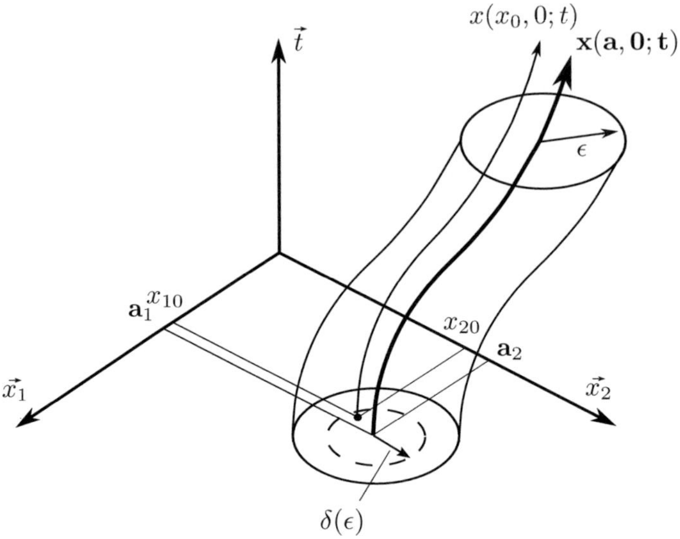

Figure 27: The LIAPOUNOV stability definition

As an example, consider a brief investigation of the stability of various solutions of the governing differential equation for the mathematical pendulum

$$\ddot{x} + \omega_0^2 \sin x = 0 \tag{313}$$

which may also be written in the form

$$\begin{aligned} \dot{x}_1 &= \omega_0 x_2, \\ \dot{x}_2 &= -\omega_0 \sin x_1. \end{aligned} \tag{314}$$

A brief scrutiny of the phase diagram, Figure 28, indicates that $\boldsymbol{x} \equiv \mathbf{0}$ (the lower equilibrium position of the pendulum) is a stable solution. For this example, the function $\delta(\epsilon)$ may be constructed in accordance with Figure 28. For a given $\epsilon > 0$ a phase trajectory contained entirely within the circle $|\epsilon| = \epsilon$ is chosen; a possible choice for δ then is the radius of any circle contained entirely within this phase trajectory.

The instability of the solution $\boldsymbol{x}^{\mathsf{T}} \equiv (\pi, 0)$ is equally obvious: in every arbitrarily small neighborhood of this point in the phase plane, there always exist initial conditions leading to solutions which may 'move far away from this point'. Now, what about the stability of any of the periodic solutions

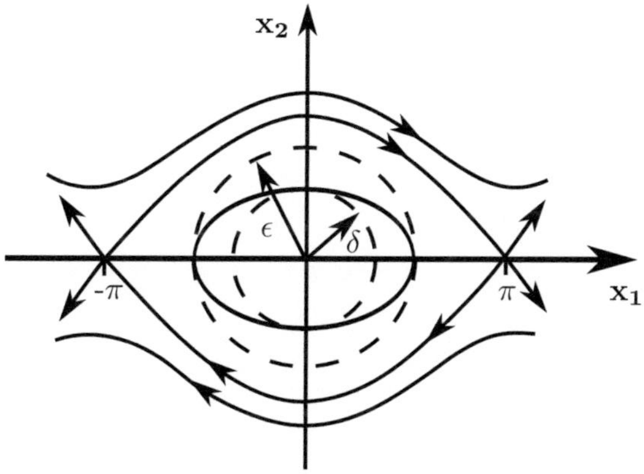

Figure 28: Stability of the static equilibrium of the pendulum

which correspond to the closed phase trajectories? All of the periodic solutions of (314) are unstable! This is apparent from Figure 29. Consider, for example, the stability of the solution which passes through the point A for $t = 0$. A small change in the initial conditions from A to B, obviously in the nonlinear system produces a slight change not only in the amplitude of oscillation but also in the period of oscillation.

The point on the phase trajectory which passes through B moves around more slowly, that is, for some time $t > 0$ it will happen that the points of the two motions occupy positions which are diametrically opposed to one another with respect to **o**; then the distance between these two points is $|CD|$. It is *not* possible to make this distance arbitrarily small by choosing B sufficiently close to A! From this, one concludes instability. For similar reasons, the motion of an artificial satellite about the earth is unstable since the square of the time of revolution is proportional to the third power of the length of the semi-major axis, in accordance with KEPLER's third law. A small change in the initial conditions may result in only a small change in the time of revolution; however, every change in the time of revolution, no matter how small, means that the 'disturbed' and 'undisturbed' motion will eventually be a large distance from one another in a sufficiently large time interval. In contrast hereto, all of the solutions of the linearized system (314),

$$\begin{aligned} \dot{x}_1 &= \omega_0 x_2, \\ \dot{x}_2 &= -\omega_0 x_1, \end{aligned} \tag{315}$$

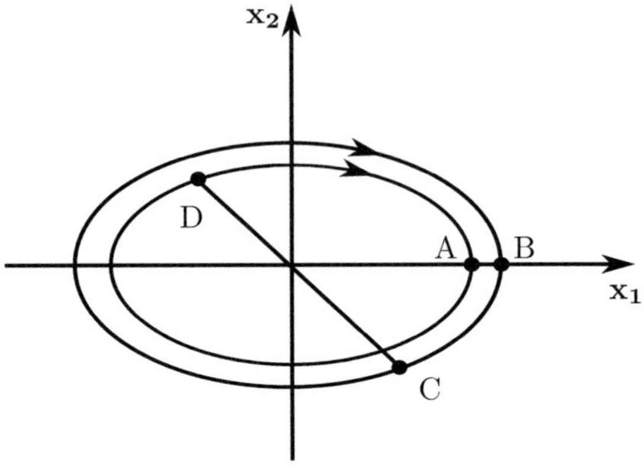

Figure 29: The stability of the periodic solutions of (314)

are stable.

Thus, in spite of the fact that the new (disturbed) phase trajectory corresponding to a small change in the initial conditions remained everywhere close to the original phase trajectory (Figure 29) it turned out that the periodic solutions of (314) were unstable. This phenomenon is present in many of the applications from mathematical physics. Thus, although a motion may be unstable in the LIAPOUNOV sense, it is useful to further distinguish that case for which the phase trajectories remain close to each other. This is the purpose of the concept of a stable trajectory or of orbital stability:

The solution $\boldsymbol{x}(\boldsymbol{a}, t_0; t)$ has a stable trajectory (or is orbitally stable) if for every (arbitrarily small) $\epsilon > 0$ there exist a $\delta(\epsilon) > 0$ a function $t_1(t)$ such that

$$|\boldsymbol{x}_0 - \boldsymbol{a}| < \delta \implies |\boldsymbol{x}(\boldsymbol{x}_0, t_0; t) - \boldsymbol{x}(\boldsymbol{a}, t_0; t_1)| < \epsilon, \forall t \geq t_0. \tag{316}$$

Expressed differently, the 'ϵ-tube' about $\boldsymbol{x}(\boldsymbol{a}, t_0; t)$ is now introduced in the phase space (in $\mathbb{R}$ as implied by $x_1, x_2, \ldots, x_n$) not in the augmented state space as before. If for every $\epsilon > 0$ there exists a δ-sphere about a such that all solutions which begin in this sphere at $t = t_0$, never leave this tube, then

$\boldsymbol{x}(\boldsymbol{a}, t_0; t)$ is orbitally stable[15]. Obviously, the periodic oscillations of the pendulum are orbitally stable.

A solution $\boldsymbol{x}(\boldsymbol{a}, t_0; t)$ is attractive if there exists an $\epsilon > 0$ such that

$$|\boldsymbol{x}_0 - \boldsymbol{a}| < \epsilon \implies \lim_{t \to \infty} |\boldsymbol{x}(\boldsymbol{x}_0, t_0; t) - \boldsymbol{x}(\boldsymbol{a}, t_0; t_1)| = 0. \tag{317}$$

A solution which is both stable and attractive is called asymptotically stable. It may very well be that a solution $\boldsymbol{x}(\boldsymbol{a}, t_0; t)$ is attractive without being stable (see Hahn 1967, p. 191 ff).

For the investigation of the stability of $\boldsymbol{x}(\boldsymbol{a}, t_0; t)$, it is often useful to introduce the coordinate transformation

$$\boldsymbol{y} = \boldsymbol{x} - \boldsymbol{x}(\boldsymbol{a}, t_0; t). \tag{318}$$

The differential equation (311) then takes on the form

$$\begin{aligned} \dot{\boldsymbol{y}} &= \dot{\boldsymbol{x}} - \dot{\boldsymbol{x}}(\boldsymbol{a}, t_0; t) = \boldsymbol{f}(\boldsymbol{x}) - \boldsymbol{f}(\boldsymbol{a}, t_0; t) = \\ &= \boldsymbol{f}(\boldsymbol{y} + \boldsymbol{x}(\boldsymbol{a}, t_0; t)) - \boldsymbol{f}(\boldsymbol{x}(\boldsymbol{a}, t_0; t)), \end{aligned} \tag{319}$$

that is,

$$\dot{\boldsymbol{y}} = \boldsymbol{g}(\boldsymbol{y}, t) \tag{320}$$

since $\boldsymbol{x}(\boldsymbol{a}, t_0; t)$ is assumed known. The solution $\boldsymbol{x}(\boldsymbol{a}, t_0; t)$ of (311) now corresponds to the trivial solution $\boldsymbol{y} = \boldsymbol{0}$ of (320) and the stability of this solution corresponds exactly to that of $\boldsymbol{x}(\boldsymbol{a}, t_0; t)$.

As an example, consider again the stability of a periodic solution $\boldsymbol{x}^*(t) = \boldsymbol{x}(\boldsymbol{a}, t_0; t)$ of (314). With $\boldsymbol{y} = \boldsymbol{x} - \boldsymbol{x}^*$ (314) becomes

$$\begin{aligned} \dot{y}_1 &= \omega_0 y_2, \\ \dot{y}_2 &= -\omega_0 (x_1^* + y_1) + \omega_0 \sin x_1^*; \end{aligned} \tag{321}$$

the system (321), however, is no longer autonomous but contains the time t explicitly! For the investigation of the motions in a neighborhood of the trivial solution $\boldsymbol{y} = \boldsymbol{0}$ of (320), it is often useful to develop the right-hand side in a Taylor series with respect to y, where the coefficients corresponding to the solution $\boldsymbol{x}^*$ are periodic in t. The periodic pendulum oscillations from (321) then become

$$\begin{aligned} \dot{y}_1 &= \omega_0 y_2, \\ \dot{y}_2 &= -\omega_0 y_1 \cos x_1^* + \frac{\omega_0}{2} y_1^2 \sin x_1^* + \dots . \end{aligned} \tag{322}$$

[15]In analytical mechanics, there also are other definitions of 'orbital' stability which turn out to be useful.

Another example which will be used a number of times concerns the free rotation of a rigid body supported by a frictionless ball-joint at its center of gravity. Let A, B, and C be the three principal moments of inertia and let p, q, and r be the projections of the angular velocity ω onto the corresponding principal axes; then the EULER equations yield the system

$$\begin{aligned} A\dot{p} &= (B-C)qr, \\ B\dot{q} &= (C-A)rp, \\ C\dot{r} &= (A-B)pq \end{aligned} \tag{323}$$

with the particular solutions $p = \omega$, $q = r = 0$; $p = 0$, $q = \omega$, $r = 0$; $p = q = 0$, $r = \omega$, where ω is an arbitrary constant in each case. Each of these solutions corresponds to the rotation of the rigid body about one of the three principal axes with angular velocity ω. To investigate the stability of the first solution, it is useful to introduce the transformation $y_1 = p - w$, $y_2 = q$, $y_3 = r$, and to write (323) in the form

$$\begin{aligned} A\dot{y}_1 &= \frac{B-C}{A} y_2 y_3, \\ B\dot{y}_2 &= \frac{C-A}{B}\left(y_1+\omega\right) y_3, \\ C\dot{y}_3 &= \frac{A-B}{C}\left(y_1+\omega\right) y_2. \end{aligned} \tag{324}$$

The trivial solution $y_1 = 0$, $y_2 = 0$, $y_3 = 0$ of (324) then corresponds to the solution $p = w$, $q = r = 0$ of (323).

The transformation which was carried out in both examples always makes it possible to reduce the concept of the stability of a motion to a treatment of the special case of the stability of an equilibrium position (or of a 'critical point'). Although the definition of the stability of a motion naturally includes that of the stability of an equilibrium position, the latter will be repeated here. Assume that the differential equation

$$\dot{\boldsymbol{y}} = \boldsymbol{g}\left(\boldsymbol{y}, t\right) \tag{325}$$

has the trivial solution $y = 0$, that is $\boldsymbol{g}(\mathbf{0}, t) \equiv \mathbf{0}$.

This trivial solution is stable if for every $\epsilon > 0$ there exists a $\delta(\epsilon) > 0$ such that

$$|\boldsymbol{y}_0| < \delta(\epsilon) \implies |\boldsymbol{y}\left(\boldsymbol{y}_0, t_0; t\right)| < \epsilon, \forall t \geq t_0. \tag{326}$$

Figure 27 now is replaced by Figure 30.

Other than the LIAPOUNOV stability definition there are a number of sometimes related concepts which - as was seen in the case of orbital stability

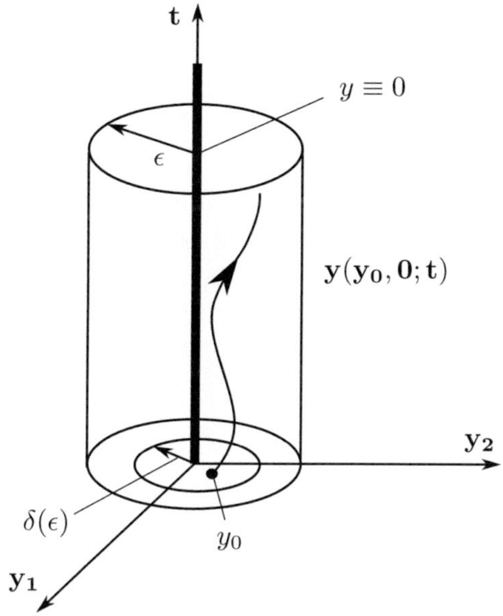

Figure 30: The stability of the trivial solution of (325)

- often play a significant role in applications. For example, an important question concerns the influence of additional 'small' terms on the right-hand side of (311) on the behavior of the solutions. The LIAPOUNOV definition allows only for disturbances in the initial conditions without producing any change in the differential equation itself. Although extensive investigations thereof are available, nothing more about this topic will be discussed here (see Hahn 1952, for example).

In applications, it is sometimes a disadvantage that LIAPOUNOV theory deals with *infinitesimal* disturbances. Thus, it may happen that a solution is L-unstable ('unstable in the sense of LIAPOUNOV') but that it may be considered to be 'stable' for all practical purposes. In this context, consider the differential equation

$$\dot{x} = x\left(a^2 - x^2\right), \tag{327}$$

where a is a constant. It has the trivial solution $x = 0$ and two further critical points $x = a$ and $x = -a$. All other solutions $x(x_0, t_0; t)$ are given by

$$a^2 (t - t_0) = \ln \left\{ \frac{x}{x_0} \sqrt{\left(\frac{a^2 - x_0^2}{a^2 - x^2} \right)} \right\} \tag{328}$$

(Figure 31). The solution $x = 0$ is unstable whereas the solutions $x = +|a|$, $x = -|a|$ are stable (even asymptotically stable). If a is a very small number, however, then $x = 0$ may still be called *practically stable* for certain applications.

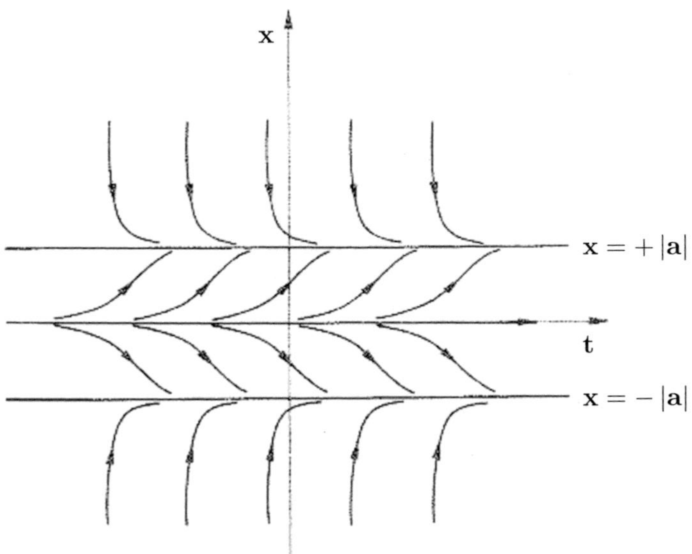

Figure 31: The stability of the solutions of (327)

Outside of the fact that a solution is stable, it thus is also important to know how far the 'disturbed' solutions will separate from the motion being investigated. Naturally, one may similarly have *practical instability* simultaneous with LIAPOUNOV stability. In Figure 32, the example of a particle is again used to illustrate this possibility. Just what constitutes practical stability or practical instability in an actual situation naturally depends on the orders of magnitude involved in the technical or physical problem under investigation.

The stability behavior of any given solution of (311) may be determined without difficulty, if the general solution is known. However, for non-linear

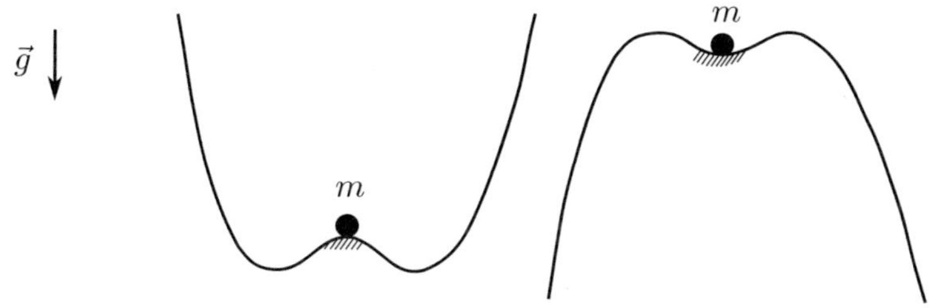

Figure 32: The concept of practical stability. left: L-unstable, eventually 'practically stable'; right: L-stable, eventually 'practically unstable'.

differential equations this is almost never the case; with few exceptions, one generally knows only certain particular, usually stationary or periodic solutions whose stability then is of interest[16]. It thus becomes necessary to search for means to clarify the stability behavior without solving the differential equations. Generally, the transformation (318) is introduced and one then has available at least two different methods for attacking the problem, both of which were developed by LIAPOUNOV. The right-hand side of eqn (320) may be developed in a Taylor series with respect to $\boldsymbol{y}$,

$$\dot{\boldsymbol{y}} = \boldsymbol{A}(t)\boldsymbol{y} + \boldsymbol{n}(\boldsymbol{y}, t), \tag{329}$$

where $\boldsymbol{n}(\boldsymbol{y}, t)$ stands for all terms of higher than first order, that is, for all of the non-linear terms in eqn (320). Since a general solution of (329) is usually not available, one is led to try to deduce results from the linearized differential equation

$$\dot{\boldsymbol{y}} = \boldsymbol{A}(t)\boldsymbol{y}. \tag{330}$$

It is a good deal easier to investigate the stability of the trivial solution of (330) rather than the stability of the solutions $\boldsymbol{y} = \boldsymbol{0}$ of (320) since the theory of linear differential equations is more complete. LIAPOUNOV was the first to obtain conditions subject to which the stability results obtained for (330) are also valid for eqn (320); this result is also called the method of first approximation. More generally, LIAPOUNOV*'s first method* consists of the

[16]Such stationary or periodic solutions often provide the basis for extensive investigations about the general character of the solutions. As noted by Poincaré, they are thus of great importance: 'Ce qui nous rend ces solutions périodiques si précieuses, c'est qu'elles sont, pour ainsi dire, la seule brèche par où nous puissons essayer de pénétrer dans une place jusqu'ici réputée inabordable.'

direct investigation of the solutions of (329). Usually, such a solution must be sought in series form. Another approach is given by what is now called LIAPOUNOV*'s second* or *direct method.* It is based on a direct estimate of the solutions of eqn (320). Consider the example:

$$\begin{aligned} \dot{y}_1 &= -y_2 + a y_1^3, \\ \dot{y}_2 &= y_1 + a y_2^3, \end{aligned} \tag{331}$$

where a is a constant. The stability of the trivial solution $\boldsymbol{y} = \boldsymbol{0}$ may be investigated, even though the general solution of equations (331) is not immediately available. Toward this purpose, the time derivative of $\frac{1}{2}|\boldsymbol{y}(t)|^2$ for an (unknown) solution $\boldsymbol{y}(t)$ of (331) is computed. The result is

$$\begin{aligned} \frac{1}{2}\frac{\mathrm{d}}{\mathrm{d}t}\left(y_1^2(t) + y_2^2(t)\right) &= y_1(t)\dot{y}_1(t) + y_2(t)\dot{y}_2(t) \\ &= y_1(t)\left\{-y_2(t) + a y_1^3(t)\right\} + y_2(t)\left\{y_1(t) + a y_2^3(t)\right\} \\ &= a\left(y_1^4(t) + y_2^4(t)\right). \end{aligned} \tag{332}$$

For $a > 0$ the norm of all of the (non-trivial) solutions grows beyond all bounds so that the solution $\boldsymbol{y} = 0$ is unstable; for $a < 0$, however, it is asymptotically stable. For $a = 0$, $\boldsymbol{y} = \boldsymbol{0}$ is stable but not asymptotically stable (this is also called *weakly stable*).

If one considers the linearized problem

$$\begin{aligned} \dot{y}_1 &= -y_2, \\ \dot{y}_2 &= y_1 \end{aligned} \tag{333}$$

with the general solution

$$\begin{aligned} y_1(t) &= y_{10}\cos t - y_{20}\sin t, \\ y_2(t) &= y_{10}\sin t + y_{20}\cos t, \end{aligned} \tag{334}$$

then it is apparent that the corresponding trivial solution is always weakly stable (eqns (333) are those of the linear oscillator). Thus, the linearized equations here yield no valid information concerning the stability of the non-linear system. In many other cases, however, the deductions for the linear system are equally valid for the complete non-linear system.

3.2 The Direct Method of LIAPOUNOV

The direct method may be used to investigate the stability of the trivial solution of

$$\dot{\boldsymbol{x}} = \boldsymbol{f}(\boldsymbol{x}, t) \tag{335}$$

without knowing the general solution or even an approximation thereof. For this purpose, one needs functions $V(\boldsymbol{x},t) : \mathbb{R}^{n+1} \to \mathbb{R}$, $V(\boldsymbol{x},t) \in C^1$ (in other words: V is a mapping from the $(n+1)$-dimensional real Euclidean space into the reals; V has continuous first partial derivatives with respect to all of its variables), with the property $V(0,t) \equiv 0$, defined in a neighborhood of the point $\boldsymbol{x} = \mathbf{o}$ in phase space for all $t \geq t_0$.

First, *time-independent* functions of this type will be considered. In the following, some properties of $V(\boldsymbol{x})$ will be used frequently and it is convenient to list the corresponding definitions.

The function $V(\boldsymbol{x})$ is positive definite iff there exists an $h > 0$ such that $V(\boldsymbol{x}) > 0$ for $0 < |\boldsymbol{x}| \leq h$.

The function $V(\boldsymbol{x})$ is positive semi-definite iff there exists an $h > 0$ such that $V(\boldsymbol{x}) \geq 0$ for $0 < |\boldsymbol{x}| \leq h$ (in every arbitrarily small neighborhood of $\boldsymbol{x} = 0$ there may be points such that $V = 0$).

Negative definiteness is analogously defined.

The function $V(\boldsymbol{x})$ is indefinite iff it is neither definite nor semi-definite. In that case, every arbitrarily small neighborhood of $\boldsymbol{x} = 0$ contains points for which V takes on positive values, as well as points for which V takes on negative values.

Consider some examples for n = 3:

$V(\boldsymbol{x}) = x_1^2 + x_2^2 + x_3^4$	positive definite (h is arbitrary);
$V(\boldsymbol{x}) = x_1^2 + 2x_1x_2 + 3x_2^2 + x_3^2$	positive definite (h is arbitrary);
$V(\boldsymbol{x}) = x_1^2 + x_2^2 + x_3^2 - x_3^3$	positive definite (for sufficiently small h);
$V(\boldsymbol{x}) = x_1^2 + x_2^2 + 2x_1x_2 + x_3^2$	positive semi-definite;
$V(\boldsymbol{x}) = x_1^2 + x_2^2$	positive semi-definite;
$V(\boldsymbol{x}) = x_1$	indefinite; and
$V(\boldsymbol{x}) = x_1^2 + x_2^2 - x_3^4$	indefinite.

To check the definiteness or indefiniteness of a function, one would like to have as general criteria as possible. Unfortunately, criteria exist for only fairly simple functions $V(\boldsymbol{x})$.

If one has

$$V(\lambda \boldsymbol{x}) = \lambda^m V(\boldsymbol{x}) \tag{336}$$

for arbitrary λ and $\boldsymbol{x}$, then $V(\boldsymbol{x})$ is called a 'form of order m'. It is immediately clear that a form is definite or indefinite on all of $\mathbb{R}^n$ as long as the same is valid in only a neighborhood of $x = 0$. Furthermore, a form with odd order certainly is always indefinite, whereas a form of even order may be definite, indefinite, or semi-definite. A simple criterion exists only for quadratic forms which are, of course, given by

$$V(\boldsymbol{x}) = \frac{1}{2}\boldsymbol{x}^\mathsf{T}\boldsymbol{A}\boldsymbol{x}, \qquad \boldsymbol{A}^\mathsf{T} = \boldsymbol{A} = (a_{ij})\,. \tag{337}$$

Here, a theorem due to SYLVESTER applies: *A necessary and sufficient condition for the positive definiteness of the form* (337) *is*

$$a_{11} > 0, \begin{vmatrix} a_{11} & a_{12} \\ a_{21} & a_{22} \end{vmatrix} > 0, \ldots, \begin{vmatrix} a_{11} & a_{12} & \ldots & a_{1n} \\ a_{21} & a_{22} & \ldots & a_{2n} \\ \vdots & & & \vdots \\ a_{n1} & a_{n2} & \ldots & a_{nn} \end{vmatrix} > 0.$$

This theorem is proven in Bellman (1969), for example.

Now, let $V(\boldsymbol{x})$ be a positive definite (indefinite) form of order m. Then it is easy to prove the following theorem:

There exists a number $a > 0$, depending only on $V(\boldsymbol{x})$, such that

$$V(\boldsymbol{x}) + W(\boldsymbol{x}) \tag{338}$$

is also positive definite (indefinite), where $W(\boldsymbol{x})$ is an arbitrary function $\mathbb{R}^n \to \mathbb{R}$, which satisfies the inequality

$$|W(\boldsymbol{x})| < a|\boldsymbol{x}|^m \mathit{in}\ 0 < |\boldsymbol{x}| \leq h \tag{339}$$

and for which $W(\mathbf{0}) = 0$.

The proof will be given only for the definite case. For this purpose, introduce $U = V + W$ and note that

$$U(\boldsymbol{x}) = |\boldsymbol{x}|^m V\left(\frac{\boldsymbol{x}}{|\boldsymbol{x}|}\right) \geq a|\boldsymbol{x}|^m + W(\boldsymbol{x}) \tag{340}$$

where $a = \min_{|\boldsymbol{x}|=1} V(\boldsymbol{x})$; from condition (339) it then follows that $U = V + W$ is also definite.

This theorem has the immediate corollary: *If $V(\boldsymbol{x})$ is a definite (indefinite) form, then the same is true for*

$$U(\boldsymbol{x}) = V(\boldsymbol{x}) + W(\boldsymbol{x})$$

provided $W(\boldsymbol{x})$ is a form of the same order with sufficiently small coefficients. Finally, consider an arbitrary function $V(\boldsymbol{x})$ which has a Taylor series expansion about $\boldsymbol{x} = \boldsymbol{0}$. Write this expansion as the sum

$$V(\boldsymbol{x}) = V_m(\boldsymbol{x}) + V^*(\boldsymbol{x}) \tag{341}$$

where $V_m(\boldsymbol{x})$ is a form of order m representing the non-vanishing terms of lowest order in the expansion of V, whereas all of the terms of higher order are contained in $V^*(\boldsymbol{x})$. Then one may consider $V^*(\boldsymbol{x})$ itself as an mth order form whose coefficients are themselves continuous functions of $\boldsymbol{x}$ vanishing for $\boldsymbol{x} = \boldsymbol{0}$. From this then follows the theorem:

If V_m is definite (indefinite), then $V = V_m + V^*$ is also definite (indefinite).

The definiteness or indefiniteness of V thus depends on the terms of lowest order in the expansion of V, *if these do not consist of a semi-definite form.*

Consider some examples for n = 2:

$V(\boldsymbol{x}) = x_1^2 + x_2^2 + x_1x_2^2 + x_2^3$ positive definite;

$V(\boldsymbol{x}) = x_1^2 - x_2^2 + x_1x_2^2 + x_2^3$ indefinite;

$V(\boldsymbol{x}) = x_1^2$ positive semi-definite;

$V(\boldsymbol{x}) = x_1^2 - 2x_1x_2^2$ indefinite;

$V(\boldsymbol{x}) = x_1^2 - 2x_1x_2^2 + x_2^4 + x_1^4 = \left(x_1 - x_2^2\right)^2 + x_1^4$ positive definite;

$V(\boldsymbol{x}) = x_1^2 - 2x_1x_2^2 + x_2^4 + x_1^4 + x_1x_2^5$ indefinite.

On the parabola $x_1 = x_2^2$ the last of the functions $V(\boldsymbol{x})$ above takes on values which are given by $V = x_2^7 + x_2^8$ and it changes sign on this curve at the origin of coordinates. It thus is apparent that the addition of terms of higher order may eliminate the definiteness or indefiniteness of a function $V(\boldsymbol{x})$, if $V(\boldsymbol{x})$ is not a form.

The theorem about the definiteness of $V = V_m + V^*$ still is valid even if one does not assume that V^* begins with the terms of order $m + 1$, but instead demands that there exist numbers $a > 0$, $\alpha > 0$ such that

$$|V(\boldsymbol{x})| < a|\boldsymbol{x}|^{m+\alpha}$$

holds in a neighborhood $|\boldsymbol{x}| \leq h$ of $\boldsymbol{x} = \boldsymbol{0}$.

If $V(\boldsymbol{x})$ is positive definite, then, for sufficiently small c, $V(\boldsymbol{x}) = c$ corresponds to a one-parameter family of hypersurfaces which contain the point $\boldsymbol{x} = \mathbf{0}$ in their interior (see Figure 33).

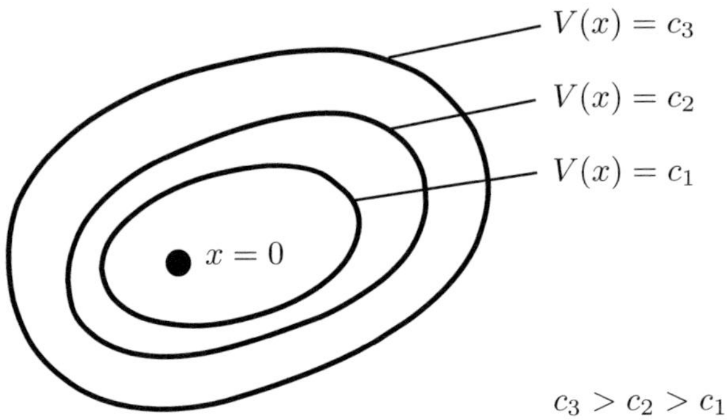

Figure 33: The geometric interpretation of the positive definite function $V(\boldsymbol{x})$.

Until now, just functions depending only on $\boldsymbol{x}$ and not on t have been considered. Now, a dependence on t is also admitted[17].

The function $V(\boldsymbol{x}, t)$ *is positive definite if there exists a positive definite function* $W(\boldsymbol{x})$, *independent of* t, *such that the inequality* $V(\boldsymbol{x}, t) \geq W(\boldsymbol{x})$, $\forall t \geq t_0$ *is satisfied for* $|\boldsymbol{x}| < h$..

The function $V(\boldsymbol{x}, t) = \mathrm{e}^{-t}|\boldsymbol{x}|^2$ is not positive definite, since it tends to zero for $t \to \infty$; the function $V(\boldsymbol{x}, t) = (2 + \sin t)\,|\boldsymbol{x}|^2$, however, is positive definite. The geometric interpretation of a positive definite function $V(\boldsymbol{x}, t)$ is obvious.

The function $V(\boldsymbol{x}, t)$ is positive semi-definite iff it takes on no negative values in the neighborhood $|\boldsymbol{x}| < h$, for sufficiently small h, and for $t \geq t_0$. t_0. Thus, semi-definiteness here is defined in the same manner as in the time-independent case; the same holds for indefiniteness. The definitions for negative definiteness and of negative semi-definiteness are analogous.

A further definition which was unnecessary in the time-independent case now is needed:

[17]With this it then is possible to cite the theorems for the general case where the time appears explicitly on the right-hand side of (335). The examples, however, will generally be restricted to the treatment of autonomous (that is, time-independent) problems. Simple, non-autonomous problems are treated in the exercises.

The function $V(\boldsymbol{x}, t)$ *is uniformly small if there exists a positive definite function* $W(\boldsymbol{x})$ *such that the inequality* $|V(\boldsymbol{x}, t)| \leq W(\boldsymbol{x})$, $\forall t \geq t_0$ *is satisfied for* $|\boldsymbol{x}| \leq h$.

Thus, the function $V(\boldsymbol{x}, t) = |\boldsymbol{x}| \sin t$ is uniformly small; the function $V(\boldsymbol{x}, t) = \sin\{t(x_1 + \cdots + x_n)\}$, however, is not.

These functions, previously discussed, may now be used to formulate the LIAPOUNOV stability and instability theorems for the stability of the trivial solution of (335).

First stability theorem. *If there is a positive definite function* $V(\boldsymbol{x}, t)$ *such that*

$$\dot{V}(\boldsymbol{x}, t) = \frac{\partial V}{\partial t} + \mathrm{grad} V \cdot \boldsymbol{f}(\boldsymbol{x}, t)$$

is negative semi-definite or vanishes identically, then the trivial solution of (335) *is stable.*

Second stability theorem. *If there is a positive definite and uniformly small function* $V(\boldsymbol{x}, t)$ *for* (335) *such that* $\dot{V}(\boldsymbol{x}, t)$ *is negative definite, then the trivial solution of* (335) *is asymptotically stable.*

LIAPOUNOV's instability theorem. *If there is a uniformly small function* $V(\boldsymbol{x}, t)$ *for* (335) *such that* $\dot{V}(\boldsymbol{x}, t)$ *is positive definite, whereas* $V(\boldsymbol{x}, t)$ *may take on positive values for arbitrarily small* $\boldsymbol{x}$ *and arbitrarily large* $t > t_0$, *then the trivial solution is unstable.*

All three of these theorems can be proven by simple geometric means (for example, see Malkin (1959)). In particular, in the autonomous case the second theorem is based on the fact that the trivial solution certainly is asymptotically stable if all of the solution trajectories $\boldsymbol{x}(t)$ penetrate the trajectories $V(\boldsymbol{x}) = c$ of Figure 33 from outside to inside. Functions which satisfy the theorems above are called LIAPOUNOV functions.

► Some applications

(1) LAGRANGE*'s theorem concerning the stability of equilibrium.* Consider a conservative holonomic mechanical system with kinetic energy $T(\boldsymbol{q}, \dot{\boldsymbol{q}}) = \frac{1}{2}\dot{\boldsymbol{q}}^\mathsf{T}\boldsymbol{A}(\boldsymbol{q})\dot{\boldsymbol{q}}$ and with potential energy $U(\boldsymbol{q})$ and assume that the system has an equilibrium position for $\boldsymbol{q} = \boldsymbol{0}$. With the Lagrangian $L = T - U$ the equations of motion follow from LAGRANGE's equations

$$\frac{d}{dt}\frac{\partial L}{\partial \dot{q}_i} - \frac{\partial L}{\partial q_i} = 0, i = 1, 2, \ldots, n \tag{342}$$

and may always be written in the normal form

$$\ddot{\boldsymbol{q}} = \boldsymbol{f}(\boldsymbol{q}, \dot{\boldsymbol{q}}) \tag{343}$$

since $\boldsymbol{A}(\boldsymbol{q})$ is assumed to be a symmetric positive definite matrix for all $\boldsymbol{q}$. The system (343) may now be written as a first-order system. Prior thereto, it is advantageous to introduce different coordinates. Instead of the column matrix $\dot{\boldsymbol{q}}$ of the generalized velocities, the generalized momenta

$$\boldsymbol{p} = \frac{\partial T}{\partial \dot{\boldsymbol{q}}} = \boldsymbol{A}(\boldsymbol{q})\dot{\boldsymbol{q}} \tag{344}$$

are introduced. The kinetic energy is expressed in terms of $\boldsymbol{p}$ rather than $\boldsymbol{q}$ and the Hamiltonian $H(\boldsymbol{q}, \boldsymbol{p})$ is defined as the sum of the kinetic and of the potential energy:

$$H(\boldsymbol{q}, \boldsymbol{p}) = \frac{1}{2}\boldsymbol{p}^{\mathsf{T}}\boldsymbol{A}^{-1}(\boldsymbol{q})\boldsymbol{p} + U(\boldsymbol{q}). \tag{345}$$

The equations of motion now are given by

$$\begin{aligned} \dot{\boldsymbol{q}} &= \frac{\partial H}{\partial \boldsymbol{p}}, \\ \dot{\boldsymbol{p}} &= \frac{\partial H}{\partial \boldsymbol{q}}. \end{aligned} \tag{346}$$

Obviously, the function $H(\boldsymbol{q}, \boldsymbol{p})$ is positive definite if the potential energy, given to within an arbitrary additive constant, is set equal to zero for $\boldsymbol{q} = \boldsymbol{0}$ and if it has a minimum there. Furthermore, one has

$$\dot{H} = \left(\frac{\partial H}{\partial \boldsymbol{q}}\right)^{\mathsf{T}} \dot{\boldsymbol{q}} + \left(\frac{\partial H}{\partial \boldsymbol{p}}\right)^{\mathsf{T}} \dot{\boldsymbol{p}} = 0 \tag{347}$$

along the solutions of eqns (346) so that $H(\boldsymbol{q}, \boldsymbol{p})$ is a *first integral* of the system (346). The function H satisfies all of the hypotheses of the first stability theorem of LIAPOUNOV and from this it follows that the static equilibrium position $(\boldsymbol{q}, \boldsymbol{p}) = (\boldsymbol{0}, \boldsymbol{0})$ is stable. In stability theory this result is known as the LAGRANGE-DIRICHLET Theorem.

Consider briefly two extensions of this theorem: the case of dissipative systems and the case of conservative systems with gyroscopic forces. If the system is dissipative, eqns (342) are replaced by

$$\frac{d}{dt}\frac{\partial L}{\partial \dot{q}_i} - \frac{\partial L}{\partial q_i} = Q_i(\boldsymbol{q}, \dot{\boldsymbol{q}}), i = 1, 2, \ldots, n \tag{348}$$

where one has $\sum_{i=1}^{n} Q_i(\boldsymbol{q}, \dot{\boldsymbol{q}})\dot{q}_i \leq 0$. In fact, if $\sum_{i=1}^{n} Q_i(\boldsymbol{q}, \dot{\boldsymbol{q}})\dot{q}_i$ is negative definite with respect to $\dot{\boldsymbol{q}}$, then one has *complete dissipation (complete damping)*. All motions of the system then experience damping (complete dissipation implies the *pervasive damping* as defined previously, not every system with pervasive damping, however, has complete dissipation!). If one again chooses $V = T + U$, then it is apparent that the second stability theorem is not applicable, since $\dot{V}$ is negative definite only with respect to $\dot{\boldsymbol{q}}$ but not with respect to $\boldsymbol{q}$, $\dot{\boldsymbol{q}}$. However, asymptotic stability of the equilibrium position still follows from Krasovskii's theorem (Krasovskii (1963), page 82).

Other than terms quadratic in $\dot{\boldsymbol{q}}$ the Lagrangian occasionally also contains terms which are linear in $\dot{\boldsymbol{q}}$. The corresponding terms in the equations of motion (342) are called *gyroscopic terms.* These always occur, for example, when a cyclic coordinate is being eliminated. It may easily be checked that even then a minimum of the potential energy still provides a sufficient condition for stability - in the case of complete dissipation, even for asymptotic stability.

In elastomechanics the LAGRANGE-DIRICHLET stability theorem often is used as the definition of stability. There, an equilibrium position is stable iff the corresponding potential energy takes on a relative minimum, and it is unstable otherwise. This approach is unsatisfactory for various reasons. On the one hand, the concept of stability is best visualized in terms of motion: an equilibrium position is called stable iff all *motions* whose initial conditions are sufficiently close to the equilibrium position occur in a sufficiently small given neighborhood of the equilibrium (for this purpose, some concept of distance must be introduced). On the other hand, however, the stability of equilibrium positions for *non-conservative* systems for which it may not even be possible to define a potential energy, and that of periodic and other motions is to be investigated also.

The main reason for this approach in elastomechanics is due to the considerable difficulties which may be encountered in the use of a LIAPOUNOV stability theory extended to include applications involving partial differential equations.

(2) The instability of equilibrium for a maximum of $U(\boldsymbol{q})$. Consider again the system described by eqns (346) and choose

$$V = \boldsymbol{p}^{\mathsf{T}}\boldsymbol{q} \tag{349}$$

as a LIAPOUNOV function. Write

$$\boldsymbol{A}^{-1}(\boldsymbol{q}) = \boldsymbol{A}^{-1}(\boldsymbol{0}) + \boldsymbol{b}(\boldsymbol{q}), \tag{350}$$

where $\boldsymbol{B}(\boldsymbol{0}) = \boldsymbol{0}$ and $\boldsymbol{B}(\boldsymbol{q})$ is continuous. From (345), (349) and (350) one then obtains

$$\dot{V} = \boldsymbol{p}^\mathsf{T} \boldsymbol{A}^{-1}(\boldsymbol{0})\boldsymbol{p} + \boldsymbol{p}^\mathsf{T} \left(b_{ij}(\boldsymbol{q}) - \frac{1}{2} \sum_{k=1}^{n} q_k \frac{\partial b_{ij}}{\partial q_k} \right) \boldsymbol{p} - \boldsymbol{q}^\mathsf{T} \frac{\partial U}{\partial \boldsymbol{q}}. \qquad (351)$$

The coefficients of the second quadratic form in $\boldsymbol{p}$ tend to zero as $|\boldsymbol{q}| \to 0$, so that $\dot{V}$ is at least positive definite with respect to $\boldsymbol{p}$, since $\boldsymbol{A}^{-1}(\boldsymbol{0})$ is a positive definite matrix. $U(\boldsymbol{q})$ now is written in the form

$$U(\boldsymbol{q}) = U_m(\boldsymbol{q}) + U_{m+1}(\boldsymbol{q}) + \ldots, \qquad (352)$$

where U_k stands for the kth-order terms in the power series expansion of U. Since $\boldsymbol{q} = \boldsymbol{0}$ is to be an equilibrium position, one has $\frac{\partial U}{\partial \boldsymbol{q}}$ and $m \leq 2$ follows. By using EULER's theorem concerning homogeneous functions, (351) may be replaced by

$$\dot{V} = \boldsymbol{p}^\mathsf{T} \boldsymbol{A}^{-1}(\boldsymbol{0})\boldsymbol{p} + \boldsymbol{p}^\mathsf{T} \left(b_{ij}(\boldsymbol{q}) - \frac{1}{2} \sum_{k=1}^{n} q_k \frac{\partial b_{ij}}{\partial q_k} \right) \boldsymbol{p} - mU_m - (m+1)U_{m+1} - \ldots. \qquad (353)$$

If U_m is negative definite with respect to $\boldsymbol{q}$, then $\dot{V}$ is positive definite in $\boldsymbol{p}$, $\boldsymbol{q}$ whereas $V = \boldsymbol{p}^\mathsf{T}\boldsymbol{q}$ takes on positive and negative values in every arbitrarily small neighborhood of $(\boldsymbol{q}, \boldsymbol{p}) = (\boldsymbol{0}, \boldsymbol{0})$. This fulfills the hypotheses of the instability theorem; it has been shown that the equilibrium position is unstable if U_m is negative definite, that is, if $U(\boldsymbol{q})$ has a maximum, *and if the existence of this maximum follows from the terms of lowest order in the series expansion* $U = U_m + U_{m+1} + \ldots$.

LIAPOUNOV already proved this theorem. For an engineer, this theorem is extremely unsatisfactory since he believes that he knows from observation that the equilibrium position is unstable not only for a maximum of the potential energy but generally and without additional restrictions whenever there is no minimum. Furthermore, this LIAPOUNOV theorem may not even be used to show the instability of $q_1 = 0$, $q_2 = 0$, for

$$U = -q_1^2 - q_2^4,$$

since $U_2 = -q_1^2$ is not negative definite but only negative semi-definite. The hypothesis 'U_m negative definite' may not be omitted in the present theorem, even though this is occasionally done without comment (for example, see LaSalle and Lefschetz (1967) and Hahn (1952)).[18]

[18] Even for a system with $2T = p_1^2 + p_2^2$, $U = -\left(q_1 - q_2^2\right)^2 - \frac{3}{32} q_2^4$, $n = 2$, the function $V = \boldsymbol{p}^\mathsf{T}\boldsymbol{q}$ is not a LIAPOUNOV function. In fact, one has $\boldsymbol{q}^\mathsf{T}\left(\frac{\partial U}{\partial q}\right) = -2\left(q_1 - \frac{3}{2}q_2^2\right)^2 + \frac{1}{8}q_2^4$, and this expression is indefinite, even though U has a maximum for $\boldsymbol{q} = \boldsymbol{0}$ (the expression is positive for $\frac{7}{4}q_2^2 > q_1 > \frac{5}{4}q_2^2$ and negative for $q_1 < \frac{5}{4}q_2^2$ and $q_1 > \frac{7}{4}q_2^2$). Thus $\dot{V}$ is also indefinite.

Since the turn of the century, numerous mathematicians have concerned themselves with the question of the stability of an equilibrium position of a conservative system in the absence of a minimum of $U(\boldsymbol{q})$. Painlevé (1897) gave an example and showed that it is quite possible to have stability even in the absence of a minimum of $U(\boldsymbol{q})$! A general proof of instability for a maximum of $U(\boldsymbol{q})$ without any additional hypotheses on $U_m(\boldsymbol{q})$ was not available until given by Hagedorn (1971) (including a further, more detailed bibliography). In many practical applications, the stability problem is simplified considerably because of the often present dissipation of mechanical energy: then it is a great deal easier to prove instability!

(3) Stability of the rotational motion of a rocket. For nearly-level flight-trajectories of a projectile, it may be assumed that the center of gravity experiences approximately linear and uniform motion. The same is approximately valid (at least during short time intervals) for the power-off motion of a rocket. Let β be the angle between the rocket axis and its projection onto the vertical plane of flight and let α be the angle between this projection and the velocity vector. The equations of motion then are given by

$$\begin{aligned} A\ddot{\beta} + A\dot{\alpha}^2 \sin\beta\cos\beta - Cn\dot{\alpha}\cos\beta &= eR\sin\beta\cos\alpha\,, \\ A\ddot{\alpha}\cos\beta - 2A\dot{\alpha}\dot{\beta}\sin\beta + Cn\dot{\beta} &= eR\sin\alpha\,, \end{aligned} \tag{354}$$

where C is the moment of inertia with respect to the longitudinal axis of symmetry and n is the constant projection of the angular velocity in the direction of this axis. The moment of inertia with respect to an arbitrary diameter through the center of gravity is given by A and e denotes the distance between the center of gravity and the center of pressure (the point of action of the resultant R of the air resistance). For small angles α, β the resistance R may be taken to be constant. These equations of motion are derived, for example, in Gantmacher (1970), pages 186 ff, and they are discussed in Luré (1968), pages 238 ff. The equations (354) have the particular solution $\alpha = \beta = 0$, $\dot{\alpha} = \dot{\beta} = 0$. The stability of this trivial solution is to be investigated.

From the theorems of mechanics, two first integrals of motion are immediate: the energy integral

$$F_1\left(\alpha, \dot{\alpha}, \beta, \dot{\beta}\right) = \frac{1}{2}A\left(\dot{\beta}^2 + \dot{\alpha}^2\cos^2\beta\right) + eR(\cos\alpha\cos\beta - 1) \tag{355}$$

and the angular momentum integral

$$F_2\left(\alpha, \dot{\alpha}, \beta, \dot{\beta}\right) = A\left(\dot{\beta}\sin\alpha - \dot{\alpha}\cos\beta\sin\beta\cos\alpha\right) + Cn(\cos\alpha\cos\beta - 1). \tag{356}$$

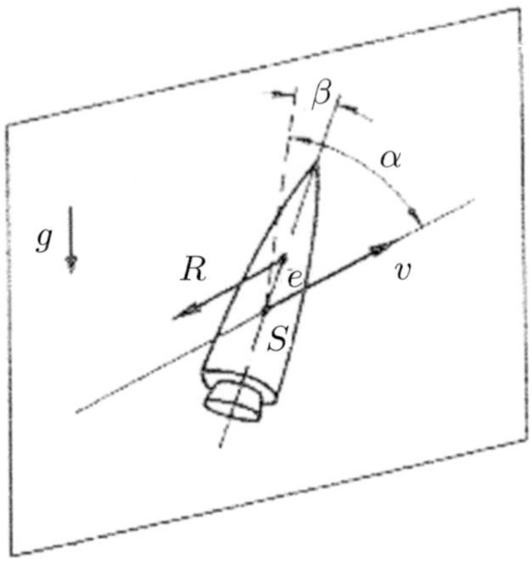

Figure 34: Stability of a rocket.

It is easy to convince oneself that F_1 and F_2 actually are first integrals of (354), that is, that one has $F_1 = 0$ and $F_2 = 0$ for all of the solutions of (354). For F_1 , for example, one has

$$\begin{aligned}\dot{F}_1 &= \frac{\partial F_1}{\partial \alpha}\dot{\alpha} + \frac{\partial F_1}{\partial \dot{\alpha}}\ddot{\alpha} + \frac{\partial F_1}{\partial \beta}\dot{\beta} + \frac{\partial F_1}{\partial \dot{\beta}}\ddot{\beta} \\ &= -eR\dot{\alpha}\sin\alpha\cos\beta + \cos\beta\left(eR\sin\alpha - Cn\beta + 2A\dot{\alpha}\dot{\beta}\sin\beta\right)\dot{\alpha} - \\ &\quad - A\dot{\alpha}^2\dot{\beta}\cos\beta\sin\beta - eR\dot{\beta}\cos\alpha\sin\beta + \dot{\beta}\,(eR\sin\beta\cos\alpha \\ &\quad - A\dot{\alpha}^2\sin\beta\cos\beta + Cn\dot{\alpha}\cos\beta) \equiv 0.\end{aligned}$$

Unfortunately, neither of these first integrals F_1 and F_2 is definite. Consider, thus, a new integral

$$V = F_1 - \lambda F_2, \tag{357}$$

where lambda is a constant which is to be chosen in has m such a way that V becomes positive definite. One has

$$\begin{aligned}V = &\frac{1}{2}\{A\dot{\alpha}^2 + (Cn\lambda - eR)\beta^2 + 2A\lambda\dot{\alpha}\beta\} + \\ &+ \frac{1}{2}\left\{A\dot{\beta}^2 + (Cn\lambda - eR)\alpha^2 - 2A\lambda\dot{\beta}\alpha\right\} + \ldots,\end{aligned} \tag{358}$$

where only the terms of second order in the expansion of V with respect to α, $\dot{\alpha}$, β, $\dot{\beta}$ have been indicated. All additional terms are at least fourth order. Clearly, V_2 is positive definite precisely when the two quadratic forms

$$Ax^2 + (Cn\lambda - eR)x_2^2 \pm 2A\lambda x_1 x_2$$

are positive definite. However, V then is also positive definite and the trivial solution is stable. From Sylvester's theorem one obtains

$$A > 0, \qquad \begin{vmatrix} A & \pm A\lambda \\ \pm A\lambda & Cn\lambda - eR \end{vmatrix} > 0 \tag{359}$$

as necessary and sufficient conditions for the positive definiteness of V_2, that is,

$$A > 0, \qquad A(Cn\lambda - eR) - A^2\lambda^2 > 0$$

or

$$A > 0, \qquad Cn\lambda - eR - A\lambda^2 > 0. \tag{360}$$

The first condition in (360) is always fulfilled. In order to be able to satisfy the second condition for at least one real value of λ, the two roots λ_1, λ_2 must be real and distinct from one another; this is the case only if one has

$$C^2n^2 - 4AeR > 0. \tag{361}$$

Whenever (361) holds, one may always choose λ in such a way that V_2 and hence V is definite with the consequent stability of the trivial solution. The inequality (361) provides a lower bound on n for which the stability of the trivial solution is assured. Later, it will be shown that the trivial solution is unstable for smaller values of n.

Note also that the trivial solution here is stable in spite of the fact that the potential energy, represented by the second term in (355), has a maximum for $\alpha = \beta = 0$. This is possible only because the equations of motion (354) contain terms which are linear in the generalized velocities. Under certain circumstances, such so called gyroscopic terms may serve to stabilize the equilibrium position of a conservative mechanical system even for a maximum of the potential energy.

(4) Free rotation of a rigid body about a fixed point. Consider again EULER's equations (323) for the rotation of a rigid body. From mechanics it is known that both energy and angular momentum are conserved. The first integrals thus are known to be

$$\begin{aligned} T &= \frac{1}{2}\left(Ap^2 + Bq^2 + Cr^2\right), \\ L &= Ap\boldsymbol{i} + Bq\boldsymbol{j} + Cr\boldsymbol{k}, \end{aligned} \tag{362}$$

where $\boldsymbol{i}$, $\boldsymbol{j}$, $\boldsymbol{k}$ are the unit vectors in the directions of the principal axes of inertia (for example, $d\boldsymbol{i}/dt = \boldsymbol{\omega} \times \boldsymbol{i} = r\boldsymbol{j} - q\boldsymbol{k}$). Naturally, the magnitude of the angular momentum is also a first integral, as is its square

$$L^2 = A^2p^2 + B^2q^2 + C^2r^2. \tag{363}$$

Eqns (324) now are used to investigate the stability of the particular solution $p = \omega$, $q = 0$, $r = 0$. In terms of these new coordinates the expressions (362) and (363) result in the two first integrals

$$\begin{aligned} F_1(y_1, y_2, y_3) &= Ay_1^2 + By_2^2 + Cy_3^2 + 2A\omega y_1, \\ F_2(y_1, y_2, y_3) &= A^2y_1^2 + B^2y_2^2 + C^2y_3^2 + 2A^2\omega y_1. \end{aligned} \tag{364}$$

With $2A\omega y_1 = F_1 - Ay_1^2 - By_2^2 - Cy_3^2$ it follows that

$$\begin{aligned} F_3(y_1, y_2, y_3) &= A^2y_1^2 + B^2y_2^2 + C^2y_3^2 - A^2y_1^2 - ABy_2^2 - ACy_3^2 \\ &= B(B - A)y_2^2 + C(C - A)y_3^2 \end{aligned} \tag{365}$$

is a first integral and, naturally, that

$$\begin{aligned} F_3(y_1, y_2, y_3) &= F_1^2 \pm F_3 \\ &= \left\{Ay_1^2 + By_2^2 + Cy_3^2 2A\omega y_1\right\}^2 \pm \\ &\pm \left\{B(B - A)y_2^2 + C(C_A)y_3^2\right\} \end{aligned} \tag{366}$$

also is. However, the function F corresponding to the upper sign is positive definite for $A < B, C$; it is positive definite with the lower sign for $A > B, C$. Thus, the rotations about the axes corresponding to the least and to the largest principal moments of inertia are stable!

3.3 Supplementary Remarks Concerning the Direct Method of Liapounov

The Liapounov theorems considered in the previous section provide sufficient conditions for stability and instability. They contain no hint as to how a function $V(\boldsymbol{x}, t)$ may be found in a particular case. In problems from mechanics or, more generally, for differential equations which describe the behavior of a physical system, it is often possible to deduce a suitable Liapounov function by using general physical principles. Thus, one often makes use of energy expressions, various first integrals, or the entropy in certain cases. It can be proven (see Krasovskii (1963), for example) that for every differential equation with the trivial solution $\boldsymbol{x} = \boldsymbol{0}$ there indeed exists a Liapounov function which satisfies the hypotheses of at least one of the three LIAPOUNOV theorems. In many cases, however, it just cannot be found. There are a multitude of procedures which have been proposed for the systematic construction of these functions (without solving the differential equation), but they are either too complicated or are suited only for special classes of differential equations. For this reason, much effort has gone into adjusting the LIAPOUNOV theorems in such a way that they may be more easily used and, consequently, there are a large number of related stability theorems. One of the most important is the following theorem due to Chetayev; a proof may be found in Malkin (1959), for example. For this instability theorem, the function $V(\boldsymbol{x}, t)$ need not be defined in a complete neighborhood of $\boldsymbol{x} = \mathbf{0}$ but only in a 'cone' whose apex is located at $\boldsymbol{x} = \mathbf{0}$. For instability it suffices to establish the existence of a single solution which moves away from $\boldsymbol{x} = \mathbf{0}$. In the case of Chetayev's theorem, this solution develops with the 'cone'. The theorem here is cited only for the autonomous case

Consider the differential equation

$$\dot{\boldsymbol{x}} = \boldsymbol{f}(\boldsymbol{x}), \qquad \boldsymbol{f}(\mathbf{0}) = \mathbf{0} \tag{367}$$

and assume that there is a function $V(\boldsymbol{x})$ such that

1. *in every arbitrarily small neighborhood of $\boldsymbol{x} = \mathbf{0}$ there exists a region in which $V > 0$ holds and on whose boundary one has $V = 0$, and*
2. *at all points of the region defined by $V > 0$ the derivative $\dot{V}$ takes on positive values,*

then, the solution $\boldsymbol{x} = \mathbf{0}$ of (367) is unstable. There is one essential disadvantage to the LIAPOUNOV theorems which have been considered. For example, if one tries to prove the asymptotic stability of the trivial solution of

$$\ddot{x} + c\dot{x} + \omega_0^2 x = 0 \tag{368}$$

by using the energy expression $E = \frac{1}{2}\left(\dot{x}^2 + \omega_0^2 x^2\right)$, then it turns out that this is not possible. In fact, one has

$$\dot{E} = \dot{x}\ddot{x} + \omega_0^2 x\dot{x} = -c\dot{x}^2 \tag{369}$$

and it follows that $\dot{E}$ is only negative semi-definite in x, $\dot{x}$. Since $E(x, \dot{x})$ is positive-definite, Theorem 1 guarantees stability but Theorem 2, concerning asymptotic stability, is not applicable. Although (369) may be solved explicitly by analytical methods and one may of course also prove asymptotic stability by using some other function, it would clearly be desirable to change Theorem 2 in such a way that asymptotic stability may also be deduced by using the energy expression. One might expect that this new theorem then may also be used with more success for other more complicated differential equations. Such a generalization of the second LIAPOUNOV theorem will be given shortly; another generalization will be taken up in the process . Namely, it is evident from the geometric significance of the direct method that not only may the functions $V(\boldsymbol{x},t)$ be used to investigate stability, but one may also determine the 'domain of attraction' of $\boldsymbol{x} = \mathbf{0}$ in the asymptotically stable case. The term 'domain of attraction' here is used to designate the region of those initial conditions $\boldsymbol{x}_0$ which lead to motions $\boldsymbol{x}(\boldsymbol{x}_0, t_0; t)$ such that $\boldsymbol{x}(\boldsymbol{x}_0, t_0; t) \to 0$ for $t \to \infty$.

The following discussion will be restricted to the autonomous differential eqn (367) and to functions $V(\boldsymbol{x})$ which are independent of t. The set

$$G = \{\boldsymbol{x}_0 | \boldsymbol{x}\,(\boldsymbol{x}_0, t_0; t) \in G, \qquad \forall t \geq t_0\}$$

(in words: G is the set of all points $\boldsymbol{x}_0$, such that if $\boldsymbol{x}$ is located in G then $\boldsymbol{x}(\boldsymbol{x}_0, t_0; t)$ is also in G for all $t \geq t_0$) is called an *invariant set* of (367). This term now is used to formulate *La Salle's theorem:*
Let $V(\boldsymbol{x}) \in C^1$,

$$\begin{aligned}
&\Omega_l = \{\boldsymbol{x} | V(\boldsymbol{x}) < l\}, && \Omega_l \textit{ bounded},\\
&V(\boldsymbol{x}) > 0 \qquad \textit{for} && \mathbf{0} \neq \boldsymbol{x} \in \Omega_l,\\
&\dot{V}(\boldsymbol{x}) \leq 0 \qquad \textit{for} && \boldsymbol{x} \in \Omega_l,\\
&L = \left\{\boldsymbol{x} \in \Omega_l | \dot{V}(\boldsymbol{x}) = 0\right\},
\end{aligned}$$

and let M *be the largest invariant set in* L. *Then, every solution* $\boldsymbol{x}(t)$ *in* Ω_l *tends to* M *for* $t \to \infty$ *(For a proof, see LaSalle and Lefschetz (1967)).*

This theorem may be used not only to investigate the stability of the trivial solution of (367) but also to investigate the stability of limit cycles. For the determination of the domain of attraction of $\boldsymbol{x} = \mathbf{0}$ it is advantageous to write La Salle's theorem in a somewhat less general form, such as may be found, for example, in Krasovskii (1963).

Krasovskii's theorem: Let $V(\boldsymbol{x}) \in C^1$, $V(\mathbf{0}) = 0$,

$$\begin{aligned}
&\Omega_l = \{\boldsymbol{x} | V(\boldsymbol{x}) \leq l\}, &&\\
&V(\boldsymbol{x}) > 0 \qquad \textit{for} && \mathbf{0} \neq \boldsymbol{x} \in \Omega_l,\\
&\dot{V}(\boldsymbol{x}) \leq 0 \qquad \textit{for} && \boldsymbol{x} \in \Omega_l.
\end{aligned}$$

If there is no solution $\boldsymbol{x}^*(t)$ *of* (367), *other than the trivial solution, for which* $\dot{V}(\boldsymbol{x}^*(t)) \equiv 0$ *and which lies completely in* Ω_l, *then every solution* $\boldsymbol{x}(\boldsymbol{x}_0, t_0; t)$, $\boldsymbol{x}_0 \in \Omega_l$ *tends to* $\boldsymbol{x} = \mathbf{0}$ *and* $\boldsymbol{x} = \mathbf{0}$ *is asymptotically stable.*

This theorem now is used on (368) with $V(\boldsymbol{x}) = E(x, \dot{x})$; here, Ω_l is the interior of the ellipse $\frac{1}{2}\left(\dot{x}^2 + \omega_0^2 x^2\right) = l$. $\dot{V}$ vanishes on the x-axis, that is, for $\dot{x} = 0$. Other than the trivial solution, there is no further solution $\boldsymbol{x}^*(t)$ for which $\dot{V} = -c\dot{x}^2$ vanishes identically. Thus, the trivial solution also is asymptotically stable and for arbitrarily large l the ellipse is located in the domain of attraction of $\boldsymbol{x} = \mathbf{0}$ which here consists of the whole phase space.[19]

A second, less trivial example, which will also be discussed later in connection with self-excited oscillations, will now be treated. Consider the differential equation

$$\ddot{x} + \epsilon\left(x^2 - 1\right)\dot{x} + x = 0 \tag{370}$$

with $\epsilon < 0$ (for the case $\epsilon > 0$ the equation is known as the Van der Pol equation).

The trivial solution is asymptotically stable and its domain of attraction is

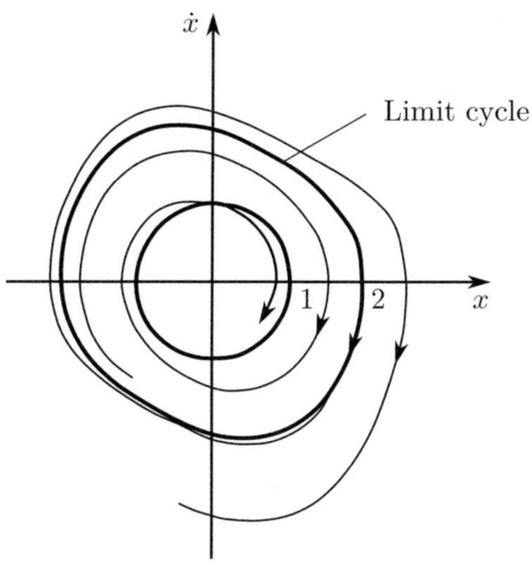

Figure 35: Phase diagram for (370) with $\epsilon < 0$.

to be estimated. The corresponding phase diagram is sketched in Figure 35. It contains an isolated (unstable) periodic solution-a so-called *limit cycle.*

[19]The asymptotic stability of the equilibrium position may be proven for (348) in the completely damped case in the same manner as long as the equilibrium position is an 'isolated' one.

Since (370) cannot be solved analytically, an estimate of the domain of attraction actually is of some importance here. The LIAPOUNOV function is chosen to be

$$V(x, \dot{x}) = \frac{1}{2}\left(\dot{x}^2 + x^2\right) \tag{371}$$

with the result

$$\dot{V}(x, \dot{x}) = \dot{x}\ddot{x} + x\dot{x} = -\epsilon\left(x^2 - 1\right)\dot{x}^2. \tag{372}$$

For $x^2 < 1$ the function $\dot{V}(x, \dot{x})$ is negative semi-definite since $\epsilon < 0$ was assumed. In order to enforce $x^2 < 1$, $l = \frac{1}{2}$ is chosen, so that Ω_l is the interior of the circle $\dot{x}^2 + x^2 = 1$. Since there are no non-trivial solutions in Ω_l for which $\dot{V}$ vanishes identically, $x = 0$, $\dot{x} = 0$ is asymptotically stable and the region $\dot{x}^2 + x^2 \leq 1$ is located totally within the domain of attraction of the trivial solution, independent of the value of $|\epsilon|$. This result may be considerably improved upon by introducing a 'suitable' coordinate transformation prior to the application of the theorem. This will be dealt with later in the case of self-excited oscillations (see Hagedorn and DasGupta (2007)).

3.4 Stability by the First Approximation (Autonomous Case)

A discussion of the 'method of the first approximation' will be given next. The method is used to obtain results concerning the stability of the trivial solution of

$$\dot{\boldsymbol{x}} = \boldsymbol{A}\boldsymbol{x} + \boldsymbol{h}(\boldsymbol{x}, t) \tag{373}$$

by making use of the linearized differential equation

$$\dot{\boldsymbol{x}} = \boldsymbol{A}\boldsymbol{x}. \tag{374}$$

Here, $\boldsymbol{A}$ is a constant real-valued matrix and there exist numbers $\beta > 1, \alpha \geq 0$, such that

$$|\boldsymbol{h}(\boldsymbol{x}, t)| \leq \alpha|\boldsymbol{x}|^{\beta} \tag{375}$$

holds in a neighborhood of $\boldsymbol{x} = \boldsymbol{0}$.

In order to proceed in the indicated manner, the stability of the linear system (374) must be investigated first and it must then be determined when these results are also valid for the non-linear system (374).

The manner of obtaining a solution of (374) is known from the theory of linear differential equations or, equivalently, from the theory of linear oscillations. The substitution $\boldsymbol{x} = \boldsymbol{a}e^{\lambda t}$ in equation (374) results in the characteristic equation

$$\det(\boldsymbol{A} - \lambda \boldsymbol{E}) = 0 \tag{376}$$

from which one may then calculate the eigenvalues of $\boldsymbol{A}$. For the case $n = 2$, it is easy to draw the phase trajectories corresponding to the various combinations of eigenvalues. The following cases arise (Figure 36):

- (a) λ_1, λ_2 real, $\lambda_1 > 0$, $\lambda_2 > 0$, 'unstable node';
- (b) λ_1, λ_2 real, $\lambda_1 > 0$, $\lambda_2 < 0$, 'saddle point';
- (c) λ_1, λ_2 real, $\lambda_1 < 0$, $\lambda_2 < 0$, 'stable node';
- (d) λ_1, λ_2 imaginary ($\lambda_1 = -i\lambda_2$, since $\boldsymbol{A}$ is real-valued), 'centre';
- (e) λ_1, λ_2 complex (conjugates, since $\boldsymbol{A}$ is real), negative real part, 'stable focus';
- (f) λ_1, λ_2 complex, positive real parts, 'unstable focus'.

In addition, there may of course be vanishing and multiple eigenvalues; in each of these cases the general solution may be stated without difficulty.

It is known from linear algebra (for example, see Pontryagin (1965) or Gantmacher (1958)) that for arbitrary real matrices $\boldsymbol{A}$ there always exists a coordinate transformation

$$\boldsymbol{y} = \boldsymbol{S}\boldsymbol{x} \tag{377}$$

which may be complex in general, such that

$$\dot{\boldsymbol{y}} = \boldsymbol{S}\dot{\boldsymbol{x}} = \boldsymbol{S}\boldsymbol{A}\boldsymbol{x} = \boldsymbol{S}\boldsymbol{A}\boldsymbol{S}^{-1}\boldsymbol{y} \tag{378}$$

and such that (374) may always be transformed to the form

$$\dot{\boldsymbol{y}} = \boldsymbol{J}\boldsymbol{y}, \tag{379}$$

where only the main diagonal and adjacent diagonal are non-zero in the JORDAN *normal form* $\boldsymbol{J} = \boldsymbol{S}\boldsymbol{A}\boldsymbol{S}^{-1}$. Here, the matrix $\boldsymbol{J}$ consists of 'elementary blocks' whose main diagonal consists of one and the same eigenvalue, say λ_i, while all of the elements of the right-adjacent diagonal are unity; all of the remaining elements of the matrix are zero. Thus, an 'elementary block' has the form

$$\begin{pmatrix} \lambda_i & 1 & 0 & \dots & 0 \\ 0 & \lambda_i & 1 & \dots & 0 \\ 0 & 0 & \lambda_i & & \vdots \\ \vdots & \vdots & \vdots & & 1 \\ 0 & 0 & 0 & \dots & \lambda_i \end{pmatrix}. \tag{380}$$

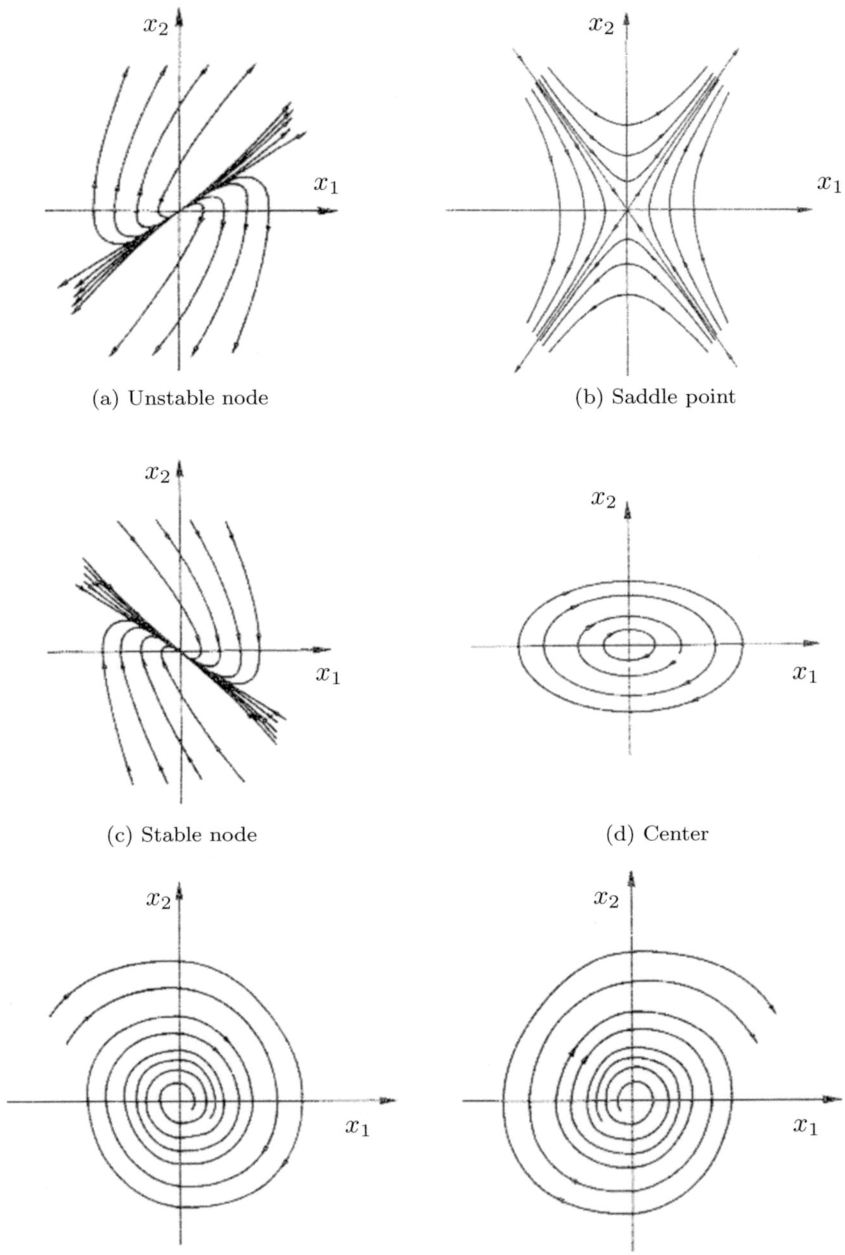

(a) Unstable node (b) Saddle point

(c) Stable node (d) Center

(e) Stable focus (f) Unstable focus

Figure 36: Phase trajectories of an autonomous linear system with $n = 2$.

Corresponding to every eigenvalue λ_i there exists at least one elementary block. For each distinct eigenvalue there always exists exactly one block of 1×1 form so that the adjacent diagonal vanishes. Such blocks may also occur for multiple eigenvalues. If all eigenvalues are distinct, then $\boldsymbol{J}$ reduces to the main diagonal form

$$\boldsymbol{J} = \begin{pmatrix} \lambda_1 & 0 & \dots & 0 \\ 0 & \lambda_2 & \dots & \vdots \\ \vdots & & & 0 \\ 0 & 0 & \dots & \lambda_n \end{pmatrix}. \tag{381}$$

A more typical structure of $\boldsymbol{J}$, however, is given by

$\boldsymbol{J} =$

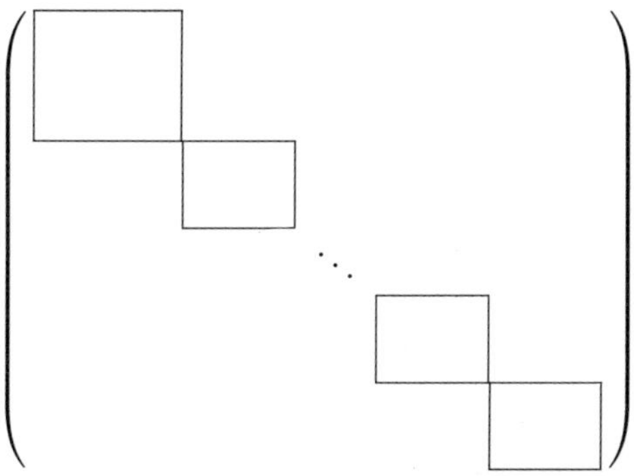

, (382)

where the squares represent elementary blocks of the type (380). The transformation $\boldsymbol{S}$ thus decouples the system (374) into as many single systems of differential equations as there exist elementary blocks. Consider, therefore, a system of order m of the form

$$\dot{\boldsymbol{z}} = \begin{pmatrix} \lambda & 1 & 0 & \dots & 0 \\ 0 & \lambda & 1 & \dots & 0 \\ 0 & 0 & \lambda & & \vdots \\ \vdots & \vdots & \vdots & & 1 \\ 0 & 0 & 0 & \dots & \lambda \end{pmatrix} \boldsymbol{z}. \tag{383}$$

Define a corresponding function

$$V(\boldsymbol{x}) = -(\gamma - \epsilon) \left[z_1 \bar{z}_1 + (\gamma - \epsilon)^2 z_2 \bar{z}_2 + \cdots + (\gamma - \epsilon)^{2(m-1)} z_m \bar{z}_m \right], \tag{384}$$

where $\lambda = \gamma + \mathrm{i}\delta$ (γ, δ real, i = unit imaginary), ϵ and γ are different real constants, and where $\bar{z}$ denotes the complex conjugate of z. From (383) and

(384) one obtains

$$\dot{V}(\boldsymbol{z}) = 2\left(\gamma - \frac{1}{\gamma-\epsilon}\right)V - \left[2z_1\bar{z}_1 + 2(\gamma-\epsilon)^2 z_2\bar{z}_2 + \cdots + \right.$$
$$+ 2(\gamma-\epsilon)^{2(m-1)} z_m\bar{z}_m + (\gamma-\epsilon)\left(z_1\bar{z}_2 + \bar{z}_1 z_2\right) +$$
$$+ (\gamma-\epsilon)^3\left(z_2\bar{z}_3 + \bar{z}_2 z_3\right) + \cdots +$$
$$\left. + (\gamma-\epsilon)^{2m-3}\left(z_{m-1}\bar{z}_m + \bar{z}_{m-1}z_m\right)\right] . \tag{385}$$

It may easily be established that the expression in brackets is always positive definite by investigating the matrix

$$\begin{pmatrix} 2 & 1 & 0 & \dots & 0 \\ 1 & 2 & 1 & \dots & . \\ 0 & 1 & 2 & \dots & . \\ \vdots & & & & . \\ \vdots & & & 2 & 1 \\ 0 & \dots & \dots & 1 & 2 \end{pmatrix} . \tag{386}$$

of the corresponding quadratic form in the new variables $(\gamma-\epsilon)^k z_{k+1}$. If $\gamma < 0$, then ϵ may be chosen to satisfy $\epsilon > \gamma - (\frac{1}{\gamma})$, resulting in a positive definite V and a negative definite $\dot{V}$. If one now adds the functions $V(\boldsymbol{z})$ defined for each block, one obtains a function $V(\boldsymbol{y})$. If all λ_i have a negative real part, then $V(\boldsymbol{y})$ also is positive definite and $\dot{V}(\boldsymbol{y})$ is negative definite. The functions V and $\dot{V}$ remain definite even when (377) is used to replace $\boldsymbol{y}$ by the real variable $\boldsymbol{x}$. It follows that the trivial solution is asymptotically stable. Instability may be proven in a similar manner if at least one eigenvalue has a positive real part.

In summary, the following statements are valid in connection with (374):

- *all EV (eigenvalues) have negative real part ⇒ asymptotic stability;*
- *at least one EV has positive real part ⇒ instability;*
- *there do not exist any EV with positive real part, but there are EV with vanishing real part of such a type that only 'singleton blocks' correspond to them in the Jacobi normal form ⇒ stability (non-asymptotic);*
- *there do not exist any EV with positive real part, but there are EV with vanishing real part of such a type that not only 'singleton blocks' correspond to them ⇒ instability.*

Confidence in the last two assertions may easily be gained by means of the following two examples. Consider the differential equation

$$\ddot{x} + \omega^2 x = 0. \tag{387}$$

If this is viewed as a first-order system, then

$$\boldsymbol{J} = \begin{pmatrix} \mathrm{i}\omega & 0 \\ 0 & \mathrm{i}\omega \end{pmatrix} \tag{388}$$

is the JORDAN normal form of $\boldsymbol{A}$ and it follows from the previous comments that the equilibrium position is stable-but not asymptotically stable. Here, this may be verified by means of the available general solution. However, a consideration of the system

$$\begin{aligned} \dot{x}_1 &= x_2, \\ \dot{x}_2 &= 0 \end{aligned} \tag{389}$$

with

$$\boldsymbol{J} = \begin{pmatrix} 0 & 1 \\ 0 & 0 \end{pmatrix}, \tag{390}$$

indicates that a '2x2 block' corresponds to the multiple eigenvalue 'zero', and instability follows. Again this is easily verified by using the general solution

$$\begin{aligned} x_1 &= x_{10} + x_{20}t, \\ x_2 &= x_{20}. \end{aligned} \tag{391}$$

Thus, a necessary and sufficient condition for the asymptotic stability of (374) is given by the requirement that all of the roots of (376) have negative real parts. Generally, it is not a simple manner to determine all of the roots of (376). One thus makes use of criteria which provide assertions about the real parts of the roots of the characteristic equation

$$a_0\lambda^n + a_1\lambda^{n-1} + a_2\lambda^{n-2} + \cdots + a_{n-1}\lambda + a_n = 0, \qquad a_0 > 0 \tag{392}$$

without having to resort to an actual solution of the equation. The most important of these criteria is a *theorem due to* HURWITZ.

Use the coefficients of (392) *to construct the minors*

$$\Delta_1 = a_1, \quad \Delta_2 = \begin{vmatrix} a_1 & a_0 \\ a_3 & a_2 \end{vmatrix}, \quad \Delta_3 = \begin{vmatrix} a_1 & a_0 & 0 \\ a_3 & a_2 & a_1 \\ a_5 & a_4 & a_3 \end{vmatrix},$$

$$\Delta_n = \begin{vmatrix} a_1 & a_0 & 0 & 0 & \dots & 0 \\ a_3 & a_2 & a_1 & a_0 & \dots & 0 \\ \vdots & & & & & \vdots \\ 0 & 0 & 0 & & \dots & a_n \end{vmatrix} = a_n\Delta_{n-1}; \tag{393}$$

then a necessary and sufficient condition assuring that all of the roots of (392) *have negative real parts is given by*

$$\Delta_1 > 0, \quad \Delta_2 > 0, \dots, \Delta_{n-1} > 0, \quad a_n > 0 \tag{394}$$

(For a proof, see Chetayev (1961), for example).

An application of HURWITZ's theorem to the quadratic equation

$$a_0 x^2 + a_1 x + a_2 = 0, \qquad a_0 > 0 \tag{395}$$

yields

$$a_1 > 0, \qquad \begin{vmatrix} a_1 & a_0 \\ 0 & a_2 \end{vmatrix} = a_1 a_2 > 0, \tag{396}$$

which may be replaced by the conditions $a_0 > 0$, $a_1 > 0$, $a_2 > 0$.

For the cubic equation

$$a_0 x^3 + a_1 x^2 + a_2 x + a_3 = 0, \qquad a_0 > 0 \tag{397}$$

one obtains

$$a_1 > 0, \qquad \begin{vmatrix} a_1 & a_0 \\ a_3 & a_2 \end{vmatrix} = a_1 a_2 - a_3 a_0 > 0, \qquad a_3 > 0, \tag{398}$$

and for the quartic equation

$$a_0 x^4 + a_1 x^3 + a_2 x^2 + a_3 x + a_4 = 0, \qquad a_0 > 0 \tag{399}$$

the criterion yields the conditions

$$a_1 > 0, \qquad \begin{vmatrix} a_1 & a_0 \\ a_3 & a_2 \end{vmatrix} > 0, \qquad \begin{vmatrix} a_1 & a_0 & 0 \\ a_3 & a_2 & a_1 \\ 0 & a_4 & a_3 \end{vmatrix} > 0, \qquad a_4 > 0 \tag{400}$$

or, equivalently,

$$\begin{aligned} & a_1 > 0, \qquad a_1 a_2 - a_0 a_3 > 0, \\ & a_3 \left(a_1 a_2 - a_0 a_3\right) - a_4 a_1^2 > 0, \qquad a_4 > 0. \end{aligned} \tag{401}$$

Some additional simplification is possible; namely, the third and the fourth of these conditions imply

$$a_3 \left(a_1 a_2 - a_0 a_3\right) > a_4 a_1^2 > 0$$

so that the second condition may be replaced by $a_3 > 0$. Thus, the conditions (401) may also be written in the form

$$\begin{aligned} & a_1 > 0, \qquad a_3 > 0, \\ & a_3 \left(a_1 a_2 - a_0 a_3\right) > 0, \qquad a_4 > 0. \end{aligned} \tag{402}$$

For larger systems, the computation of the determinants Δ_i, $i = 1, 2, \ldots, n$ in (394) is rather tedious. Hence, it becomes desirable to replace the conditions (394) by others which may be more easily evaluated. This is the case, for example, in the criterion of LIÉNARD-CHIPART, wherein the conditions (394) are replaced by

$$\begin{aligned} &a_1 > 0,\, a_2 > 0, \ldots,\, a_n > 0, \\ &\Delta_{n-1} > 0, \qquad \Delta_{n-3} > 0, \ldots. \end{aligned} \tag{403}$$

These also are necessary and sufficient conditions so that (403) is equivalent to (394) (for example, see Gantmacher (1970)). For the example (399), one thus immediately obtains (402) in place of (401).

It will now be investigated in what manner the stability behavior of (373) is determined by that of the linearized differential equation (374). If all of the real parts of the EV of $\boldsymbol{A}$ are negative, then an approach analogous to the one described in (384) may be used to construct a quadratic form which is itself positive definite and whose derivative is negative definite on the solutions of (374). This quadratic form then may also be used as a LIAPOUNOV function for (373). If one computes its time derivatives on the solutions of (373), then $\dot{V}$ is given by the negative definite quadratic form of the linear case with additional 'small terms of higher order', since $\boldsymbol{h}(\boldsymbol{x}, t)$ satisfies the inequality (375); thus, $\dot{V}$ also is negative definite on the solutions of (373)! Also, it follows that the trivial solution of the non-linear equation (373) also is asymptotically stable if all of the EV of $\boldsymbol{A}$ have negative real parts! It may even be shown that one may choose an arbitrary negative definite quadratic form $W = \boldsymbol{x}^\mathsf{T}\boldsymbol{C}\boldsymbol{x}$ in this case, so that $\dot{V} = W(\boldsymbol{x})$ on the solutions of (374), where the corresponding function $V(\boldsymbol{x}) = \boldsymbol{x}^\mathsf{T}\boldsymbol{B}\boldsymbol{x}$ always exists and is positive definite. Because of

$$\dot{V} = \dot{\boldsymbol{x}}^\mathsf{T}\boldsymbol{B}\boldsymbol{x} + \boldsymbol{x}^\mathsf{T}\boldsymbol{B}\dot{\boldsymbol{x}} = \boldsymbol{x}^\mathsf{T}\boldsymbol{A}^\mathsf{T}\boldsymbol{B}\boldsymbol{x} + \boldsymbol{x}^\mathsf{T}\boldsymbol{B}\boldsymbol{A}\boldsymbol{x} = \boldsymbol{x}^\mathsf{T}\boldsymbol{C}\boldsymbol{x} \tag{404}$$

the three matrices then are always related by the 'LIAPOUNOV' equation

$$\boldsymbol{A}^\mathsf{T}\boldsymbol{B} + \boldsymbol{B}\boldsymbol{A} = \boldsymbol{C}. \tag{405}$$

This means that the matrix equation (405) has a positive definite solution $\boldsymbol{B}$ for every negative definite $\boldsymbol{C}$ as long as all of the EV of $\boldsymbol{A}$ have negative real parts (for example, see Hahn (1967)).

In a similar manner, one may construct a LIAPOUNOV function for (374) if at least one of the EV of $\boldsymbol{A}$ has a positive real part, and then use this function to prove the instability of the trivial solution of (373). A summary of these results is given by the statements:

- *if all of the EV of* $\boldsymbol{A}$ *have negative real parts, then the trivial solution of* (373) *is asymptotically stable;*
- *if at least one EV of* $\boldsymbol{A}$ *has a positive real part, then the trivial solution of* (373) *is unstable.*

These statements are valid independent of the higher order terms; $h(\boldsymbol{x}, t)$ need only satisfy the inequality (375). The linear part of (373) then is also said to exhibit a dominant stability behavior. However, one may also show:

- *if* $\boldsymbol{A}$ *has no EV with positive real part but at least one EV with vanishing real part, then the higher order terms in* (373) *may always be chosen in such a manner that one obtains either stability or instability, just as desired.*

In this case, the question of stability may not be decided on the basis of the linearized equation, but the effect of the non-linear terms must be taken into account; this is called the *critical case.* The solution of the stability problem in the critical case is often quite difficult and may be accomplished only in the special cases where one is able to find a LIAPOUNOV function. Systematic investigations of various types of critical cases may be found in Malkin (1959) for example.

Now, some *examples* appropriate to these stability and linearization theorems.

(1) Rotation of a rigid body about a fixed point. Consider once again the force free and moment free motion of a rigid body about a point (Example (4) in section 3.2). The equations of motion for 'disturbances' of the rotation with angular speed ω about an axis with corresponding moment of inertia $\boldsymbol{A}$ were given in section 3.1 as:

$$\begin{aligned} \dot{y}_1 &= \frac{B-C}{A} y_2 y_3, \\ \dot{y}_2 &= \frac{C-A}{B} (y_1 + \omega) y_3, \\ \dot{y}_3 &= \frac{A-B}{C} (y_1 + \omega) y_2; \end{aligned} \tag{406}$$

with corresponding linearized differential equations

$$\begin{aligned} \dot{y}_1 &= 0, \\ \dot{y}_2 &= \frac{C-A}{B} \omega y_3, \\ \dot{y}_3 &= \frac{A-B}{C} \omega y_2 \end{aligned} \tag{407}$$

and with characteristic equation

$$\begin{vmatrix} -\lambda & 0 & 0 \\ 0 & -\lambda & \dfrac{C-A}{B}\omega \\ 0 & \dfrac{A-B}{C}\omega & -\lambda \end{vmatrix} = 0. \tag{408}$$

It has a root $\lambda_1 = 0$ along with

$$\lambda_{2,3} = \pm\omega\sqrt{\frac{(A-B)(C-A)}{BC}}. \tag{409}$$

For the case $A > B, C$ or for $A < B, C$, λ_2 and λ_3 are imaginary and all three EV have zero real parts: the case is a critical case, that is, the linearized differential equations yield no information concerning the stability of rotation about the principal axes corresponding to the largest and the smallest principal moment of inertia. In this case, stability has already been shown previously (Example (4) of section 3.2) by means of a suitable LIAPOUNOV function.

However, if $B > A > C$ or $C > A > B$, then λ_2 and λ_3 are real. Then there exists one EV with positive real part and it follows that the trivial solution of (407) as well as that of the non-linear (324) are unstable. Thus, rotations about the principal axis corresponding to the intermediate principal moment of inertia are unstable!

(2) The heavy symmetric top. The stability of rotation of a heavy symmetric top about its vertical axis of symmetry will now be investigated. First, the equations of motion are established. The usual manner of defining the EULER angles is illustrated in Figure 37; the $Oxyz$ axis system is fixed in space. It is assumed that the axi-symmetric top rests in a frictionless ball-joint at 0 and that the ζ-axis coincides with the axis of symmetry passing through 0 and through the center of gravity S, with C as the corresponding moment of inertia. In order to avoid any indeterminacy of the angles Φ and Ψ for the vertical position of the axis of the top, the positive y-direction is taken to be the same as that of the gravitational acceleration (the coordinates of the upright top are then given by $\Phi = 0$, $\Theta = \frac{\pi}{2}$, Ψ arbitrary).

The potential energy thus may be written as

$$U = mgl \sin\Theta \cos\Phi \tag{410}$$

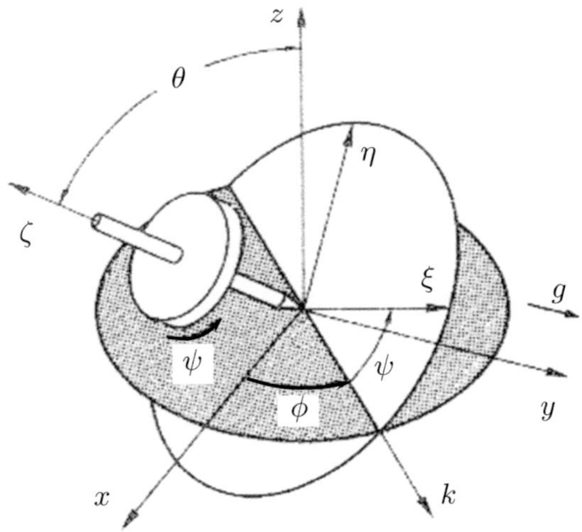

Figure 37: Definition of the EULER angles.

with $l = |OS|$, and the kinetic energy as

$$T = \frac{1}{2}A\left(\dot{\Theta}\cos\Psi + \dot{\Phi}\sin\Theta\sin\Psi\right)^2 + \frac{1}{2}B\left(\dot{\Theta}\sin\Psi - \dot{\Phi}\sin\Theta\cos\Psi\right)^2 + \\ + \frac{1}{2}C\left(\dot{\Psi} + \dot{\Phi}\cos\Theta\right)^2, \tag{411}$$

where A and B are the moment of inertia about the ξ and η axes, respectively. Because of symmetry, $A = B$ so that the Lagrangian has the form

$$L = T - U = \frac{1}{2}A\left(\dot{\Theta}^2 + \dot{\Phi}^2\sin^2\Theta\right) + \frac{1}{2}C\left(\dot{\Psi} + \dot{\Phi}\cos\Theta\right)^2 - \\ - mgl\sin\Theta\cos\Phi. \tag{412}$$

The Lagrangian equations for Θ and Φ are given by

$$A\left(\ddot{\Theta} - \dot{\Phi}^2\sin\Theta\cos\Theta\right) + C\left(\dot{\Psi} + \dot{\Psi}\cos\Theta\right)\dot{\Phi}\sin\Theta + mgl\cos\Theta\cos\Phi = 0, \tag{413}$$

$$A\left(\ddot{\Phi}\sin\Theta + 2\dot{\Theta}\dot{\Phi}\cos\Theta\right)\sin\Theta + C\frac{\mathrm{d}}{\mathrm{d}t}\left(\left(\dot{\Psi} + \dot{\Phi}\cos\Theta\right)\cos\Theta\right) - \\ - mgl\sin\Theta\sin\Phi = 0. \tag{414}$$

The angle Ψ is a cyclic coordinate from which it follows that

$$\frac{\partial L}{\partial \dot{\Psi}} = C\left(\dot{\Psi} + \dot{\Phi}\cos\Theta\right) = nC \tag{415}$$

is constant; that is, the 'eigenrotation' n (the component of the angular velocity in the direction of the axis of symmetry) is also constant. For $\sin\Theta \neq 0$, eqns (413) and (414) then take on the form

$$\begin{aligned} &A\left(\ddot{\Theta} - \dot{\Phi}^2\cos\Theta\sin\Theta\right) + Cn\dot{\Phi}\sin\Theta + mgl\cos\Theta\cos\Phi = 0, \\ &A\left(\ddot{\Phi}\sin\Theta + 2\dot{\Theta}\dot{\Phi}\cos\Theta\right) - Cn\dot{\Theta} - mgl\sin\Phi = 0 \end{aligned} \tag{416}$$

with the particular solution $\Theta = \frac{1}{2}\pi$, $\Phi = 0$, corresponding to the rotation of the upright top about its axis of symmetry. Experience indicates that this axis position is stable as long as n is sufficiently large.

Use

$$\begin{aligned} \Theta &= \frac{\pi}{2} + \bar{\Theta}, \\ \Phi &= \bar{\Phi} \end{aligned} \tag{417}$$

and linearize (416) with respect to the 'disturbances' $\bar{\Theta}$ and $\bar{\Phi}$. This results in

$$\begin{aligned} A\ddot{\bar{\Theta}} + Cn\dot{\bar{\Phi}} - mgl\bar{\Theta} &= 0, \\ A\ddot{\bar{\Phi}} - Cn\dot{\bar{\Theta}} - mgl\bar{\Phi} &= 0 \end{aligned} \tag{418}$$

with characteristic equation

$$\begin{vmatrix} A\lambda^2 - mgl & Cn\lambda \\ -Cn\lambda & A\lambda^2 - mgl \end{vmatrix} = 0 \tag{419}$$

or, equivalently,

$$A^2\lambda^4 + \left(-2Amgl + C^2n^2\right)\lambda^2 + (mgl)^2 = 0 \tag{420}$$

(it is superfluous here to write (418) as a first-order system). A necessary and sufficient condition to assure that there exist no roots with positive real part is given by

$$C^2n^2 > 4Amgl. \tag{421}$$

If this condition is violated, then the rotation of the upright top is unstable; if the condition is satisfied, then one has the critical case and the non-linear terms must be taken into consideration in order to draw any final conclusions about stability. It can be shown that the rotation about the upright axis of symmetry of the top is indeed stable as long as (421) is satisfied.

3.5 Stability by the First Approximation (Periodic Case)

In the previous section it was assumed that linearization resulted in a linear differential equation with constant coefficients. It will now be assumed that

$$\dot{\boldsymbol{x}} = \boldsymbol{A}(t)\boldsymbol{x} + \boldsymbol{h}(\boldsymbol{x}, t) \tag{422}$$

holds, where $\boldsymbol{h}(\boldsymbol{x}, t)$ is assumed small in the sense indicated in condition (375). Again, the linearized differential equation

$$\dot{\boldsymbol{x}} = \boldsymbol{A}(t)\boldsymbol{x} \tag{423}$$

is investigated first and any implications concerning the stability of the trivial solution of (422) are treated subsequently.

The square matrix $\boldsymbol{F}(t)$ which satisfies the differential equation

$$\dot{\boldsymbol{F}}(t) = \boldsymbol{A}(t)\boldsymbol{F}(t) \tag{424}$$

subject to the initial condition $\boldsymbol{F}(t_0) = \boldsymbol{E}$ (the unit matrix) is called the *fundamental matrix* of the system (423). The specific solution determined by the initial condition $\boldsymbol{x}(t_0) = \boldsymbol{x}_0$ then is given by $\boldsymbol{x}(t) = \boldsymbol{F}(t)\boldsymbol{x}_0$. The existence of the fundamental matrix of (423) follows from known existence theorems (e.g., see Pontryagin (1965)).

For a constant $\boldsymbol{A}$ the form of the fundamental matrix of (423) is easily deduced, since the ith column of $\boldsymbol{F}(t)$ is given by the solution vector $\boldsymbol{x}(t)$ corresponding to the initial conditions $x_1 = x_2 = \ldots x_{i-1} = x_{i+1} \cdots = x_n = 0$, $x_i = 1$. With $t_0 = 0$ one often writes

$$\boldsymbol{F}(t) = \mathrm{e}^{\boldsymbol{A}t} = \left(\boldsymbol{E} + \boldsymbol{A}t + \frac{1}{2!}\boldsymbol{A}^2t^2 + \frac{1}{3!}\boldsymbol{A}^3t^3 + \ldots\right); \tag{425}$$

convergence may easily be proven for arbitrary values of t.

Next, the periodic case is discussed, where $\boldsymbol{A}(t)$ is periodic with period T so that one has $\boldsymbol{A}(t+1) = \boldsymbol{A}(t)$, $\forall t \geq 0$. Periodic differential equations of this type were investigated by FLOQUET. LIAPOUNOV later showed that corresponding to every periodic matrix $\boldsymbol{A}(t)$ there always exists a non-singular continuous matrix $\boldsymbol{P}(t)$ with the same period T and with $\boldsymbol{P}(t_0) = \boldsymbol{P}(t_0 + T) = \boldsymbol{E}$, such that the transformation

$$\boldsymbol{y} = \boldsymbol{P}(t)\boldsymbol{x} \tag{426}$$

transforms the differential eqn (423) into the form

$$\dot{\boldsymbol{y}} = \boldsymbol{B}\boldsymbol{y} \tag{427}$$

where $\boldsymbol{B} = \left(\dot{\boldsymbol{P}}(t) + \boldsymbol{P}(t)\boldsymbol{A}(t)\right)\boldsymbol{P}^{-1}(t)$ is a constant matrix independent of t. The solutions of (427), however, have already been discussed in section 3.4. If the fundamental matrix of (427) for $t_0 = 0$ is written as $\mathrm{e}^{\boldsymbol{B}t}$, then the fundamental matrix of (423) may be written as

$$\boldsymbol{F}(t) = \boldsymbol{P}^{-1}(t)\mathrm{e}^{\boldsymbol{B}t} \tag{428}$$

in view of the transformation (426). It follows that the solutions of (423) are products of periodic functions with $\mathrm{e}^{\boldsymbol{B}t}$ and stability is determined by the eigenvalues of $\boldsymbol{B}$. Unfortunately, it is generally not possible to obtain the matrix $\boldsymbol{P}(t)$ explicitly. If it were, one could compute $\boldsymbol{B}$ and then use the HURWITZ criterion on $\boldsymbol{B}$, for example. In stability investigations of the linear periodic differential eqn (423), it thus is common to obtain an approximation to

$$\boldsymbol{F}(t) = \boldsymbol{P}^{-1}(T)\mathrm{e}^{\boldsymbol{B}T} = \mathrm{e}^{\boldsymbol{B}T} \tag{429}$$

by numerical or analytical means. In view of

$$\boldsymbol{x}(T) = \mathrm{e}^{\boldsymbol{B}T}\boldsymbol{x}(0) \tag{430}$$

and because of the continuity of $\boldsymbol{P}(t)$, one has:

- *the trivial solution of* (423) *is asymptotically stable iff all of the EV of the matrix* $\mathrm{e}^{\boldsymbol{B}T}$ *have a magnitude less than one;*
- *if an EV of the matrix* $\mathrm{e}^{\boldsymbol{B}T}$ *with magnitude greater than one exists, then the trivial solution of* (423) *is unstable;*
- *if the matrix* $\mathrm{e}^{\boldsymbol{B}T}$ *has no EV with magnitude greater than one but has one or more EV with magnitude one, then the trivial solution of* (423) *may be stable or unstable, depending on the structure of the* JORDAN *normal form corresponding to* $\mathrm{e}^{\boldsymbol{B}T}$.

Note than an EV of the matrix $\mathrm{e}^{\boldsymbol{B}T}$ with magnitude larger than one corresponds to an EV with positive real part for the matrix $\boldsymbol{B}$ (for example, see Knobloch and Kappel (1974)). The eigenvalues of the matrix $\boldsymbol{B}$ are often designated as the characteristic exponents of the system (423).

Thus, if one wished to solve the stability problem of the linear system, one would first determine an approximate solution of the matrix equation (423) on the interval $[0, T]$ with initial conditions $\boldsymbol{F}(0) = \boldsymbol{E}$. Thereafter, one computes the EV of $\mathrm{e}^{\boldsymbol{B}T} = \boldsymbol{F}(T)$ and checks their magnitudes. Usually, this procedure would be applied in a form which is tailored to the particular type of differential eqn (423) under investigation (for example, see Malkin (1959)).

Ultimately, one may prove linearization theorems which correspond exactly to those for systems with constant coefficients:

- *if all of the EV of e^{BT} have magnitudes smaller than one, then the trivial solution of* (422) *is asymptotically stable, independent of the form of the higher order terms;*
- *if e^{BT} has at least one EV of magnitude greater than one, then the trivial solution of* (422) *is unstable, independent of the form of the higher order terms;*
- *if e^{BT} has no EV of magnitude greater than one but has at least one EV of magnitude equal to one, then the non-linear terms $\boldsymbol{h}(\boldsymbol{x}, t)$ may always be chosen in such a way that one obtains either stability or instability of the trivial solution in* (422), *just as desired.*

Because of the complicated calculations involved in the analytical determination of the fundamental matrix by means of perturbation or other methods, only a few simple examples will be considered here:

(1) The HILL *and the* MATHIEU *differential equations.* One of the most important ordinary differential equations with periodic coefficients is HILL's differential equation

$$\ddot{x} + g(t)x = 0, \tag{431}$$

where $g(t)$ is a periodic function with period T. For example, such a differential equation describes the oscillations of a spring-mass-system with periodically changing spring stiffness, the small (linear) oscillations of a pendulum with periodically excited support, the transverse oscillations of a beam with pulsating axial load, as well as the oscillations of a loudspeaker signal coil. The most important special case arises when $g(t)$ has the form

$$g(t) = \omega_0^2(1 + \epsilon \cos \Omega t) \tag{432}$$

in which case (431) becomes the MATHIEU differential equation

$$\ddot{x} + \omega_0^2(1 + \epsilon \cos \Omega t)x = 0. \tag{433}$$

Here, parametrically excited oscillations as described by (431) with $\epsilon \ll 1$ will be treated. This terminology is founded on the fact that one does not have a periodic excitation independent of x as is usual in the treatment of forced oscillations, that is, a term of the type $P \sin \Omega t$ 'on the right-hand side', but much rather that the system parameters such as the mass or the stiffness themselves are periodic functions of the time. Properly, this type of problem belongs in the realm of linear oscillations.

The stability of the trivial solution $x(t) \equiv 0$ of (431) and (433) is to be investigated. Consider, for the moment, the more general eqn (431) and write it in the form

$$\begin{aligned} \dot{x}_1 &= x_2, \\ \dot{x}_2 &= -g(t)x_1 \end{aligned} \tag{434}$$

in which one now needs to compute the eigenvalues of the matrix $\boldsymbol{F} = \mathrm{e}^{\boldsymbol{B}T}$ (see (429)). One has

$$\boldsymbol{F}(T) = \begin{pmatrix} \bar{x}_1(T) & \bar{\bar{x}}_1(T) \\ \bar{x}_2(T) & \bar{\bar{x}}_2(T) \end{pmatrix}, \tag{435}$$

where $\bar{x}(t)$ and $\bar{\bar{x}}(t)$ are the solutions of (434) corresponding to the initial conditions $\bar{x}_1(0) = 1$, $\bar{x}_2(0) = 0$, and $\bar{\bar{x}}_1(0) = 0$, $\bar{\bar{x}}_2(0) = 1$. The solutions may be determined approximately by numerical or analytical means. The characteristic equation is obtained in the form

$$\begin{vmatrix} \bar{x}_1(T) - \rho & \bar{\bar{x}}_1(T) \\ \bar{x}_2(T) & \bar{\bar{x}}_2(T) - \rho \end{vmatrix} = 0, \tag{436}$$

which may also be written as

$$\rho^2 - 2a\rho + 1 = 0 \tag{437}$$

with $a = \frac{1}{2}\left(\bar{x}_1(T) + \bar{\bar{x}}_2(T)\right)$.[20] The product of the two roots of (437) is always equal to one so that either both roots have magnitude one or one has magnitude greater than one and the other a magnitude less than one. From $\rho = a \pm \sqrt{(a^2 - 1)}$ it follows that for $a^2 < 1$ both roots are complex with magnitude one, whereas the roots are real, one with magnitude larger and one with magnitude smaller than one, for $a^2 > 1$. For the limiting case a $a^2 = 1$, one has the double root $\rho = a$; that is, either $\rho = 1$ or $\rho = -1$ as roots of multiplicity 2. From the theorems cited, it now is apparent that the trivial solution of (431) is *unstable for* $a^2 > 1$ and is *stable* (weakly stable) *for* $a^2 < 1$. The result for $a^2 > 1$ remains valid even when 'small' non-linear terms are added in (431).

Thus, stability results for (431) may be obtained in the following manner: the solutions $\bar{x}(t)$ and $\bar{\bar{x}}(t)$ on the interval $[0, T]$ are determined first; thereafter, with $a = \frac{1}{2}\left(\bar{x}_1(T) + \bar{\bar{x}}_2(T)\right)$ instability may be deduced for $a^2 > 1$ and stability for $a^2 < 1$.

The stability of solutions of the Mathieu differential equation (433) obviously depends on the values of the parameters ω_0, Ω, and ϵ. The introduction of the dimensionless time τ, with $\Omega t = 2\tau$, in (433) results in

$$x'' + (\delta + \eta \cos 2\tau)x = 0\ , \tag{438}$$

[20] If $\boldsymbol{F}(r)$ is the fundamental matrix of $\dot{\boldsymbol{x}} = \boldsymbol{A}(t)\boldsymbol{x}$, then the Wronskian determinant is given by

$$\Delta(t) = \det \boldsymbol{F}(t) = \Delta(t_0)\exp\left[\int_{t_0}^{t_1} \sum_{i=1}^{n} a_{ii} \mathrm{d}t\right],$$

and it then follows from (434) that the ρ-independent term in (437) is equal to unity.

where $x'' = \frac{\mathrm{d}^2 x}{\mathrm{d}\tau^2}$, $\delta = \left(\frac{2\omega_0}{\Omega}\right)^2$, $\eta = \epsilon\left(\frac{2\omega_0}{\Omega}\right)^2$ so that the stability of the trivial solution depends only on δ and η or, equivalently, on $\frac{\Omega}{\omega_0}$ and ϵ. In the parameter plane of δ, η or $\frac{\Omega}{\omega_0}$, ϵ, respectively, there exist regions in which (438) is unstable and others in which (438) is stable (because of linearity, one may speak of the stability of a differential equation, since all of the solutions exhibit the same stability behavior).

Approximate solutions for $\bar{x}(\tau)$ and $\bar{\bar{x}}(\tau)$ in the form $x(\tau) = x_0(\tau) + \eta x_1(\tau) + \eta^2 x_2(\tau) + \ldots$ may be obtained in the by now familiar manner; this is relatively easy, since only solutions on the interval $[0, \pi]$ are to be approximated so that the secular terms here do not introduce the difficulties which are encountered for approximations which are to be valid for 'secular times' (large values of τ) also.

The stability investigation is simplified still further if one realizes that the stability[21] and instability regions in the δ, η-plane are separated by curves whose corresponding parameter values result in periodic solutions. Thus, the boundaries of the stability regions in the parameter plane may be obtained by determining those parameters δ, η which result in periodic solutions. This will be done by using perturbation methods based on LINDSTEDT's procedure. The treatment will be limited to small values of δ and η. With

$$\begin{aligned} \delta &= n^2 + \eta\delta_1 + \eta^2\delta_2 + \ldots, \\ x(\tau) &= x_0 + \eta x_1 + \eta^2 x_2 + \ldots \end{aligned} \tag{439}$$

equation (438) may be used to deduce

$$\begin{aligned} x_0'' + n^2 x_0 &= 0, \\ x_1'' + n^2 x_1 &= -\left(\delta_1 + \cos 2\tau\right) x_0, \\ x_2'' + n^2 x_2 &= -\left(\delta_1 + \cos 2\tau\right) x_1 - \delta_2 x_0, \ \ldots . \end{aligned} \tag{440}$$

It can be shown that periodic solutions are possible only for $n = 1, 2, \ldots$ (with the assumption of another value for n it becomes apparent after a computation of the first approximation that the resulting solution cannot be periodic). Periodic solutions with initial conditions $x(0) = 1$, $x'(0) = 0$, and $x(0) = 0$, $x'(0) = 1$ are to be determined, and a convenient way to satisfy these conditions is to impose them on x_0 in the calculations of the

[21] More precisely: those regions in the parameter plane whose points correspond to parameter values for which the trivial solution of (433) is stable.

first approximation. Consequently, one has

$$x_0 = \cos n\tau \tag{441a}$$

or, respectively,

$$x_0 = \frac{1}{n}\sin n\tau. \tag{441b}$$

The first approximations corresponding to $n = 0, 1, 2$ are now computed. For $n = 0$ with $x_0 \equiv 1$ the expression (441a)-(a) results in

$$x_1'' = -\delta_1 - \cos 2\tau; \tag{442}$$

for periodic solutions, one must choose $\delta_1 = 0$ so that

$$x_1 = \frac{1}{4}\cos 2\tau + C. \tag{443}$$

This yields

$$x_2'' = -\delta_2 - \frac{1}{8} - C\cos 2\tau - \frac{1}{8}\cos 4\tau, \tag{444}$$

and $\delta_2 = -\frac{1}{8}$ is chosen. Based on (439), one now has

$$\delta = -\frac{1}{8}\eta^2 + \mathcal{O}\left(\eta^3\right). \tag{445}$$

For $n = 0$, the expression (441b) must be replaced by $x_0 \equiv 0$, and it is impossible to construct a non-trivial periodic solution from this initial formulation!

For $n = 1$, the procedure is begun with $x_0(\tau) = \cos\tau$ which results in

$$x_1'' + x_1 = -\left(\delta_1 + \frac{1}{2}\right)\cos\tau - \frac{1}{2}\cos 3\tau, \tag{446}$$

with $\delta_1 = -\frac{1}{2}$ and $x_1(\tau) = \frac{1}{16}(\cos 3\tau - \cos\tau)$ as a consequence. From

$$x_2'' + x_2 = -\left(\frac{1}{32} + \delta_2\right)\cos\tau + \frac{1}{16}\cos 3\tau - \frac{1}{32}\cos 5\tau \tag{447}$$

one then concludes $\delta_2 = -\frac{1}{32}$ and obtains

$$\delta = 1 - \frac{1}{2}\eta - \frac{1}{32}\eta^2 + \mathcal{O}\left(\eta^3\right). \tag{448}$$

If the process had been begun with $x_0(\tau) = \sin\tau$, the result would have been

$$\delta = 1 + \frac{1}{2}\eta - \frac{1}{32}\eta^2 + \mathcal{O}\left(\eta^3\right). \tag{449}$$

For $n = 2$, the same approach yields the two equations

$$\delta = 4 + \frac{5}{48}\eta^2 + \mathcal{O}\left(\eta^3\right) \tag{450}$$

and

$$\delta = 4 - \frac{1}{48}\eta^2 + \mathcal{O}\left(\eta^3\right). \tag{451}$$

A sketch of the curves $\delta(\eta)$ in terms of the variables $\frac{\Omega}{\omega_0}$ and ϵ in accordance with (448) through (451) yields the wedge-shaped, shaded regions emanating from the $\frac{\Omega}{\omega_0}$-axis as shown in Figure 38; although not calculated above, the curves for $n = 3, 4$ have also been included. Parameter combinations

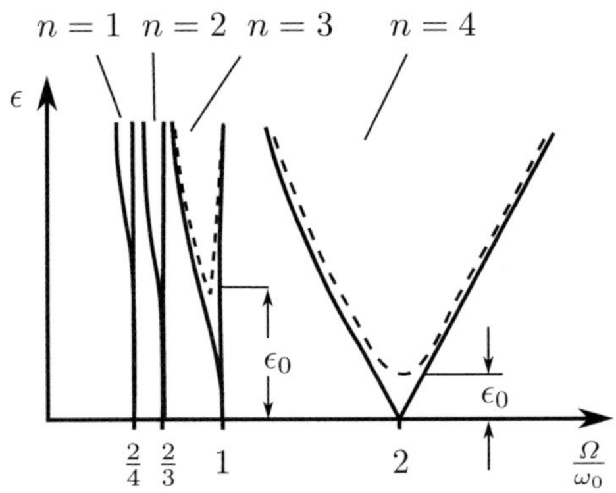

Figure 38: The instability regions of the MATHIEU differential equation.

corresponding to points on these curves result in periodic solutions. For the shaded regions in between, one has $a^2 > 1$ resulting in an unstable trivial solution. For parameter combinations which are located outside the shaded regions, the trivial solution is stable. The instability regions emanate from the points $\frac{\Omega}{\omega_0} = \frac{2}{n}$ and they accumulate at $\frac{\Omega}{\omega_0} = 0$. If one, futhermore, includes linear damping, (433) must be replaced by

$$\ddot{x} + \beta\dot{x} + \omega_0^2(1 + \epsilon\cos\Omega t)x = 0. \tag{452}$$

The instability regions then are bounded by the dashed lines in Figure 38. For sufficiently small values of the excitation (small values of ϵ), the trivial solution always remains asymptotically stable and corresponding to every instability region there exists a 'threshold' $\epsilon = \epsilon_0$ which must be exceeded before instability can occur. The linear differential eqn (452) has a dominant stability behavior so that non-linear terms can no longer eliminate the stability or instability of the trivial solution.

(2) The MATHIEU *differential equation with additional non-linear terms.* Consider again the MATHIEU differential equation with additional linear and non-linear damping terms as well as a non-linear restoring force so that the equation now has the form

$$\ddot{x} + \omega_0^2(1 + \epsilon \cos \Omega t)x + \beta \dot{x} + \delta \dot{x}^2 \operatorname{sgn} \dot{x} + \alpha x^3 = 0. \tag{453}$$

With α, β and δ small and of the same order of magnitude as ϵ, eqn (453) may also be written as

$$\ddot{x} + \omega_0^2(1 + \epsilon \cos \Omega t)x + \epsilon \left(e\dot{x} + D\dot{x}^2 \operatorname{sgn} \dot{x} + f x^3\right) = 0. \tag{454}$$

From Figure 38 it is apparent that the first-order instability region, emanating from $\frac{\Omega}{\omega_0} = 2$ for the undamped case, is the most important region in any applications; thus, only this region will be investigated here. The method of slowly changing phase and amplitude will be used and special consideration will be given to the non-trivial stationary solutions and their stability.

The time is transformed with the substitution $\Omega t = \tau$; the derivative with respect to the transformed time is designated by a prime with $\dot{x} = \Omega x'$. The stability of the solutions in a neighborhood of the critical frequency $\Omega_1 = 2\omega_0$ is of particular interest, and it is thus convenient to use $\Omega = \Omega_1(1-\lambda)$ where λ is a small factor of the order of magnitude of ϵ. The use of the abbreviations $K = \frac{\omega_0}{\Omega_1}$, $A = \frac{f}{\omega_0^2}$, and $E = \frac{e}{\omega_0}$ along with the omission of terms which are small and of second order in equation (454) result in

$$\begin{aligned} x'' + K^2(1+2\lambda)x + \epsilon \big(K^2 x \cos\tau + EKx' + \\ + Dx'^2 \operatorname{sgn} x' + AK^2 x^3\big) = 0. \end{aligned} \tag{455}$$

With the transformations

$$\begin{aligned} x &= a(\tau) \sin\{K\tau + \Psi(\tau)\}, \\ x' &= Ka(\tau) \cos\{K\tau + \Psi(\tau)\}, \end{aligned} \tag{456}$$

where $a(\tau) > 0$, eqn (455) may be written as

$$\begin{aligned} a' &= -\lambda K a \sin 2q + \epsilon K a \left\{ -\frac{1}{2} \sin 2q \cos \tau - E \cos^2 q - \right. \\ &\quad \left. - D a \cos^3 q \operatorname{sgn}(\cos q) - A a^2 \sin^3 q \cos q \right\}, \\ \Psi' &= 2\lambda K \sin^2 q + \epsilon K \left\{ \sin^2 q \cos \tau + E \sin q \cos q + \right. \\ &\quad \left. + D a \sin q \cos^2 q \operatorname{sgn}(\cos q) + A a^2 \sin^4 q \right\}, \end{aligned} \tag{457}$$

where $q = K\tau + \Psi(\tau)$.

It is now assumed that $a(\tau)$ and $\Psi(\tau)$ are 'slowly' varying and that the right-hand sides of (457) may thus be replaced by their temporal means. The result of a choice of $K = \frac{\omega_0}{\Omega_1} = \frac{1}{2}$ for the first-order instability region then results in

$$\begin{aligned} a' &= -\frac{\epsilon}{8} a \sin 2\Psi - \frac{2\epsilon}{3\pi} D a^2 - \frac{\epsilon}{4} E a, \\ \Psi' &= -\frac{\epsilon}{8} \cos 2\Psi + \frac{\lambda}{2} + \frac{3\epsilon}{16} A a^2. \end{aligned} \tag{458}$$

Naturally, this system has the trivial solution $a \equiv 0$; but, in addition, there are also other stationary non-trivial solutions given by $a = a_0 = \text{const.}$, $\Psi = \Psi_0 = \text{const.}$ The non-trivial stationary solutions are investigated first.

The elimination of Ψ_0 from eqns (458) easily yields the expression for the deviation λ of the frequency Ω from the nominal value Ω_1 in terms of the stationary amplitude a_0:

$$\lambda = -\frac{3\epsilon}{8} A a_0^2 \pm \epsilon \sqrt{\frac{1}{16} - \frac{E^2}{4} - \frac{16}{9\pi^2} D^2 a_0^2 - \frac{4}{3\pi} D E a_0}. \tag{459}$$

Real values for λ are obtained when the expression under the radical is non-negative; two cases are to be distinguished . If $D = 0$, that is, no quadratic damping is present, then one obtains real values for λ as long as $E \leq \frac{1}{2}$; a_0 may take on arbitrary values in this case. For $D > 0$ there exist non-trivial stationary solutions for all those values of a_0 which satisfy the inequality

$$a_0 \leq \frac{3\pi}{16D}(1 - 2E)$$

(see Figure 39).

From (459) it is furthermore apparent that the values of λ become independent of the constants A and D for $a_0 \to 0$.

Next, the stability of the stationary solutions just treated will be considered. Toward this purpose, a and Ψ in (458) are replaced by $a = a_0 + \bar{a}$ and $\Psi = \Psi_0 + \bar{\Psi}$, where $\bar{a}$ and $\bar{\Psi}$ represent the deviations from the stationary

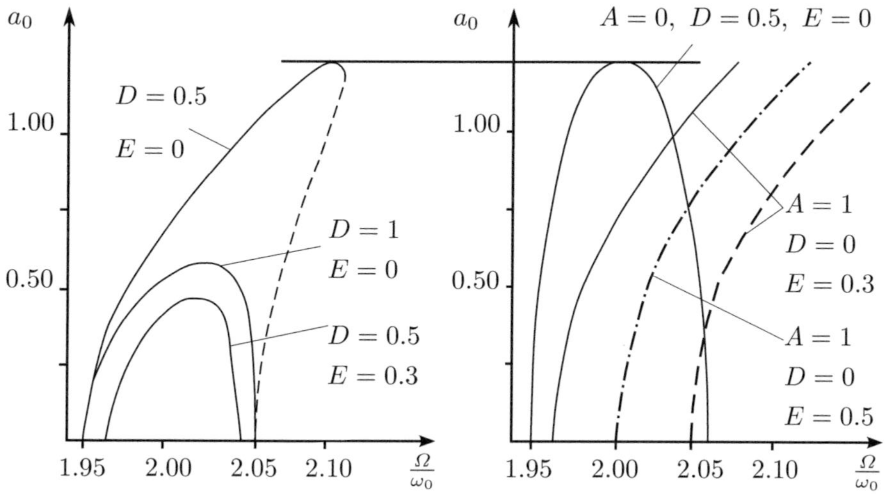

Figure 39: Instability regions of the MATHIEU-equation with linear and with quadratic damping and with cubic restoring term. Left with A=1 and $\epsilon = 0.1$, right with $\epsilon = 0.1$.

solution. Their substitution in (458) and the subsequent linearization of the equation with respect to $\bar{a}$ and $\bar{\Psi}$ results in

$$\begin{aligned} \bar{a}' &= \epsilon \left\{ -\frac{1}{8} \sin 2\Psi_0 - \frac{4}{3}\frac{D}{\pi} a_0 - \frac{E}{4} \right\} \bar{a} - \epsilon \left\{ \frac{a_0}{4} \cos 2\Psi_0 \right\} \bar{\Psi}, \\ \bar{\Psi}' &= \epsilon \frac{3A}{8} a_0 \bar{a} + \epsilon \left\{ \frac{1}{4} \sin 2\Psi_0 \right\} \bar{\Psi}. \end{aligned} \tag{460}$$

These equations may be used to investigate the stability of the stationary solutions a_0, Ψ_0 of (458) by means of the method of the first approximation. An application of the HURWITZ criterion then yields the stability conditions

$$\epsilon \left[\frac{2D}{\pi} a_0 + \frac{E}{2} \right] > 0, \tag{461}$$

$$A \frac{\lambda}{\epsilon} > -\frac{64}{27\pi^2} D^2 - \frac{8}{9\pi} \frac{DE}{a_0} - \frac{3}{8} A^2 a_0^2. \tag{462}$$

The inequality (461) is always satisfied as long as $\epsilon \neq 0$ and D and E are not simultaneously zero. When the values of λ resulting from (459) are substituted in (462), one obtains the condition

$$\pm A\sqrt{\frac{1}{16}-\frac{E^2}{4}-\frac{16}{9\pi^2}D^2a_0^2-\frac{4}{3\pi}DEa_0} > -\frac{64}{27\pi^2}D^2-\frac{8\,DE}{9\pi\,a_0}. \tag{463}$$

For the upper sign, this inequality is always satisfied, that is, the corresponding branch of the resonance curve $a_0(\lambda)$ is always stable; it corresponds to the left (increasing) part of the resonance curve in Figure 39. Generally, the condition (463) is also satisfied for the lower sign for certain values of a_0; the stability boundary is however difficult to obtain explicitly. Eventually, one arrives at the following stability results:

- For $D = 0$, $0 < E < \frac{1}{2}$, $\epsilon > 0$ the right-hand part of the resonance curve in Figure 39 is always unstable.
- For $A = 0$, all of the resonance curve is stable, as long as $D > 0$ and $\epsilon > 0$ are valid.
- For $E = 0$, the right-hand part of the resonance curve is stable for all a_0 which satisfy the inequality

$$a_0 > \frac{3\pi}{4D}\sqrt{\frac{1}{16}-\frac{4096\,D^4}{729\pi^4\,A^2}}. \tag{464}$$

It follows that all of the right-hand part of the resonance curve is stable for $E = 0$ as long as $D^2 > A\frac{27\pi^2}{256}$ the case.

The resonance curves corresponding to several values of A, D, E and ϵ have been depicted in Figure 39. From the figure as well as from (459), it is apparent that the quadratic damping has no influence on the width of the interval which is cut out of the $\frac{\Omega}{\omega_0}$-axis by the resonance curve. The width of this interval clearly depends only on E. An essential difference between the resonance curves corresponding to the case of quadratic damping and that with only linear damping lies in the fact that with increasing values for D and decreasing values for A, the right-hand part of the curve also becomes stable. This is true up to a limiting value of a_0 where the curve has a vertical tangent perpendicular to the $\frac{\Omega}{\omega_0}$-axis; this may easily be deduced from (459) together with (463). A further difference consists of the fact that the resonance curves according to the first approximation now are closed, something which was not the case for linear damping.

Finally, the stability of the trivial solution will be investigated. With the substitution

$$a\sin\Psi = y \quad \text{and} \quad a\cos\Psi = z$$

equations (458) result in

$$y' = -\frac{\epsilon}{4}Ey + \left(-\frac{\epsilon}{8} + \frac{\lambda}{2}\right) z - \frac{2\epsilon}{3\pi} Dy\sqrt{y^2 + z^2} + \\ + \frac{3\epsilon}{16} Az\left(y^2 + z^2\right), \tag{465}$$
$$z' = \left(-\frac{\epsilon}{8} - \frac{\lambda}{2}\right) y - \frac{\epsilon}{4}Ez - \frac{2\epsilon}{3\pi} Dz\sqrt{y^2 + z^2} - \frac{3\epsilon}{16} Ay\left(y^2 + z^2\right).$$

The HURWITZ-criterion may now be used to investigate the stability of the zero position by means of the first method of LIAPOUNOV, where only the linear part of eqns (465) need be considered. This results in the conditions $E > 0$ and $\lambda^2 > \frac{\epsilon^2}{16}\left(1 - 4E^2\right)$. It follows that the null position is always stable as long as $E > \frac{1}{2}$ and that for $0 < E < \frac{1}{2}$ the null position is stable for

$$\lambda > \frac{\epsilon}{4}\sqrt{1 - 4E^2} \quad \text{and} \quad \lambda < -\frac{\epsilon}{4}\sqrt{1 - 4E^2}$$

and is unstable for

$$\frac{\epsilon}{4}\sqrt{1 - 4E^2} > \lambda > -\frac{\epsilon}{4}\sqrt{1 - 4E^2}$$

(see Figure 40).

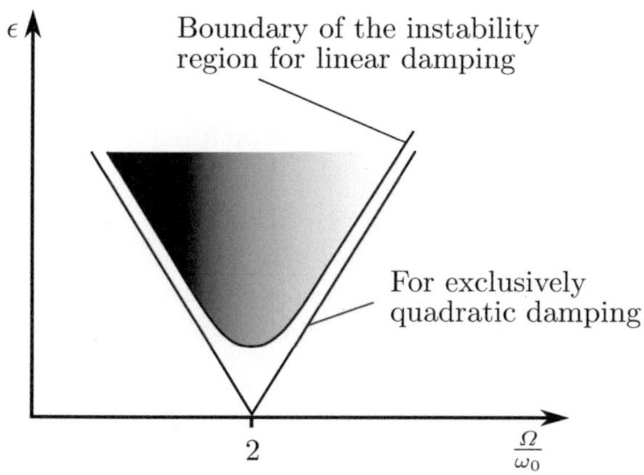

Figure 40: Boundaries of the instability region of first order.

Only the linear damping enters into this result, that is, for $E > 0$ quadratic damping has no influence on the width of the instability region of first order.

The case $E = 0$ still needs to be investigated. The linearization of (465) with $E = 0$ yields

$$\begin{aligned} y' &= \left(-\frac{\epsilon}{8} + \frac{\lambda}{2}\right) z, \\ z' &= \left(-\frac{\epsilon}{8} - \frac{\lambda}{2}\right) y. \end{aligned} \tag{466}$$

The instability of the trivial solution for $-\frac{\epsilon}{4} < \lambda < \frac{\epsilon}{4}$ follows immediately.

For $\lambda > \frac{\epsilon}{4}$ and $\lambda < -\frac{\epsilon}{4}$ no conclusions based on (466) can be drawn, since the trivial solution of (466) is only stable but not asymptotically stable. This corresponds to the critical case, and stability therefore must be investigated by means of LIAPOUNOV's second method or the direct method of LIAPOUNOV.

For $D > 0$ the LIAPOUNOV function $V = \frac{1}{2}y^2 + \frac{1}{2}Bz^2$ may be employed successfully, where B is chosen so as to assure a vanishing of the second-order terms in V'. The use of (465) then yields

$$V' = -\frac{2D}{3\pi}\epsilon\left(y^2 + Bz^2\right)\sqrt{y^2 + z^2} + \frac{3A}{16}\epsilon yz\left(y^2 + z^2\right)(1 - B) \tag{467}$$

with $B = \frac{-\frac{\epsilon}{8}+\frac{\lambda}{2}}{\frac{\epsilon}{8}+\frac{\lambda}{2}} > 0$. For $D > 0$ there always exists a neighbourhood of the point $y = z = 0$ in which (467) is negative definite. It follows that in this case the null position is stable for $\lambda > \frac{\epsilon}{4}$ and for $\lambda < -\frac{\epsilon}{4}$, that is, quadratic damping does not influence the stability of the null position even when $E = 0$.

3.6 Stability by the First Approximation (Aperiodic Case)

In sections 3.4 and 3.5, it was shown that the asymptotic stability of the linear autonomous or periodic system implied the asymptotic stability of the trivial solution of the complete non-linear system. That this is no longer the case when the coefficients of the linear system are arbitrary functions of time is shown by the following counter-example.

Consider the linear system

$$\begin{aligned} \dot{x}_1 &= -bx_1, \\ \dot{x}_2 &= (\sin(\ln t) + \cos(\ln t) - 2b)\, x_2 \end{aligned} \tag{468}$$

for $0 < t_0 \leq t$. It consists of two uncoupled differential equations with the solution

$$\begin{aligned} x_1(t) &= x_{10}\mathrm{e}^{-b(t-t_0)}, \\ x_2(t) &= x_{20}\mathrm{e}^{-(t_0 \sin(\ln t_0) - 2bt_0)}\mathrm{e}^{(t \sin(\ln t) - 2bt)}. \end{aligned} \tag{469}$$

As long as $b > \frac{1}{2}$ the trivial solution of (468) is obviously asymptotically stable.
If one now adds a non-linear term in (468) in such a way that one obtains

$$\begin{aligned} \dot{x}_1 &= -bx_1, \\ \dot{x}_2 &= (\sin(\ln t) + \cos(\ln t) - 2b)\, x_2 + x_1^2 \end{aligned} \tag{470}$$

then the general solution has the form

$$\begin{aligned} x_1(t) &= x_{10}\mathrm{e}^{-b(t-t_0)}, \\ x_2(t) &= \mathrm{e}^{t\sin(\ln t)-2b(t-t_0)} \cdot \Bigl(x_{20}\mathrm{e}^{-t_0\sin(\ln t_0)} + \\ &\quad + x_{10}^2 \int_{t_0}^{t} \mathrm{e}^{-t_1\sin(\ln t_1)}\mathrm{d}t_1\Bigr), \end{aligned} \tag{471}$$

a result which may easily be checked. The trivial solution of (470) however, is unstable, since

$$\mathrm{e}^{t\sin(\ln t)-2bt} \int_{t_0}^{t} \mathrm{e}^{-t_1\sin(\ln t_1)}\mathrm{d}t_1$$

is unbounded as $t \to \infty$. For $t_n = \mathrm{e}^{\left(2n+\frac{1}{2}\right)\pi}$, n an integer, with $t_n\mathrm{e}^{-\pi} > t_0$, one has:

$$\int_{t_0}^{t_n} \mathrm{e}^{-\bar{t}\sin\left(\ln \bar{t}\right)}\mathrm{d}\bar{t} > \int_{t_n\mathrm{e}^{-\pi}}^{t_n\mathrm{e}^{-\frac{2\pi}{3}}} \mathrm{e}^{-\bar{t}\sin\left(\ln \bar{t}\right)}\mathrm{d}\bar{t} = \int_{\mathrm{e}^{\left(2n-\frac{1}{2}\right)\pi}}^{\mathrm{e}^{\left(2n-\frac{1}{6}\right)\pi}} \mathrm{e}^{-\bar{t}\sin\left(\ln \bar{t}\right)}\mathrm{d}\bar{t} >$$

$$\int_{\mathrm{e}^{\left(2n-\frac{1}{2}\right)\pi}}^{\mathrm{e}^{\left(2n-\frac{1}{6}\right)\pi}} \mathrm{e}^{-\bar{t}\sin\left(-\frac{\pi}{6}\right)}\mathrm{d}\bar{t} > \int_{t_n\mathrm{e}^{-\pi}}^{t_n\mathrm{e}^{-\frac{2\pi}{3}}} \mathrm{e}^{-t_n\mathrm{e}^{-\pi}\sin\left(-\frac{\pi}{6}\right)}\mathrm{d}\bar{t} =$$

$$= \mathrm{e}^{\frac{1}{2}t_n\mathrm{e}^{-\pi}} \left\{\mathrm{e}^{-\frac{2\pi}{3}} - \mathrm{e}^{-\pi}\right\} t_n.$$

For $t = t_n$ there thus follows:

$$\mathrm{e}^{t_n \sin(\ln t_n) - 2bt_n} \int\limits_{t_0}^{t_n} \mathrm{e}^{-\bar{t}\sin\left(\ln \bar{t}\right)} \mathrm{d}\bar{t} > \mathrm{e}^{t_n \sin(\ln t_n) - 2bt_n} .$$

$$\cdot \left\{ \mathrm{e}^{-\frac{2\pi}{3}} - \mathrm{e}^{-\pi} \right\} t_n \mathrm{e}^{\frac{1}{2} t_n \mathrm{e}^{-\pi}} =$$

$$= \left\{ \mathrm{e}^{-\frac{2\pi}{3}} - \mathrm{e}^{-\pi} \right\} t_n \mathrm{e}^{t_n \left(\sin(\ln t_n) - 2b + \frac{\mathrm{e}^{-\pi}}{2} \right)}$$

and this expression grows beyond all bounds as $n \to \infty$, as long as one has $1 + \frac{1}{2}\mathrm{e}^{-\pi} > 2b$!

For $\frac{1}{2} < b < \frac{1}{2} + \frac{1}{4}\mathrm{e}^{-\pi}$ one thus has the non-linear system (470) with unstable trivial solution and such that its linearized system is one with an asymptotically stable equilibrium position! It follows that the linearization theorems, as derived for autonomous and periodic systems, may not be generalized to the case with arbitrary time dependence without some additional assumptions!

However, even for systems of linear equations with arbitrary time dependent coefficients, one may define parameters which correspond to the real parts of the characteristic exponents defined in section 3.5, although with reversed sign: they are the *characteristic numbers* of LIAPOUNOV. These characteristic numbers may be used to describe the growth behavior of a function $f(t)$. The *characteristic number* z of a function $f(t)$ is defined in the following manner: *the number z is that number for which the two conditions*

$$\begin{aligned} &\overline{\lim_{t\to\infty}} |f(t)| \mathrm{e}^{(z+\epsilon)t} = +\infty, \qquad && \forall \epsilon > 0 \\ &\lim_{t\to\infty} f(t) \mathrm{e}^{(z-\epsilon)t} = 0, \qquad && \forall \epsilon > 0 \end{aligned} \tag{472}$$

are satisfied.[22] It can be shown that (472) determines the characteristic number of a function $f(t)$ uniquely and that z may also be calculated by means of the expression

$$z = -\overline{\lim_{t\to\infty}} \frac{\ln |f(t)|}{t}. \tag{473}$$

This definition may now be used to define the characteristic numbers of the solutions and consequently the characteristic numbers of a linear system and to obtain a number of stability results (see Malkin (1959)).

As it turns out, one can obtain stronger linearization results for asymptotic stability than for instability. As indicated by Krasovskii (1963), this

[22] $\overline{\lim}$ means 'lim sup'

is due to the fact that *uniform* asymptotic stability is a *structural property* whereas instability is not.

A brief discussion of the concept of uniform stability follows. In the stability definition (312), δ generally does not only depend on ϵ but also on t_0. If it is possible to obtain a function $\delta(\epsilon)$ such that the inequality (312) is satisfied for all t_0, the solution $\boldsymbol{x}(\boldsymbol{a}, t_0; t)$ is said to be *uniformly stable.* The solution $\boldsymbol{x}(\boldsymbol{a}, t_0; t)$ then is said to be *uniformly asymptotically stable* if $\boldsymbol{x}(\boldsymbol{a}, t_0; t)$ is uniformly stable and if one has $\lim_{t\to\infty} |\boldsymbol{x}(\boldsymbol{x}_0, t_0; t) - (\boldsymbol{a}, t_0; t)| = 0$ uniformly for all t_0 and for every $\boldsymbol{x}_0$, $|\boldsymbol{x}| < \delta$.

With this in mind, one may formulate the following linearization theorem, proven in Halanay (1966):

If the trivial solution of

$$\dot{\boldsymbol{x}} = \boldsymbol{A}(t)\boldsymbol{x}$$

is uniformly asymptotically stable, if $\boldsymbol{A}(t)$ *is bounded and if* $|\boldsymbol{h}(y,t)| < c|y|$ *(c is a sufficiently small positive constant which may depend on* $\boldsymbol{A}(t)$*), then the trivial solution of*

$$\dot{\boldsymbol{x}} = \boldsymbol{A}(t)\boldsymbol{x} + \boldsymbol{h}(\boldsymbol{x}, t)$$

is also uniformly asymptotically stable.

In the counter-example (equations (468) - (471)) at the beginning of this section, the conditions for *uniform* asymptotic stability were not satisfied. There are numerous additional theorems concerning linearization and *structural properties* to be found in the literature. However, this interesting and by no means exhausted area will be pursued no further here.

Bibliography

K. Anderson and P. Hagedorn. On the Energy Dissipation in Space Dampers in Bundeled Conductors of Overhead Transmission Lines. *J. of Sound and Vibration*, 180(4):539–556, 1995.

R. Bellman. *Stability theory of differential equations.* Dover, New York, 1969.

N. G. Chetayev. *The stability of motion.* Pergamon Press, New York, 1961.

F. Gantmacher. *Matrizenrechnung [Matrix algebra].* VEB Deutscher Verlag der Wissenschaften, Berlin, 1958.

F. Gantmacher. *Lectures in analytical mechanics.* MIR Publishers, Moscow, 1970.

P. Hagedorn. Die Umkehrung der Stabilitätssätze von Lagrange-Dirichlet und Routh [Converse theorems for the stability theorems of Lagrange-Dirichlet and Routh]. *Arch. rational Mech. Anal.*, 42:281–316, 1971.

P. Hagedorn and A. DasGupta. *Vibrations and Waves in Continuous Mechanical Systems.* 2007.

W. Hahn. *Theorie und Anwendung der direkten Methode von Ljapunow [Theory and application of Liapounov's direct method].* Springer, Berlin, 1952.

W. Hahn. *Stability of motions.* Springer, Berlin, 1967.

A. Halanay. *Differential equations, stability, oscillations, time lags.* Academic Press, New York, 1966.

H.W. Knobloch and F. Kappel. *Gewöhnliche Differentialgleichungen [Ordinary differential equations].* Teubner, Stuttgart, 1974.

N.N. Krasovskii. *Stability of motion.* Stanford University Press, Stanford, California, 1963.

J. LaSalle and S. Lefschetz. *Die Stabilitätstheorie von Ljapunow [Liapounow stability theory].* Bibliographisches Institut, Mannheim, 1967.

L. Luré. *Mecanique analytique [Analytical mechanics].* Libraire Universitaire, Louvain, 1968.

J.G. Malkin. *Theorie der Stabilität einer Bewegung [Theory of the stability of motion].* Akademie-Verlag, Berlin, 1959.

P. Painlevé. Sur la réciproque de la théorème de Lagrange [On the converse of Lagrange's theorem]. *Comptes Rendus*, 25:1021–4, 1897.

L.S. Pontryagin. *Gewöhnliche Differentialgleichungen [Ordinary differential equations].* VEB Deutscher Verlag der Wissenschaften, Berlin, 1965.

S.P. Timoshenko and J.N. Goodier. *Theory of Elasticity*, volume 3e. McGrw-Hill Book Co., Singapore, 1970.

Variational Principles in Mechanics and Control

Gottfried Spelsberg-Korspeter
TU Darmstadt, Darmstadt, Germany

Abstract The chapter 'Variational principles in mechanics and control' summarizes the material presented in six lectures in the CISM course no. 418. The first part considers the derivation of equations of motion for discrete and continuous systems. Founding on the basics of the calculus of variations the relation of the principle of virtual work to Lagrange's equations and Hamilton's principle is discussed. The concepts are useful for an efficient modeling of control systems. In the second part variational methods are used to introduce basics of optimal control and control system design.

1 Introduction

One of the major tasks in mastering structural control problems is to find efficient models for the systems under investigation. For mechanical systems this means that appropriate degrees of freedom have to be defined in order to keep the models managable and efficient. As a consequence we have to establish appropriate methods to set up equations of motion for discrete and continuous systems and combinations of those. Once appropriate models have been derived they can in many cases be linearized and brought into the standard form

$$\dot{\boldsymbol{x}} = \boldsymbol{A}\boldsymbol{x} + \boldsymbol{B}\boldsymbol{u} \tag{1}$$

on which a good part of linear control theory is based. For the analysis of equations of motion and corresponding control problems it is not enough just to study the equations of motion. Often further reasoning is needed for example energetic considerations or the investigation and optimization of objective functions which are functionals of the degrees of freedom. The link between the analysis of functionals and the governing differential equations for structural control problems are variational principles. Using the concept of calculus of variations the relations between energy functionals and the

P. Hagedorn, G. Spelsberg-Korspeter (Eds.), *Active and Passive Vibration Control of Structures*, CISM International Centre for Mechanical Sciences
DOI 10.1007/ 978-3-7091-1821-4_2 © CISM Udine 2014

equations of motion are established. The extension of the mathematical reasoning to control problems is the basis of the theory of optimal control.

A profound understanding of the equations is necessary to gain insight into the mechanical properties of a structure and is essential in order to set up appropriate controls. Often deep insight can already be gained from the linearized equations of motion of which the behavior is governed by eigenvalues and eigenvectors. Therefore, after establishing methods for the set up of equations of motion, we will investigate structural properties of the eigenvalue problems corresponding to the linearized equations of motion and their dependence on design parameters. The insights gained will be very helpful for example in passive control approaches for stabilizing rotors which will be studied as a technical example.

2 Contributing and non contributing forces in the principle of virtual work

The basic principle for deriving equations of motion for mechanical systems is Newton's law. If kinematical constraints such as a rigid body assumption are utilized in addition the balance equations for the moment of momentum have to be used. In order to take into account the kinematic relationships between the different bodies constraints have to be fulfilled. For obtaining the equations of motion in the form of ordinary differential equations the constraint forces have to be eliminated, which can be quite tedious.

2.1 Introductory example

As an example let us consider a beam hoist shown in figure 1 consisting of two particles. In order to apply Newton's law each of the particles has to be considered separately. The corresponding equations of motion read

$$\begin{aligned} m_P \ddot{x}_1 &= Z_1 + Z_3 \sin\varphi, \\ m_P \ddot{x}_2 &= m_p g + Z_2 + Z_3 \cos\varphi, \\ m_Q \ddot{x}_3 &= -Z_3 \sin\varphi, \\ m_Q \ddot{x}_4 &= -Z_3 \cos\varphi + m_Q g, \end{aligned}$$

where

$$\begin{aligned} \cos\varphi &= \frac{L + x_4 - x_2}{\sqrt{(L + x_4 - x_2)^2 + (x_3 - x_1)^2}}, \\ \sin\varphi &= \frac{x_3 - x_1}{\sqrt{(L + x_4 - x_2)^2 + (x_3 - x_1)^2}}. \end{aligned}$$

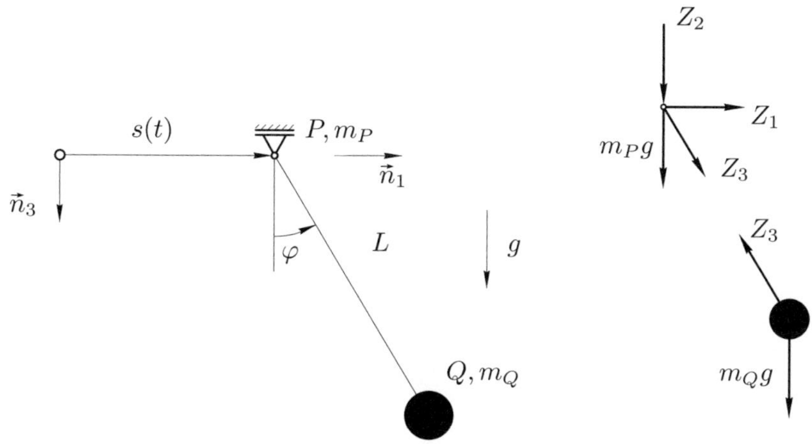

Figure 1. beam hoist

Since x_1, x_2, x_3 and x_4 are not independent we have consider the kinematical constraints

$$\begin{aligned} {}^{N_0}\vec{p}^{\,P} - s(t)\vec{n}_1 &= 0, && \text{(2a)} \\ \left| {}^{N_0}\vec{p}^{\,P} - {}^{N_0}\vec{p}^{\,Q} \right| - L &= 0. && \text{(2b)} \end{aligned}$$

In (2) we use the notation of Kane Kane and A. (1985), where ${}^{N_0}\vec{p}^{\,P}$ is the position vector from the origin N_0 to the point P. Equations (2) finally yield seven equations with the 7 unknowns x_1, x_2, x_3, x_4, Z_1, Z_2, Z_3 which clearly is not comfortable to deal with.

The first step to simplify the setup of equations of motion is the introduction of generalized coordinates. By definition generalized coordinates are independent quantities which represent the configuration of the system and therefore automatically fulfill the holonomic constraints of a system.

For the beam hoist considered here the angle φ is an intuitive choice for the generalized coordinate which in the following will be denoted by q. In order to set up the equations of motion we need the accelerations of points

P and Q which read

$$
\begin{aligned}
{}^{N_0}\vec{p}^{\,P} &= s\,\vec{n}_1, && \text{(3a)}\\
{}^{N}\vec{v}^{\,P} &= \dot{s}\,\vec{n}_1, && \text{(3b)}\\
{}^{N}\vec{a}^{\,P} &= \ddot{s}\,\vec{n}_1, && \text{(3c)}\\
{}^{N_0}\vec{p}^{\,Q} &= (s + L\sin q)\vec{n}_1 + L\cos q\,\vec{n}_3, && \text{(3d)}\\
{}^{N}\vec{v}^{\,Q} &= (\dot{s} + L\dot{q}\cos q)\vec{n}_1 - L\dot{q}\sin q\,\vec{n}_3, && \text{(3e)}\\
{}^{N}\vec{a}^{\,Q} &= (\ddot{s} + L\ddot{q}\sin q - L\dot{q}^2\sin q)\vec{n}_1 + (-L\ddot{q}\sin q - L\dot{q}^2\cos q)\vec{n}_3. && \text{(3f)}
\end{aligned}
$$

In (3) according to Kane's notation Kane and A. (1985) position vectors are indicated by $\vec{p}$, velocities by $\vec{v}$ and accelerations by $\vec{a}$. For velocity and acceleration vectors the first superscript always indicates which respect to which frame the time derivatives have been taken and the second superscript indicates the point. For example ${}^{N}\vec{a}^{\,Q}$ is the acceleration of particle Q with respect to the Newtonian frame N. From Newton's law we obtain the equations of motion in vector form

$$
\begin{aligned}
m_P{}^{N}\vec{a}^{\,P} &= (Z_1 + Z_3\sin q)\vec{n}_1 + (Z_3\cos q + Z_2 m_P g)\vec{n}_3, && \text{(4a)}\\
m_Q{}^{N}\vec{a}^{\,Q} &= -Z_3\sin q\,\vec{n}_1 + (m_Q g - Z_3\cos q)\vec{n}_3, && \text{(4b)}
\end{aligned}
$$

which yield four scalar equations

$$
\begin{aligned}
& m_P\ddot{s} = Z_1 + Z_3\sin q, && \text{(5a)}\\
& 0 = Z_3\cos q + Z_2, && \text{(5b)}\\
& m_Q(\ddot{s} + L\ddot{q}\sin q - L\dot{q}^2\sin q) = -Z_3\sin q, && \text{(5c)}\\
& -m_Q(-L\ddot{q}\sin q - L\dot{q}^2\cos q) = -m_Q g + Z_3\cos q, && \text{(5d)}
\end{aligned}
$$

from which the constraint forces have to be eliminated. In this case the equation of motion can be simply obtained by adding (5c) multiplied by $\cos q$ and (5d) multiplied by $\sin q$

$$
m_Q(\ddot{s}\cos q + L\ddot{q}) = -m_Q g\sin q. \tag{6}
$$

However especially for larger systems a formulation without constraint forces is much more efficient and will now be introduced through the principle of virtual work.

2.2 Partial velocities

The velocity of a particle in a holonomic system with n degrees of freedom is given by

$$ {}^N\vec{v}^P = \frac{{}^N d}{dt}{}^{N_0}\vec{p}^P = \sum_{s=1}^{n} \frac{{}^N\partial^{N_0}\vec{p}^P}{\partial q_s}\dot{q}_s + \frac{{}^N\partial^{N_0}\vec{p}^P}{\partial t} \tag{7} $$

and is always linear in the generalized velocities $\dot{q}_s$. In a shorter form (7) can be written as

$$ {}^N\vec{v}^P = \sum_{r=1}^{n} {}^N\vec{v}_r^P\dot{q}_r + {}^N\vec{v}_0^P , $$

where ${}^N\vec{v}_r^P = {}^N\vec{v}_r^P(q_1,\ldots,q_n,t)$ is called the rth partial velocity of P in N and ${}^N\vec{v}_0^P = {}^N\vec{v}_0^P(q_1,\ldots,q_n,t)$ is a remaining term that only depends on the generalized coordinates and time. For the beam hoist example we obtain from (3b)

$$ \begin{aligned} {}^N\vec{v}_1^P &= \vec{0}, \\ {}^N\vec{v}_0^P &= \dot{s}\vec{n}_1, \end{aligned} $$

and from (3e)

$$ \begin{aligned} {}^N\vec{v}_1^Q &= L\cos q\,\vec{n}_1 - L\sin q\,\vec{n}_3, \\ {}^N\vec{v}_0^Q &= \dot{s}\vec{n}_1. \end{aligned} $$

Visualizing the partial velocities in figure 2 we observe that $\vec{Z}_3$ is orthogonal to ${}^N\vec{v}_1^Q$ and therefore the virtual work $\vec{Z}_3 \cdot {}^N\vec{v}_1^Q = 0$ of $\vec{Z}_3$ vanishes. The same is true for the virtual work of $\vec{Z}_1$ and $\vec{Z}_2$ since ${}^N\vec{v}_1^P = 0$. As we shall show in the next section this observation is not specific for the particular example but quite general for forces ensuring constraints. Therefore we can make use of this fact and formulate the equations of motion without constraint forces.

Consider a system with K particles and n degrees of freedom. Newton's

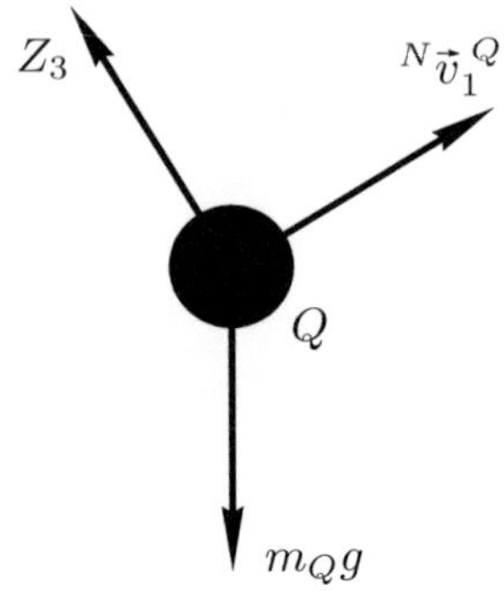

Figure 2. Virtual velocities of the beam hoist

law directly yields K equations in vector form

$$m_1 {}^N\vec{a}^{P_1} = \sum_k \vec{F}_{1k} + \sum_k \vec{Z}_{1k},$$

$$\vdots$$

$$m_i {}^N\vec{a}^{P_i} = \sum_k \vec{F}_{ik} + \sum_k \vec{Z}_{ik},$$

$$\vdots$$

$$m_K {}^N\vec{a}^{P_K} = \sum_k \vec{F}_{Kk} + \sum_k \vec{Z}_{Kk},$$

where $\vec{Z}_{jk}$ are constraint forces and $\vec{F}_{jk}$ are all other forces. Every particle P_i has n partial (virtual) velocities ${}^N\vec{v}_r^{P_i}$, $\quad r = 1, \ldots, n$. Multiplication of the ith equation with ${}^N\vec{v}_r^{P_i}$ and addition yields

$$\begin{aligned}
&(-m_1 {}^N\vec{a}^{P_1} + \sum_k \vec{F}_{1k} + \sum_k \vec{Z}_{1k}) \cdot {}^N\vec{v}_r^{P_1} \\
\vdots& \\
+ \quad &(-m_i {}^N\vec{a}^{P_i} + \sum_k \vec{F}_{ik} + \sum_k \vec{Z}_{ik}) \cdot {}^N\vec{v}_r^{P_i} \\
\vdots& \\
+ \quad &(-m_K {}^N\vec{a}^{P_K} + \sum_k \vec{F}_{Kk} + \sum_k \vec{Z}_{Kk}) \cdot {}^N\vec{v}_r^{P_K} = 0.
\end{aligned}$$

This generates n scalar equations of the form

$$\sum_{i=1}^{K}(-m_i{}^N\vec{a}^{P_i}\cdot{}^N\vec{v}_r^{P_i}+\sum_k\vec{F}_{ik}\cdot{}^N\vec{v}_r^{P_i})=0 \qquad r=1,\cdots,n, \tag{8}$$

which do not contain constraint forces since their virtual work vanishes. Forces with vanishing virtual work are also called non-contributing forces because they do not occur in the equations of motion. The equations of motion can therefore be set up by considering kinematics and contributing forces only without using free body diagrams containing constraint forces. This is also referred to as the principle of virtual work. For the example of the beam hoist we have

$$0=-m_P({}^N\vec{a}^P\cdot{}^N\vec{v}_1^P)+m_Pg(\vec{n}_3\cdot{}^N\vec{v}_1^P)-m_Q({}^N\vec{a}^Q\cdot{}^N\vec{v}_1^Q)+m_Pg(\vec{n}_3\cdot{}^N\vec{v}_1^Q),$$

which using (3) directly yields

$$-m_Q(\ddot{s}\cos q+\ddot{q}L)-m_QgL\sin q=0. \tag{9}$$

In order to make efficient use of the principle of virtual work it remains to be specified how to identify contributing and non contributing forces.

2.3 Constraints leading to ideal constraint forces

In this section we will consider four very general cases of kinematical constraints in which only ideal constraint forces namely non contributing forces arise.

Particles rigidly connected through massless joints

Probably the most common constraint is the enforcement of a prescribed distance between particles. It can be seen for example as the basis for the concept of rigid bodies which can be thought of a set of rigidly constrained particles. The velocities of two particles P_i, and P_j on a rigid body are not independent but given by the relation

$$\begin{aligned}{}^N\vec{v}^{P_j} &= {}^N\vec{v}^{P_i}+{}^N\vec{\omega}^B\times{}^{P_i}\vec{p}^{P_j},\\ {}^N\vec{v}^{P_i} &= {}^N\vec{v}^{P_j}+{}^N\vec{\omega}^B\times{}^{P_j}\vec{p}^{P_i}.\end{aligned}$$

Using that actio equals reactio the contribution of the forces acting between the particles on the virtual work can therefore be written as

$$\vec{F}_{ij}\cdot{}^N\vec{v}_r^{P_i}+\vec{F}_{ji}\cdot{}^N\vec{v}_r^{P_j}\ =\vec{F}_{ij}\cdot({}^N\vec{v}_r^{P_i}-{}^N\vec{v}_r^{P_j})=\vec{F}_{ij}\cdot({}^N\vec{\omega}_r^B\times{}^{P_i}\vec{p}^{P_j}).$$

In case of $\vec{F}_{ij}\|{}^{P_i}\vec{p}^{P_j}$ the contribution of the constraint forces to the virtual work is zero. However, it is easy to think of cases where the contact forces

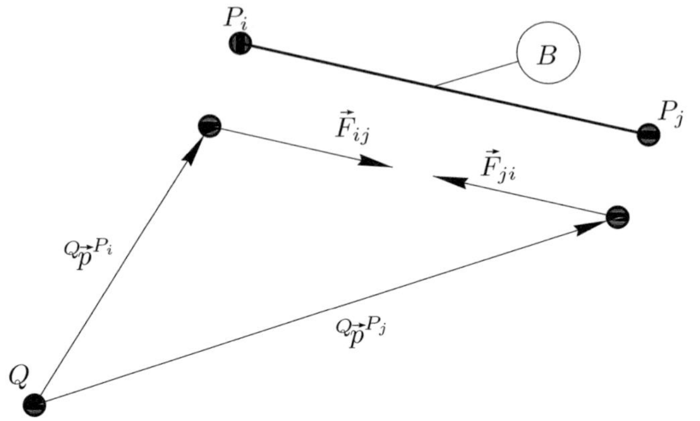

Figure 3. Two particles on a rigid body

do not lie on the lines between the particles. On the other hand, at least theoretically, there are many ways of physically ensuring the constraint of keeping the distance between particles constant. One possibility is to use a statically determined truss in which the nodes are defined as massless particles. Since the dynamical behavior of the structure cannot depend on the way the constraint is enforced, the virtual work of all constraint forces vanishes by the same argument.

Velocity of a particle prescribed as a function of time

For the next case consider a particle of which the position is prescribed as a function of time. Then

$$ {}^N\vec{v}^{P_i} = {}^N\vec{v}^{P_i}(t) = {}^N\vec{v}_0^{P_i} $$

and all partial velocities vanish. Therefore the virtual work of the forces which enforce the constraint vanishes.

Particle sliding on a frictionless rigid surface

For the next case consider a particle P which is sliding on a rigid body without penetrating it. Then the velocity of P is given by

$$ {}^N\vec{v}^P = {}^N\vec{v}^Q + {}^B\vec{v}^P, $$

where Q is the contact point on the rigid body and ${}^B\vec{v}^P$ is the relative velocity between the particle P and the surface of the rigid body. Since no penetration occurs the velocity of the particle P is orthogonal to the normal

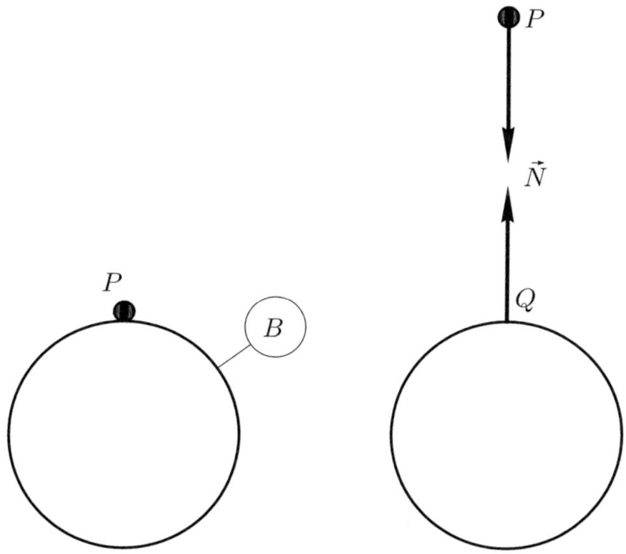

Figure 4. Particles sliding on a rigid body without penetration

vector of the surface $\vec{n}$, i.e.

$$^N\vec{v}^P \cdot \vec{n} = 0$$

and hence

$$^B\vec{v}^P = \sum {}^B\vec{v}_r^P \dot{q}_r + {}^B\vec{v}_0^P$$

multiplied by $\vec{n}$ yields

$$\sum \left({}^B\vec{v}_r^P \cdot \vec{n}\right)\dot{q}_r + \left({}^B\vec{v}_0^P \cdot \vec{n}\right) = 0.$$

Since the generalized velocities $\dot{q}_r$ are independent by definition it follows

$$\left({}^B\vec{v}_r^P \cdot \vec{n}\right) = 0, \qquad r = 1, \ldots, n$$
$$\left({}^B\vec{v}_0^P \cdot \vec{n}\right) = 0.$$

Therefore the virtual work of the contact force which is directed along $\vec{n}$ vanishes.

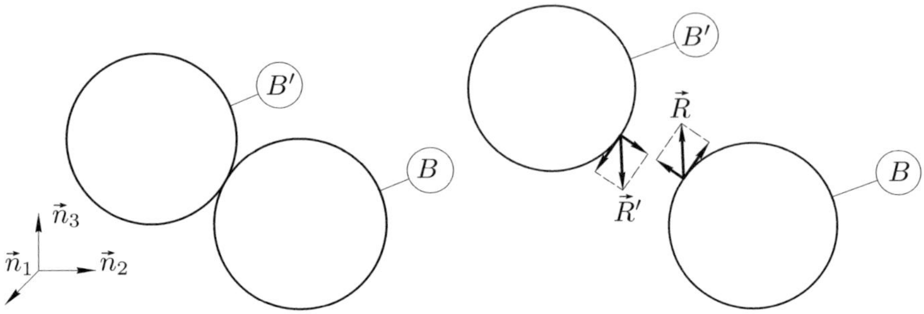

Figure 5. Two bodies rolling on each other without slip

Two rigid bodies rolling on each other

Finally consider the case of a rigid body B' rolling on another body B without sliding. Mathematically this means that the velocity of the contact point P' on B' equals the velocity of the contact point P on B, i.e.

$$^N\vec{v}^P = {}^N\vec{v}^{P'},$$

and therefore the virtual velocity of the contact forces read

$$\vec{R} \cdot {}^N\vec{v}_r^P + \vec{R}' \cdot {}^N\vec{v}_r^{P'} = (\vec{R} - \vec{R}') \cdot {}^N\vec{v}_r^P = 0$$

due to actio equals reactio $\vec{R} = \vec{R}'$.

Equations of motion for a system of particles

Summarizing we have seen that for a system of particles the equations of motion read

$$-\sum_{i=1}^{K} m_i({}^N\vec{a}^{P_i} \cdot {}^N\vec{v}_r^{P_i}) + \sum_{i=1}^{K}(\vec{F}_i \cdot {}^N\vec{v}_r^{P_i}) = 0, \qquad (10)$$

which means that the only input for the equations of motion is

- kinematics,
- mass and mass distribution,
- contributing forces and the velocities of the points where they act.

The equations of motion can therefore be obtained without considering constraint forces. It is therefore possible to set up the equations of motion completely analytically without considering any kind of free body diagram. The principle of virtual work was developed by Lagrange, therefore the equations (10) are often referred to Lagrange's equations of first kind. In the following we will show that there is a strong relation between the equations of motion and energy relations.

2.4 Relation of the principle of virtual work to Lagrange's equations of second kind

Consider the product of the acceleration and the virtual velocity of a particle in (10) which can be expressed as

$$\begin{aligned}\vec{a}\cdot\vec{v}_r &= \frac{d}{dt}(\vec{v}_r\cdot\vec{v}) - \frac{d\vec{v}_r}{dt}\cdot\vec{v}\\ &= \frac{d}{dt}(\frac{\partial\vec{v}}{\partial\dot{q}_r}\cdot\vec{v}) - \frac{d\vec{v}_r}{dt}\\ &= \frac{d}{dt}\frac{1}{2}\frac{\partial\vec{v}^2}{\partial\dot{q}_r} - \frac{1}{2}\frac{\partial\vec{v}^2}{\partial q_r}.\end{aligned} \tag{11}$$

The calculation makes use of the fact that

$$\begin{aligned}\frac{\partial\vec{v}}{\partial q_r} &= \frac{\partial}{\partial q_r}(\sum_{s=1}^{n}\vec{v}_s\dot{q}_s + \vec{v}_o)\\ &= \sum_{s=1}^{n}(\frac{\partial\vec{v}_s}{\partial q_r}\dot{q}_s + \vec{v}_s\frac{\partial\dot{q}_s}{\partial q_r}) + \frac{\partial\vec{v}_o}{\partial q_r},\end{aligned}$$

which considering that $\frac{\partial\dot{q}_s}{\partial q_r} = 0$ and

$$\begin{aligned}\frac{\partial\vec{v}_s}{\partial q_r} &= \frac{\partial}{\partial q_r}(\frac{\partial\vec{p}}{\partial q_s}) = \frac{\partial}{\partial q_s}(\frac{\partial\vec{p}}{\partial q_r}) = \frac{\partial\vec{v}_r}{\partial q_s},\\ \frac{\partial\vec{v}_r}{\partial t} &= \frac{\partial}{\partial t}(\frac{\partial\vec{p}}{\partial q_r}) = \frac{\partial}{\partial q_r}(\frac{\partial\vec{p}}{\partial t}) = \frac{\partial\vec{v}_o}{\partial q_r}\end{aligned}$$

yields

$$\frac{\partial\vec{v}}{\partial q_r} = \sum_{s=1}^{n}\frac{\partial\vec{v}_r}{\partial q_s}\dot{q}_s + \frac{\partial\vec{v}_r}{\partial t} = \frac{d\vec{v}_r}{dt}.$$

Using (11) for a system of k particles with the kinetic energy

$$T = \frac{1}{2}\sum_{i=1}^{k} m_i({}^N\vec{v}^{P_i})^2,$$

we have

$$\sum_{i=1}^{k} -m_i^N\vec{a}_i^{P_i}\cdot^N\vec{v}_r^{P_i} = -(\frac{d}{dt}\frac{\partial T}{\partial\dot{q}_r} - \frac{\partial T}{\partial q_r})$$

which means that the equations of motion can be written as

$$\frac{d}{dt}\frac{\partial T}{\partial \dot{q}_r} - \frac{\partial T}{\partial q_r} = \sum \vec{F}^{P_i} \cdot \frac{\partial^O \vec{p}^{P_i}}{\partial q_r}. \tag{12}$$

We now divide the contributing forces $\vec{F}^{P_i}$ into forces which come from a potential and nonconservative forces

$$\vec{F}^{P_i} = \underbrace{\vec{P}_i}_{\text{conservative}} + \underbrace{\vec{N}_i}_{\text{non conservative}} .$$

For the conservative forces we have

$$\vec{P}_i = -\nabla U = -(\frac{\partial U}{\partial x_i}\vec{e}_x + \frac{\partial U}{\partial y_i}\vec{e}_y + \frac{\partial U}{\partial z_i}\vec{e}_z),$$

where $U = \sum U_i = U(q_1, \ldots, q_n, t)$ is the sum of all potentials depending on the generalized coordinates. Using

$$x_i =^O \vec{p}^{P_i} \cdot \vec{e}_x\,, \qquad y_i =^O \vec{p}^{P_i} \cdot \vec{e}_y\,, \qquad z_i =^O \vec{p}^{P_i} \cdot \vec{e}_z$$

for the virtual work of the potential forces we obtain

$$\sum_{i=1}^{k} \vec{P}_i \cdot \frac{\partial^O \vec{p}^{P_i}}{\partial q_r} = -\sum_{i=1}^{k} (\frac{\partial U_i}{\partial x_i}\frac{\partial x_i}{\partial q_r} + \frac{\partial U_i}{\partial y_i}\frac{\partial y_i}{\partial q_r} + \frac{\partial U_i}{\partial z_i}\frac{\partial z_i}{\partial q_r}) \tag{13}$$

$$= -\sum_{i=1}^{k} \frac{\partial U_i}{\partial q_r} = -\frac{\partial U}{\partial q_r}. \tag{14}$$

From (12) and (13) it follows by defining the Lagrange function as

$$L = T - U,$$

that the equations of motion can be written as

$$\frac{d}{dt}\frac{\partial L}{\partial \dot{q}_r} - \frac{\partial L}{\partial q_r} = Q_r, \qquad Q_r = \sum_{i=1}^{k} \vec{N}_i \cdot \frac{\partial^O \vec{p}^{P_i}}{\partial q_r} \tag{15}$$

which are the well known Lagrange equations of second kind. All non-potential forces have to be taken into account through the generalized forces Q_r.

In the derivation of Lagrange's equations of second kind we have established the connection between equations of motion and energy expressions

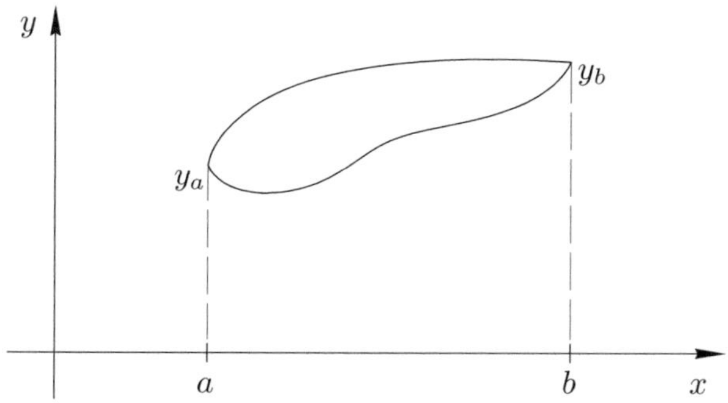

Figure 6. Different trajectories from a to b

starting from Newton's law. In the following we will see that Lagrange's equations can also be derived form the energy expressions using the concept of the calculus of variations, which is also very important in control theory as we will see. In the following paragraph we will establish the basics.

3 Introduction to calculus of variations and Hamilton's principle

As a first step towards the introduction of the calculus of variations we need to introduce the the meaning of a functional.

Definition. The Integral

$$J[y] = \int_a^b F(x, y, y')\,\mathrm{d}x$$

is called a functional. A functional is a function of functions defined on an interval.

One can now search for functions $y(x)$ extremizing $J[y]$ assuming that $y(a) = y_a, y(b) = y_b$ are prescribed. A necessary condition for an extremum of $J[y]$ is that y makes $J[y]$ stationary. This implies that the first variation defined as

$$\delta J[y] = \frac{d}{d\epsilon} J[y + \epsilon\eta]\bigg|_{\epsilon=0} = 0, \qquad \epsilon \in \mathbb{R}$$

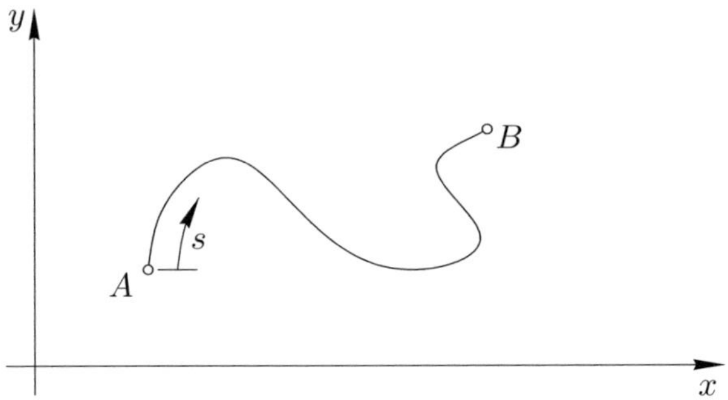

Figure 7. Path from A to B

has to vanish for arbitrary test functions $\eta(x)$ which are required to be twice continuously differentiable on the interval $[a, b]$ and vanish at the boundaries i.e.

$$\begin{aligned} &\eta(x) \in C^2[a, b], \\ &\eta(a) = 0, \qquad \eta(b) = 0. \end{aligned}$$

In the following we name the test functions as

$$\eta(x) = \delta y(x).$$

As an example consider the somehow trivial problem of finding the shortest way between two points in a flat plane. Whereas the result is intuitive the mathematical formulation of this exercise reads

$$\min \int_A^B \mathrm{d}s, \qquad \mathrm{d}s = \sqrt{\mathrm{d}x^2 + \mathrm{d}y^2}, \qquad \frac{\mathrm{d}s}{\mathrm{d}x} = \sqrt{1 + y'^2}.$$

A necessary condition for an optimum is that the first variation of the functional vanishes, i.e.

$$\delta \int_{x_a}^{x_b} \sqrt{1+y'^2}\,\mathrm{d}x = 0, \tag{16a}$$

$$\frac{\mathrm{d}}{\mathrm{d}\epsilon} \int_{x_a}^{x_b} \sqrt{1+(y'+\epsilon\delta y')^2}\,\mathrm{d}x \bigg|_{\epsilon=0} = 0, \tag{16b}$$

$$\int_{x_a}^{x_b} \frac{y'}{\sqrt{1+y'^2}}\delta y'\,\mathrm{d}x = 0. \tag{16c}$$

In order to be able to draw conclusions from (16c) we use integration by parts to obtain

$$-\int_{x_a}^{x_b} \left(\frac{y'}{\sqrt{1+y'^2}}\right)' \delta y\,\mathrm{d}x + \frac{y'}{\sqrt{1+y'^2}}\delta y \bigg|_{x_a}^{x_b} = 0.$$

Since $\delta y(x_a) = \delta y(x_b) = 0$ and since $\delta y(x)$ is arbitrary (from main theorem of variational calculus) it follows

$$\left(\frac{y'}{\sqrt{1+y'^2}}\right)' = 0,$$

which is a simple ordinary differential equation that can be integrated analytically. For the result it follows that

$$y(x) = b + cx$$

where b, c are constants determined by $y(x_a) = y_a$ and $y(x_a) = y_a$.

We now consider the more general case of a functional

$$\delta J[y] = \delta \int_a^b F(x, y, y')\,\mathrm{d}x = 0$$

depending on an unknown function $y(x)$ and its first derivative satisfying the boundary conditions $\delta y(a) = \delta y(b) = 0$. Carrying out the variations similar to the previous example it follows taking into account the boundary conditions

$$
\begin{aligned}
0 = \delta J[y] &= \frac{\mathrm{d}}{\mathrm{d}\epsilon} \int_a^b F(x, y + \epsilon\delta y, y' + \epsilon\delta y')\,\mathrm{d}x \bigg|_{\epsilon=0}, \\
&\int_a^b \frac{\partial F}{\partial y}\delta y + \frac{\partial F}{\partial y'}\delta y'\,\mathrm{d}x = 0, \\
&\int_a^b \frac{\partial F}{\partial y}\delta y - \frac{\mathrm{d}}{\mathrm{d}x}\frac{\partial F}{\partial y'}\delta y\,\mathrm{d}x = 0.
\end{aligned}
$$

From the main theorem of variational calculus we conclude

$$
\frac{\mathrm{d}}{\mathrm{d}x}\frac{\partial F}{\partial y'} - \frac{\partial F}{\partial y} = 0 \tag{17}
$$

which is known as the Euler equation. We observe that the structure of (17) is very similar to (15). We use this observation to state Hamilton's principle Meirovitch (2001); Hagedorn (1989). For a conservative mechanical system with the Lagrange function

$$
L = T - U = L(q_1, ..., q_n, \dot{q}_1, ..., \dot{q}_n, t)
$$

the equations of motion can be derived from

$$
\begin{aligned}
&\delta \int_{t_1}^{t_2} L\,\mathrm{d}t = 0 \\
&\delta q_i(t_1) = 0, \qquad \delta q_i(t_2) = 0.
\end{aligned}
$$

Indeed an easy calculation yields

$$
\begin{aligned}
&\int_{t_1}^{t_2} \left(-\frac{\mathrm{d}}{\mathrm{d}t}\frac{\partial L}{\partial \dot{q}_1} + \frac{\partial L}{\partial q_1}\right)\delta q_1 + ... + \left(-\frac{\mathrm{d}}{\mathrm{d}t}\frac{\partial L}{\partial \dot{q}_n} + \frac{\partial L}{\partial q_n}\right)\delta q_n\,\mathrm{d}t = 0, \\
&\Rightarrow \frac{\mathrm{d}}{\mathrm{d}t}\frac{\partial L}{\partial \dot{q}_i} - \frac{\partial L}{\partial q_i} = 0,
\end{aligned}
$$

which shows that Lagrange's equations are the Euler equations for Hamilton's principle. Of course non potential forces can be considered through their virtual work as previously derived for Lagrange's equations.

4 Continuous systems

In the last section we have studied different methods in order to set up the equations of motion for discrete systems with a finite number of degrees of freedom which we regarded as a set of possibly connected particles. The major benefit we found from variational methods namely the principle of virtual work and Hamilton's principle was that we did not have to eliminate constraint forces from the equations of motion. In this section we will see that continuous systems can be treated in a very similar way. The major difference to the discrete systems where the degrees of freedom are scalar valued functions of time is that for continuous systems the degrees of freedom are functions of space and time. The continuous distribution of material parameters also leads to the challenge that stiffness, mass and other material parameters cannot be separated as for discrete systems. However it is still possible to simplify equations of motion through kinematical assumptions which means that many of the nice properties of the variational approaches form the last section can also be used in the continuous case. In order to show this we start by shortly reviewing the basic equations of the theory of elasticity and then turn to the introduction of simplified structural models like rods, beams, plates shells, etc.. As a reference for this chapter the book Washizu (1974) is recommended where detailed derivations of different structural models can be found.

4.1 Kinematics of deformation

In order to analyse continuous systems the first step is to describe their deformation mathematically. The three dimensional elastic body is thought of as a collection of infinitesimal parallelipipeds (c.f. figure 8) which deform under external forcing. There are two most common ways to analyse the deformation. In fluid problems the usual approach is to take an Eulerian point of view i.e. to turn the view to a control volume fixed in space and to investigate what happens at each particular point fixed in space. It is clear that this means that at the particular point different material points pass by. The common description for structural problems is to give each material point a coordinate and to describe what happens to the particular material point. This point of view is associated with Lagrange. The deformation of the body is expressed in terms of material coordinates x_1, x_2, x_3 which determine the position of material points of the body in the undeformed configuration. The position vector of a material point in the undeformed

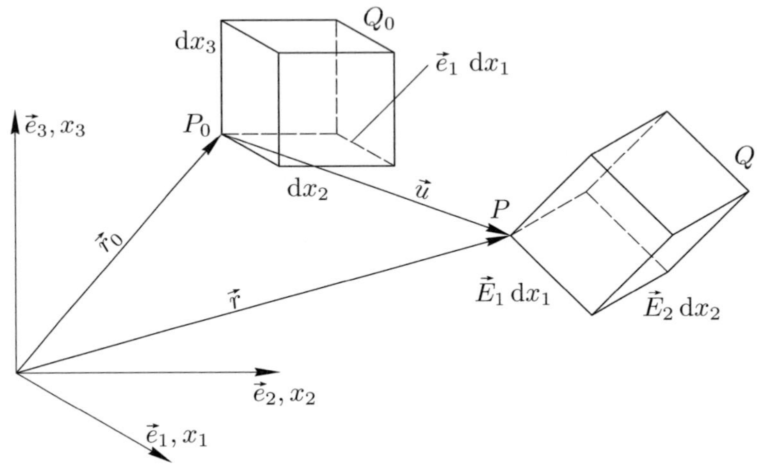

Figure 8. Kinematics of an infinitesimal parallelipiped

configuration is denoted by $\vec{r}_0$ and the position of a point in the deformed configuration is denoted by $\vec{r}$ (c.f. figure 8). These position vectors in a natural way define coordinates associated to the boundaries of the infinitesimal parallelipiped under consideration. A lattice of the parallelipiped that was parallel to a vector $\vec{e}_1$ of the global coordinate system is stretched and directed in a different direction $\vec{E}_1$ in the deformed configuration (c.f. figure 8), where

$$\vec{e}_i = \frac{d\vec{r}_0}{dx_i}, \qquad \vec{E}_i = \frac{d\vec{r}}{dx_i}.$$

It follows that the position vector of a material point in the deformed configuration is given by

$$\vec{r} = \vec{r}_0 + \vec{u},$$

where $\vec{u}$ is the displacement of the point. In order to measure how the paralelliped deforms we consider a vector on its diagonal given by

$$d\vec{r}_0 = dx_1\vec{e}_1 + dx_2\vec{e}_2 + dx_3\vec{e}_3 = dx_i\vec{e}_i.$$

The square of the length of the vector is therefore given by

$$d\vec{r}_0 \cdot d\vec{r}_0 = \delta_{ij}\, dx_i\, dx_j,$$

where δ_{ij} is the well known Kronecker delta. In the deformed configuration the vector $\vec{r}_0$ deforms to

$$\mathrm{d}\vec{r} = \vec{E}_1\,\mathrm{d}x_1 + \vec{E}_2\,\mathrm{d}x_2 + \vec{E}_3\,\mathrm{d}x_3$$

and the length of the deformed vector is calculated as

$$\mathrm{d}\vec{r}\cdot\mathrm{d}\vec{r} = E_{ij}\,\mathrm{d}x_i\,\mathrm{d}x_j, \qquad E_{ij} = \vec{E}_i\cdot\vec{E}_j.$$

A measure for the deformation of the parallelipiped is therefore given by

$$(\,\mathrm{d}\vec{r}\cdot\mathrm{d}\vec{r}) - (\,\mathrm{d}\vec{r}_0\cdot\mathrm{d}\vec{r}_0) = (E_{ij} - \delta_{ij})\,\mathrm{d}x_i\,\mathrm{d}x_j,$$

which motivates the definition of the Green-Lagrange strain tensor as

$$e_{ij} = \frac{1}{2}(E_{ij} - \delta_{ij}) = e_{ji}.$$

From $\vec{r} = \vec{r}_0 + \vec{u}$ it follows that

$$e_{ij} = \frac{1}{2}\left[\left(\underbrace{\frac{\partial\vec{r}_0}{\partial x_i}}_{\vec{e}_i} + \frac{\partial\vec{u}}{\partial x_i}\right)\left(\underbrace{\frac{\partial\vec{r}_0}{\partial x_j}}_{\vec{e}_j} + \frac{\partial\vec{u}}{\partial x_j}\right) - \delta_{ij}\right].$$

Substituting

$$\vec{u} = u_1\vec{e}_1 + u_2\vec{e}_2 + u_3\vec{e}_3$$

it follows

$$e_{ij} = \frac{1}{2}(u_{i,j} + u_{j,i} + u_{k,i}u_{k,j}), \qquad u_{i,j} = \frac{\partial u_i}{\partial x_j}, \tag{18}$$

which is the common expression of the Green-Lagrange strain tensor, defining the strain in terms of the displacement field. For small deformations (18) can be linearized as

$$\epsilon_{ij} = \frac{1}{2}(u_{i,j} + u_{j,i}). \tag{19}$$

Based on the description of strain one can define constitutive laws for different materials. Before we do so we however turn to the analysis of stress.

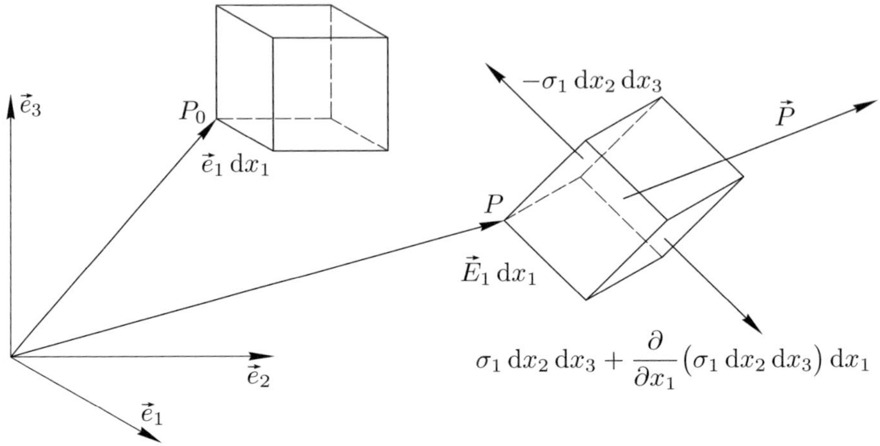

Figure 9. Force balance at an infitesimal parallelipiped

4.2 Analysis of stress

As for the discrete case the basis for the analysis of continuous systems is Newton's law. The 3-dimensional elastic body is thought of a collection of infenitesimal parallelipipeds (c.f. figure 9) which deform under external forcing. As explained in the previous section the deformation of the body is expressed in terms of material coordinates x_1, x_2, x_3 which determine the position of material points of the body in the undeformed configuration. A lattice of the parallelipiped that was parallel to a vector $\vec{e}_1$ of the global coordinate system is stretched and directed in a different direction $\vec{E}_1$ in the deformed configuration. The surfaces of the parallelipiped are loaded by six forces given by

$$\begin{aligned}
&-\vec{\sigma}_1\,\mathrm{d}x_2\,\mathrm{d}x_3, &&\vec{\sigma}_1\,\mathrm{d}x_2\,\mathrm{d}x_3 + \frac{\partial}{\partial x_1}(\vec{\sigma}_1\,\mathrm{d}x_2\,\mathrm{d}x_3)\,\mathrm{d}x_1,\\
&-\vec{\sigma}_2\,\mathrm{d}x_3\,\mathrm{d}x_1, &&\vec{\sigma}_2\,\mathrm{d}x_3\,\mathrm{d}x_1 + \frac{\partial}{\partial x_2}(\vec{\sigma}_2\,\mathrm{d}x_3\,\mathrm{d}x_1)\,\mathrm{d}x_2,\\
&-\vec{\sigma}_3\,\mathrm{d}x_1\,\mathrm{d}x_2, &&\vec{\sigma}_3\,\mathrm{d}x_1\,\mathrm{d}x_2 + \frac{\partial}{\partial x_3}(\vec{\sigma}_3\,\mathrm{d}x_1\,\mathrm{d}x_2)\,\mathrm{d}x_3
\end{aligned}$$

where higher order terms are neglected. Newton's law formulated at the parellelipiped yields

$$\rho\ddot{\vec{r}} = \vec{\sigma}_{i,i} + \vec{P},$$

where $\vec{P}$ denotes the volume forces and

$$\vec{\sigma}_i = \sigma_{ij}\vec{E}_j,$$

using index notation and σ_{ij} are the components of the stress vector in direction of the lattice vectors. A momentum balance on the parallelipiped yields

$$\vec{0} = \vec{\sigma}_1 \, \mathrm{d}x_2 \, \mathrm{d}x_3 \times \vec{E}_1 \, \mathrm{d}x_1 + \vec{\sigma}_2 \, \mathrm{d}x_3 \, \mathrm{d}x_1 \times \vec{E}_2 \, \mathrm{d}x_2 + \vec{\sigma}_3 \, \mathrm{d}x_1 \, \mathrm{d}x_2 \times \vec{E}_3 \, \mathrm{d}x_3,$$

where terms higher order are neglected. Since

$$\begin{aligned}\vec{0} &= \vec{\sigma}_i \times \vec{E}_i = \sigma_{ij}\vec{E}_j \times \vec{E}_i \\ &= (\sigma_{21} - \sigma_{12})(\vec{E}_1 \times \vec{E}_2) + (\sigma_{32} - \sigma_{23})(\vec{E}_2 \times \vec{E}_3) + (\sigma_{13} - \sigma_{31})(\vec{E}_3 \times \vec{E}_1)\end{aligned}$$

it follows

$$\sigma_{ij} = \sigma_{ji},$$

which implies that the stress tensor is symmetric.

4.3 Stress strain relations

Having introduced the stress and strain tensor the only thing missing to be able to formulate equations of motion for continuous systems are constitutive material laws, i.e. relations between stress and strain. The properties of the materials to be studied enter to the governing equations through the stress strain relations

$$\sigma_{ij} = \sigma_{ij}(e_{11}, \cdots, e_{33}), \tag{20}$$

which indicate how stresses are related to the deformation of the material. Although we assume in (20) that the stresses depend only on the strains it is very well possible that they also depend on other quantities such as the time derivatives of the strains, etc. The easiest and most common material law is Hooke's law

$$\sigma_{ij} = c_{ijkl} \, e_{kl} \tag{21}$$

which assumes a linear dependence between stresses and strains. For isotropic material it can be shown that the c_{ijkl} depend only on two independent parameters Becker and Gross (2002); Timoshenko and Goodier (1951). In the engineering literature the independent parameters are usually the modulus of elasticity E and the Poisson ratio ν. Using E and ν as parameters the stress strain relations for isotropic linear elastic material are written as

$$Ee_{11} = \sigma_{11} - \nu(\sigma_{22} + \sigma_{33}), \tag{22a}$$
$$Ee_{22} = \sigma_{22} - \nu(\sigma_{11} + \sigma_{33}), \tag{22b}$$
$$Ee_{33} = \sigma_{33} - \nu(\sigma_{11} + \sigma_{22}), \tag{22c}$$

and

$$e_{12} = \frac{1+\nu}{E}\sigma_{12} = \frac{1}{2G}\sigma_{12}, \tag{22d}$$
$$e_{23} = \qquad = \frac{1}{2G}\sigma_{23}, \tag{22e}$$
$$e_{13} = \qquad = \frac{1}{2G}\sigma_{13}, \tag{22f}$$

where the shear modulus is defined as

$$G = \frac{E}{2(1+\nu)}. \tag{23}$$

In more compact form the equations can be written as

$$\sigma_{ij} = \frac{E}{1-2\nu}e\delta_{ij} + 2G(e_{ij} - e\delta_{ij}),$$
$$e = \frac{1}{3}(e_{11} + e_{22} + e_{33}), \qquad G = \frac{E}{2(1+\nu)},$$

or

$$e_{ij} = \frac{1-2\nu}{E}\sigma\delta_{ij} + \frac{1}{2G}(\sigma_{ij} - \sigma\delta_{ij}), \tag{24a}$$
$$\sigma = \frac{1}{3}(\sigma_{11} + \sigma_{22} + \sigma_{33}). \tag{24b}$$

4.4 Basic equations of the theory of elasticity

We have now stated all ingredients to set up the basic equations of the theory of elasticity. A force balance at the parallelipiped yields

$$-\rho\ddot{\vec{r}} + \vec{\sigma}_{i,i} + \vec{P} = 0, \tag{25}$$

where

$$\vec{\sigma}_{i,i} = \frac{\partial}{\partial x_i}\vec{\sigma}_i = \frac{\partial}{\partial x_i}\sigma_{ij}\vec{r}_{,j}$$

and

$$\vec{r}_{,j} = \vec{E}_j = \frac{\partial \vec{r}}{\partial x_j}.$$

Substituting the stress strain relations and the Green Langrange strain tensor

$$e_{ij} = \frac{1}{2}\left(u_{i,j} + u_{j,i} + u_{k,i}u_{k,j}\right)$$

into (25) one obtains a system of partial differential equations, which with appropriate boundary conditions form the basic equations of elasticity. In most cases they need to be solved with numerical methods which can be very challenging.

Since the solution of problems from 3D elasticity takes a lot of computation time it is in many cases very helpful to introduce simplifying assumptions. These are usually of kinematical nature and are used to derive simplified models. Important examples are rods, beams, plates shells, etc.. In order to be able to derive these models one needs to find ways to deal with constraint forces. As in the discrete case the appropriate way to do this will be variational principles which allow a formulation of the equations of motion without constraint forces.

4.5 The principle of virtual work for continuous systems

Consider a body sketched in figure 10 which is subjected to distributed external volume forces $\vec{P}$ in the interior and surface forces $\vec{F}_E$ on the free boundaries S_1. Assume that there are geometric boundary conditions for the displacement on the surface S_2 of the body. In the interior of the body, as previously discussed, the equations of motion formulated on an infinitesimal parallelipiped read

$$-\rho\ddot{\vec{r}} + \vec{\sigma}_{i,i} + \vec{P} = \vec{0}. \tag{26}$$

On the boundary S_1 external and internal forces have to be in equilibrium yielding

$$\vec{F} = \vec{F}_E = 0 \text{ on } S_1. \tag{27}$$

As for discrete systems we can define a variation of the degrees of freedom, which in the continuous case are vector valued functions, that can for example be defined as $\vec{r} = u(x_1, x_2, x_3, t)\vec{e}_1 + v(x_1, x_2, x_3, t)\vec{e}_2 + w(x_1, x_2, x_3, t)\vec{e}_3$. The variation then reads

$$\delta\vec{r} = \frac{d}{d\epsilon}\left[(u + \epsilon\eta_u)\vec{e}_1 + (v + \epsilon\eta_v)\vec{e}_2 + (w + \epsilon\eta_w)\vec{e}_3\right]\Bigg|_{\epsilon=0} = \vec{0}, \qquad \epsilon \in \mathbb{R} \tag{28}$$

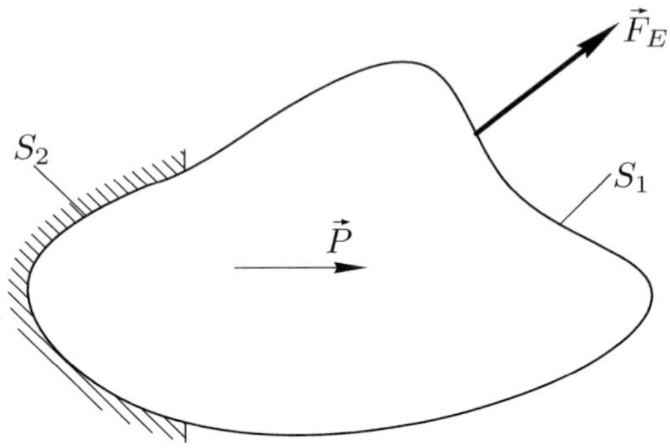

Figure 10. Body under forcing

where η_u, η_v, η_w are arbitrary twice continuously differentiable functions depending on x_1, x_2, x_3 and t which satisfy the geometric boundary conditions. We now take the scalar product of the governing equations for the interior of the continuum (26) and the governing equations for the boundary (27) with $\delta\vec{r}$ and integrate over the volume of the continuum. Since the equations are homogeneous we can add them to obtain

$$-\iiint\limits_V (-\rho\ddot{\vec{r}} + \vec{\sigma}_{i,i} + \vec{P}) \cdot \delta\vec{r}\,\mathrm{d}V + \iint\limits_{S_1} (\vec{F} - \vec{F}_E) \cdot \delta\vec{r}\,\mathrm{d}S = 0. \tag{29}$$

Using the theorem of Gauß we can write the second term as

$$-\iiint\limits_V \vec{\sigma}_{i,i} \cdot \delta\vec{r}\,\mathrm{d}V = \iiint\limits_V \vec{\sigma}_i \cdot \delta\vec{r}_{,i}\,\mathrm{d}V - \iint\limits_{S_1+S_2} \vec{F} \cdot \delta\vec{r}\,\mathrm{d}S. \tag{30}$$

Since $\delta\vec{r}$ vanishes on the geometric boundary conditions S_2 we can write (29) as

$$\iiint\limits_V \left[\rho\ddot{\vec{r}} \cdot \delta\vec{r} + \vec{\sigma}_i \cdot \delta r_{,i} - \vec{P} \cdot \delta\vec{r}\right] \mathrm{d}V - \iint\limits_{S_1} \vec{F}_E \cdot \delta\vec{r}\,\mathrm{d}S = 0. \tag{31}$$

Taking into account the definition of $\vec{\sigma}_i$ and $\vec{r}_{,i}$ we can write

$$\vec{\sigma}_i \delta\vec{r}_{,i} = \sigma_{ij} r_{,j} \delta\vec{r}_{,i} - \frac{1}{2}\sigma_{ij}\delta E_{ij} = \sigma_{ij}\delta e_{ij} \tag{32}$$

since $\delta E_{ij} = 0$. It follows

$$\iiint\limits_V \left[\rho\ddot{\vec{r}} \cdot \delta\vec{r} + \sigma_{ij}\delta e_{ij}\right] \mathrm{d}V - \iiint\limits_V \vec{P} \cdot \delta\vec{r}\,\mathrm{d}V - \iint_{S_1} \vec{F}_E \cdot \delta\vec{r}\,\mathrm{d}S = 0. \tag{33}$$

As in the discrete case the virtual work of all ideal constraint forces vanishes. The derivations in section 2.3 stay valid for the continuous case as well if one considers infinitely many partial velocities. This is valid because in an appropriate function space any test function $\eta(\vec{x}, t)$ from (28) can be expressed by an possibly infinite linear combination of base functions of the space, the weight factors serving as the degrees of freedom. Hence we obtain

$$\iiint_V \left[\rho\left(\frac{\mathrm{d}^2}{\mathrm{d}t^2}\vec{r}\right)\cdot\delta\vec{r} + \sigma_{ij}\delta e_{ij}\right]\mathrm{d}V = \sum \vec{F}\cdot\delta\vec{r}, \tag{34}$$

where we introduce the notation $\sum \vec{F}\cdot\delta\vec{r}$ for the virtual work of all contributing forces. As in the discrete case there is a strong relation between the principle of virtual work and the energy expression of a body. If we assume that the variations $\delta\vec{r}$ vanish at two arbitrary fixed points in time, we can integrate by parts and write

$$\int_{t_1}^{t_2}\iiint_V \rho\ddot{\vec{r}}\cdot\delta\vec{r}\,\mathrm{d}V\,\mathrm{d}t = -\int_{t_1}^{t_2}\iiint_V \rho\dot{\vec{r}}\cdot\delta\dot{\vec{r}}\,\mathrm{d}V\,\mathrm{d}t = -\delta\int_{t_1}^{t_2}\iiint_V \frac{1}{2}\rho\dot{\vec{r}}^2\,\mathrm{d}V\,\mathrm{d}t, \tag{35}$$

which is the variation of the kinetic energy. Additionally we can introduce the specific potential energy as

$$\overline{u}_{\mathrm{el}} = \int_0^e \sigma_{ij}(e_{ij})\,\mathrm{d}e_{ij}, \tag{36}$$

which yields

$$U = \iiint_V \overline{u}_{\mathrm{el}}\,\mathrm{d}V. \tag{37}$$

For linear elastic material we therefore have

$$\delta U = \delta\iiint_V \frac{1}{2}c_{ijkl}e_{kl}e_{ij}\,\mathrm{d}V = \iiint_V \sigma_{ij}\delta e_{ij}\,\mathrm{d}V. \tag{38}$$

and hence (34) assuming linear elastic material can be written as

$$-\delta\int_{t_1}^{t_2} T - U\,\mathrm{d}t = \int_{t_1}^{t_2}\sum \vec{F}\cdot\delta\vec{r}\,\mathrm{d}t, \tag{39}$$

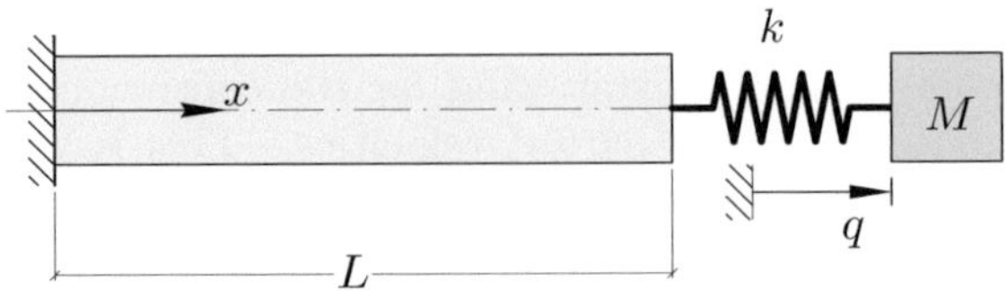

Figure 11. Rod with concentrated spring and concentrated mass

which is the formulation of Hamilton's principle for the continuous case. Having established Hamilton's principle for 3D elastic bodies specialized models can be derived by introducing kinematical constraints. Similarly also combinations of discrete and continuous systems can be analysed where the discrete systems can be thought of constrained elastic systems. In the following sections we will treat simple intuitive examples.

4.6 Derivation of structural models from 3D elasticity

As probably the simplest example of a continuous structural model in this section we discuss the elastic rod and the assumptions made in the derivation of the model. Consider a slender body and assume that we are only interested in longitudinal vibrations. If such a structure is only loaded in axial direction it is reasonable to introduce simplifying assumptions which are in fact kinematical constraints on the elastic body. For the model of a rod we assume that the crossections stay planar and can only move in x-direction along the length axis of the rod, i.e.

$$u(x, y, z) = u(x), \qquad v(x, y, z) = 0, \qquad w(x, y, z) = 0.$$

To show that we can treat discrete and continuous models in the same way we attach a spring and a point mass at the end of the rod as is shown in figure 4.6. Introducing the kinematic constraint the kinetic energy of the rod then simplifies to

$$T = \frac{1}{2} \int_0^l \rho A \dot{u}^2 \, \mathrm{d}x + \frac{1}{2} M \dot{q}^2, \tag{40}$$

where the kinetic energy of the particle at the end is taken care of in the second term with a separate degree of freedom q. Due to the kinematical assumption the expression for the specific potential energy simplifies to

$$\bar{u} = \int_0^{\epsilon} \sigma \,\mathrm{d}\epsilon = \frac{1}{2}\mathrm{E}\epsilon^2 \tag{41}$$

since $\sigma = \mathrm{E}\epsilon$, where $\epsilon = u'$. Therefore the total potential energy of the rod can be written as

$$U = \frac{1}{2}\int_0^{l} EAu'^2 \,\mathrm{d}x + \frac{1}{2}k\left(q - u(l,t)\right)^2, \tag{42}$$

where the second term is due to the kinetic energy stored in the spring. The equations of motion can now be derived using Hamilton's principle

$$\delta \int_{t_1}^{t_2} L \,\mathrm{d}t = 0, \qquad L = T - U, \tag{43}$$

where in addition to the energy expressions the geometric boundary conditions

$$u(0,t) = 0, \tag{44a}$$
$$\delta u(0,t) = 0, \tag{44b}$$

and

$$\delta u(x,t_1) = 0, \tag{44c}$$
$$\delta u(x,t_2) = 0, \tag{44d}$$

have to be considered. As in the discrete case we have to carry out the variations on the Lagrange function yielding

$$\delta \int_{t_1}^{t_2} \left[\int_0^{l} \frac{1}{2}\rho A\dot{u}^2 - \frac{1}{2}EAu'^2 \,\mathrm{d}x + \frac{1}{2}M\dot{u}^2 - \frac{1}{2}(q - u(l,t))^2 \right] \mathrm{d}t = 0, \tag{45a}$$

$$\int_{t_1}^{t_2} \left[\int_0^{l} \rho A\dot{u}\delta\dot{u} - EAu'\delta u' \,\mathrm{d}x + M\dot{q}\delta\dot{q} - k(q - u(l,t))(\delta q - \delta u(l,t)) \right] \mathrm{d}t = 0. \tag{45b}$$

In order to be able to apply the main theorem of variational calculus we have to use integration by parts with respect to x and t, yielding

$$\int_{t_1}^{t_2}\left[\int_0^l(-\rho A\ddot{u}+EAu'')\delta u\,\mathrm{d}x-(M\ddot{q}+k(q-u(l,t)))\delta q+k(q-u(l,t))\delta u(l,t)\,\mathrm{d}t\right]$$
$$+\int_0^l \rho A\dot{u}\delta u\,\mathrm{d}x\Big|_{t_1}^{t_2}-\int_{t_1}^{t_2}EAu'\delta u\,\mathrm{d}t\Big|_0^l+M\dot{q}\delta q\Big|_{t_1}^{t_2}=0.$$

Since the variations $\delta u(x,t)$ and δq are arbitrary, each term of the sum has to vanish independently. Using the fact that variations vanish at t_1 and t_2 and that they satisfy the geometric boundary conditions, i.e.

$$\delta u(x,t_1)=0,\quad \delta u(x,t_2)=0,\quad \delta q(t_1)=0,\quad \delta q(t_2)=0,\quad \delta u(0,t)=0,$$

we conclude

$$\rho A\ddot{u}-(EAu')'=0, \tag{46a}$$
$$M\ddot{q}+k(q-u(l,t)=0, \tag{46b}$$
$$EAu'=k(q-u(l,t)), \tag{46c}$$

which together with the geometric boundary condition

$$u(0,t)=0 \tag{46d}$$

defines a linear boundary value problem in x and t that can be solved uniquely.

4.7 Discussion of properties of continuous systems

There are many interesting properties and insights that can directly be derived from (39). For example it can be seen that the energy expressions only contain first order derivatives whereas the derived boundary value problems in fact have boundary conditions of higher order because of the use of integration by parts or Gauß's theorem respectively. Boundary conditions occurring in (39) are referred to as geometric boundary conditions, the additional boundary conditions arising in the process of integration by parts are called natural boundary conditions. In order to solve a physical problem one can either directly solve the variational problem or try to solve the corresponding boundary value problem. Since functions occurring in the variational problem have to fulfill less continuity and differentiability conditions the variational problem is often referred to as a weak formulation yielding so called weak solutions. Solutions of the boundary value problem are referred to as exact solutions. Whereas quite general existence and uniqueness theorems for weak solutions exist this is not the case for exact solutions. From the symmetry of the energy expressions for the conservative case one can derive orthogonality of eigenfunctions and max-min-properties for the eigenvalues which can be directly formulated for the continuous case. For the sake of brevity we however directly turn to the discretization of the equations of motion which will allow the use of the corresponding theorems for the discretized case.

4.8 Approximation Methods

Whereas for discrete systems we had to deal with ordinary differential equations and a finite number of degrees of freedom meaning that mathematically the underlying space was $\mathbb{R}^n$, for continuous systems the spaces to be considered are function spaces, which for example have to be valid for the formulation of boundary conditions. In appropriate function spaces (separable Hilbert spaces) functions can be expressed as linear combinations of other functions

$$f(x) = \sum_{i=1}^{\infty} \alpha_i W_i(x)$$

if $W_i, \quad i = 1, ..., \infty$ span the space, where the sum is infinite. A truncated series is an approximation for f. A common example for an approximation of functions in a function space is the Fourier series given by

$$f(t) = \frac{a_0}{2} + \sum_{k=1}^{\infty} a_k \cos k\omega t + b_k \sin k\omega t,$$

which is defined for periodic functions $f(t)$ on the function space L^2. The idea of expanding functions in terms of other functions is also the basis for the Ritz and the Galerkin method. However the underlying spaces are more complicated since boundary conditions have to be considered which is not possible in L^2. For many problems the appropriate spaces are Sobolev spaces however we do not go into details here. In order to approximate a function $w(x,t)$ depending on space and time we write

$$\tilde{w}(\boldsymbol{x}, t) = \sum_{i=1}^{N} W_i(\boldsymbol{x}) q_i(t) \tag{47}$$

where $W_i(x)$ are given linearly independent shape functions. If the functions $W_i(x)$ span the space and we take enough shape functions the truncated series in (47) will be a reasonably good approximation for $w(\boldsymbol{x}, t)$, provided

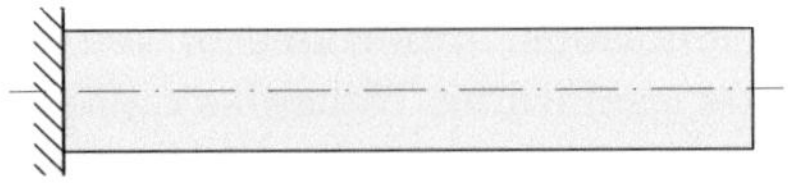

Figure 12. Elastic rod

the $q_i(t)$ are calculated appropriately. Approximation methods differ in the form the shape functions are chosen and in the way the $q_i(t)$ are determined.

Ritz method

The idea behind the Ritz method is to substitute (47) into the energy expression in Hamilton's principle and to carry out the variations with respect to the $q_i(t)$. This means that the solution of the variational problem is limited to a subspace spanned by the shape functions $W_i(x)$ which have to satisfy the geometric boundary conditions. Consider the example of the elastic rod shown in figure 12 for which as before the energy expressions read

$$T = \frac{1}{2}\int_0^l \rho A(x)\dot{u}^2 \,\mathrm{d}x,$$

$$U = \frac{1}{2}\int_0^l EA(x)u'^2 \,\mathrm{d}x.$$

Substitution of

$$u(x,t) \approx \tilde{u}(x,t) = \sum_{i=1}^{N} U_i(x)q_i(t)$$

where $U_i(0) = 0$ and $U_i(x)$ are linearly independent yields the following equations in Hamilton's principle

$$\delta\int_{t_1}^{t_2}\left[\int_0^l \frac{1}{2}\rho A\left(\sum_{i=1}^{N} U_i\dot{q}_i\right)\left(\sum_{i=1}^{N} U_j\dot{q}_j\right)\mathrm{d}x - \int_0^l EA\left(\sum_{i=1}^{N} U_i'q_i\right)\left(\sum_{i=1}^{N} U_j'q_j\right)\mathrm{d}x\right]\mathrm{d}t = 0 \tag{48}$$

with boundary conditions

$$q_i(t_1) = q_i(t_2) = 0.$$

Carrying out the variations or quicker by application of Lagrange's equations it follows

$$\sum_{i=1}^{N} \int_{0}^{L} \rho A U_i U_j \,\mathrm{d}x \,\ddot{q}_i + \sum_{i=1}^{N} \int_{0}^{L} EAU_i' U_j' \,\mathrm{d}x \, q_i = 0, \quad \text{for} \quad j = 1, ..., N,$$

which in matrix form can be written as

$$\boldsymbol{M}\ddot{\boldsymbol{q}} + \boldsymbol{K}\boldsymbol{q} = 0$$

with

$$m_{ij} = \int_{0}^{L} \rho A U_i U_j \,\mathrm{d}x,$$

$$k_{ij} = \int_{0}^{L} EAU_i' U_j' \,\mathrm{d}x.$$

Galerkin method

In the Galerkin method we directly substitute the ansatz (47) into the boundary value problem (b.v.p.). For the example considered in figure 12 we obtain

$$\rho A \sum_{i=1}^{N} U_i(x)\ddot{q}_i(t) - \left[EA \sum_{i=1}^{N} U_i'(x) q_i(t) \right]' = e(x,t),$$

which yields an error $e(x,t)$. Note that in contrast to the Ritz method the shape functions have to satisfy all boundary conditions

$$U_i(0) = 0, \qquad \text{(geometric b.c.)}$$
$$U_i'(L) = 0 \qquad \text{(natural b.c.)}.$$

In order to minimize the error $e(x,t)$ we require that the projection on the shape functions vanishes, i.e.

$$\int_{0}^{L} \left(\rho A \sum U_i \ddot{q}_i - \left[EA \sum U_i' q_i \right]' \right) U_1 \,\mathrm{d}x = 0,$$

$$\vdots$$

$$\int_{0}^{L} \left(\rho A \sum U_i \ddot{q}_i - \left[EA \sum U_i' q_i \right]' \right) U_N \,\mathrm{d}x = 0.$$

In matrix notation the equations read

$$\boldsymbol{M}\ddot{\boldsymbol{q}} + \boldsymbol{K}\boldsymbol{q} = 0, \tag{49}$$

where

$$m_{ij} = \int_0^L \rho A U_i U_j \,\mathrm{d}x, \tag{50a}$$

$$k_{ij} = -\int_0^L [EAU_i']' \, U_j \,\mathrm{d}x. \tag{50b}$$

Since all boundary conditions have to be satisfied one can use integration by parts on the stiffness matrix which shows that the matrices are formally identical to those obtained by the Ritz method. The Ritz method however converges also if the geometric boundary conditions are not satisfied, whereas the Galerkin method with matrices of the form (50) does not.

Discussion

It can be clearly seen that for both Ritz and Galerkin method the matrices $\boldsymbol{M}$ and $\boldsymbol{K}$ are positive definite, which will generally be the case if conservative problems of linear elasticity are considered. This means that all nice properties of discrete systems like the orthogonality of eigenvectors and the max-min-properties of the eigenvalues carry over also to the continuous case. This can be seen on the one hand from the convergence of the approximation methods but can also be derived directly for the continuous problems Hagedorn (1989).

5 Aspects of design and optimization for active an passive control systems

In the last chapters we have focused on the question how to derive efficient mechanical models for control structures. We have seen that both for discrete and continuous systems the mechanical models can be represented by ordinary differential equations of the form

$$\dot{\boldsymbol{x}} = \boldsymbol{f}(\boldsymbol{x}, \boldsymbol{u}, \boldsymbol{t}), \tag{51}$$

which are written here in first order form. For many practical applications the linearized equations

$$\dot{\boldsymbol{x}} = \boldsymbol{A}\boldsymbol{x} + \boldsymbol{B}\boldsymbol{u}, \tag{52}$$

which are in many cases autonomous, dominate the system's behavior. A very common task in control theory is of course to bring a system from an initial state $\boldsymbol{x}_0$ to a final state $\boldsymbol{x}_f$ using the possible controls $\boldsymbol{u}$. The first question is therefore always, whether this is possible, namely whether the systems is controllable, meaning it can be brought from $\boldsymbol{x}_0$ to $\boldsymbol{x}_f$ in arbitrary finite time. This can be checked by the Kalman criterion according to which the system is controllable if and only if the controllability matrix has full rank Unbehauen (2000), i.e.

$$\text{rank}\{\left[\boldsymbol{B}\,|\,\boldsymbol{A}\boldsymbol{B}\,|\ldots|\,\boldsymbol{A}^{n-1}\boldsymbol{B}\right]\} = n. \tag{53}$$

Usually it is helpful to transform the system matrix $\boldsymbol{A}$ to Jordan normal form, i.e. by writing

$$\boldsymbol{x} = \boldsymbol{Q}\boldsymbol{y},$$

where $\boldsymbol{Q}$ consists of the eigenvectors of $\boldsymbol{A}$ and the corresponding vectors from the Jordan chains, in case multiple eigenvalues with degenerate eigenspaces occur. With $\boldsymbol{J} = \boldsymbol{Q}^{-1}\boldsymbol{A}\boldsymbol{Q}$ the transformed equations read

$$\dot{\boldsymbol{y}} = \boldsymbol{J}\boldsymbol{y} + \boldsymbol{Q}^{-1}\boldsymbol{B}\boldsymbol{u}. \tag{54}$$

From (54) it is for example easy to see that the system cannot be controllable if (54) contains any decoupled equations not influenced by the controls u_i. As we stated above controllability means the system can be brought from $\boldsymbol{x}_0$ to $\boldsymbol{x}_f$ in arbitrary finite time. For practical problems this is however usually not enough. First of all, due to the limitations of the actuators the controls are also limited. Second, in many cases one wants to optimize the controls such that for example energy consumption is low or other requirements are met. In addition to the need for the optimization of controls the system needs to be designed in order to be robust against disturbances. In the next section we address the question how this can be achieved in an active or a passive manner. Afterwards we will show how variational approaches can be used to optimize controls.

5.1 Structural design and optimization of control systems

As mentioned above it is usually not enough to calculate an appropriate control to bring a system from one state to another. In addition to designing an appropriate control $\boldsymbol{u}$ one has to make sure that the corresponding solution of the system stays approxmimately valid also under small disturbances which mathematically formulated means that the solution is stable. Having found a control $\boldsymbol{u}^*$ which can be the result of an optimization we

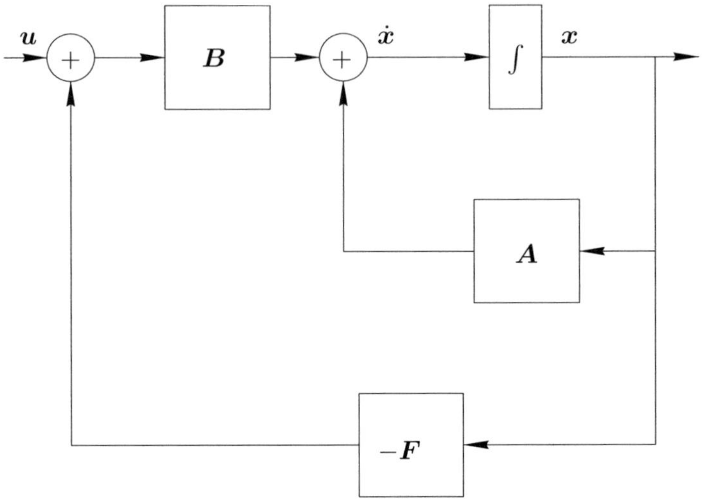

Figure 13. Block diagram of a feedback loop

have to make sure that the corresponding solution $\boldsymbol{x}^*$ is stable. For a linear autonomous system the corresponding disturbed equations read

$$\dot{\boldsymbol{x}}^* + \Delta\dot{\boldsymbol{x}} = \boldsymbol{A}(\boldsymbol{x}^* + \Delta\boldsymbol{x}) + \boldsymbol{B}\boldsymbol{u}. \tag{55}$$

Since $\boldsymbol{x}^*$ is a solution of (52) equation (55) simplifies to

$$\Delta\dot{\boldsymbol{x}} = \boldsymbol{A}\Delta\boldsymbol{x}, \tag{56}$$

a linear system of equations which determines whether disturbances in the initial conditions will grow. It is well known from stability theory (see for example Hagedorn (1988)) that the disturbances will decay when all eigenvalues of $\boldsymbol{A}$ have a negative real part. Therefore prior to designing optimal controls one has to stabilize the control path or even better to optimize the control path in order to achieve desired system properties. This can be achieved through active and passive measures.

Feedback control

Probably the most common active approach to stabilize a control system is to introduce a feedback loop around the control path which is added to the control in addition to $\boldsymbol{u}$ and proportional to the state of the system $\boldsymbol{x}$. The corresponding block diagram is given in figure 13 and the corresponding equations are given by

$$\dot{\boldsymbol{x}} = (\boldsymbol{A} - \boldsymbol{BF})\boldsymbol{x} + \boldsymbol{Bu}, \tag{57}$$

where $\boldsymbol{F}$ denotes the feedback matrix. If the system (52) is controllable the eigenvalues of the closed loop system namely the eigenvalues of the matrix $(\boldsymbol{A} - \boldsymbol{BF})$ can be placed arbitrarily Föllinger (1994); Unbehauen (2000). However it is not easy to see what placement of the eigenvalues is favorable. Therefore, in section 5.2 we will use an objective function in order to design an optimal linear control system Unbehauen (2000).

In practice it is often not possible to measure all states of the system such that the procedure described is not directly applicable. The feedback control methods have however been extended for much more general systems by developing observers for the states which are not observable Föllinger (1994); Unbehauen (2000).

Passive structural optimization

We have seen in the context of (55) that in order to successfully apply controls to a system it is desireable to have a stable control path which for linear constant coefficient systems means that all eigenvalues have a negative real part. Of course it is usually beneficial to achieve this passively without the need of a controller. For mechanical systems the common approach to achieve this is to introduce damping. In the linear case for systems with symmetric positive definite mass and stiffness matrices this is always helpful although it might not be easy to implement. One needs to insure that damping is pervasive, a task that can be analysed in close analogy to the concept of controllability Müller (1977); Hagedorn and Otterbein (1987). If the system under investigation is not conservative the question of how to stabilize it becomes more involved. In this case most of the time one needs to deal with asymmetric parts in the velocity proportional matrix and or in the stiffness matrix. For representative examples we refer to Hagedorn and Hochlenert (1987) and to section 5.3 where we study possibilities to passively stabilize a rotor in frictional contact by changing the stiffness characteristics.

5.2 Variational approaches in control theory

In section 3 we have seen how the equations of motion of a system can be derived from making a functional stationary using the calculus of variations. The same mathematical technique is also very useful in control theory. Generally we study systems of the form

$$\dot{\boldsymbol{x}} = \boldsymbol{f}(\boldsymbol{x}, \boldsymbol{u}, t), \tag{58}$$

where $\boldsymbol{x}$ is the vector of the states of the system and $\boldsymbol{u}$ is the vector of the controls. Note that in many cases the states are limited mathematically stated by $\boldsymbol{u} \in U$, where U is the set of admissible controls. The system (58) can for example be a mechanical system written as first order differential equations where the controls are external forces. The general problem in optimal control is to bring a system from an initial state $\boldsymbol{x}_0$ to a desired final state $\boldsymbol{x}_f$. This is possible in many different ways, provided the system is controllable. In order to evaluate the different controls a cost functional is implemented as

$$I = \int_{t_0}^{t_f} L(\boldsymbol{x}, \boldsymbol{u}, t)\mathrm{d}t, \tag{59}$$

which is to be minimized. Note that if we set $L(\boldsymbol{x}, \boldsymbol{u}, t) = 1$ the cost functional minimizes the time to bring the system from the initial to the final state. Mathematically the optimization problem to be solved is

$$\begin{aligned} \mathrm{Min}\, I &= \int_{t_0}^{t_f} L(\boldsymbol{x}, \boldsymbol{u}, t)\mathrm{d}t, \\ \text{s.t.} & \\ \dot{\boldsymbol{x}} &= \boldsymbol{f}(\boldsymbol{x}, \boldsymbol{u}, t). \end{aligned}$$

In order to get rid of the boundary conditions the cost functional can be augmented by the constraints using Lagrangian multipliers $\boldsymbol{\psi}(t)$ which are also called adjoint variables. The augmented cost functional then reads

$$I = \int_{t_0}^{t_f} L(\boldsymbol{x}, \boldsymbol{u}, t) - \boldsymbol{\psi}^T(\boldsymbol{f}(\boldsymbol{x}, \boldsymbol{u}, t) - \dot{\boldsymbol{x}})\mathrm{d}t. \tag{61}$$

If the controls $\boldsymbol{u}(t)$ are sufficiently smooth and unbounded the functional (69a) can be optimized using the calculus of variations with

$$\boldsymbol{x} = \boldsymbol{x}^* + \epsilon\delta\boldsymbol{x}, \tag{62a}$$

$$\boldsymbol{u} = \boldsymbol{u}^* + \epsilon\delta\boldsymbol{u}, \tag{62b}$$

$$\boldsymbol{\psi} = \boldsymbol{\psi}^* + \epsilon\delta\boldsymbol{\psi}, \tag{62c}$$

where the optimal trajectories are marked by * and the variations vanish for the boundaries t_0 and t_f. Note that the variation of $\boldsymbol{\psi}$ ensures that the equations of motion have to be fulfilled and therefore no extra constraints are needed.

As a minimal example consider a particle which is moved on a line by a control force. At time $t_0 = 0$ the particle is at rest at the position $x(0) = 1$ and is to be transferred to the origin arriving there at $t_f = 1$. The objective is not to use high control forces which can be achieved by minimizing the square of the force. Mathematically the optimization problem reads

$$\begin{aligned}
&\mathrm{Min}\frac{1}{2}\int_0^1 u^2 \mathrm{d}t, \\
&\text{s.t.} \\
&\ddot{x} = u, \\
&x(0) = 1, \qquad x(1) = 0, \qquad \dot{x}(0) = 0, \qquad \dot{x}(1) = 0.
\end{aligned}$$

In order to obtain a description compatible with (61) we write the equation of motion as a first order system

$$\begin{bmatrix} \dot{x}_1 \\ \dot{x}_2 \end{bmatrix} = \begin{bmatrix} 0 & 1 \\ 0 & 0 \end{bmatrix} \begin{bmatrix} x_1 \\ x_2 \end{bmatrix} + \begin{bmatrix} 0 \\ 1 \end{bmatrix} u \tag{63}$$

and obtain for the augmented cost functional

$$I = \int_0^1 \frac{1}{2}u^2 + \psi_1(\dot{x}_1 - x_2) + \psi_2(\dot{x}_2 - u)\mathrm{d}t.$$

A necessary condition for an optimum is that the variation δI vanishes yielding

$$\int_0^1 u\delta u + \delta\psi_1(\dot{x}_1 - x_2) + \psi_1(\delta\dot{x}_1 - \delta x_2) + \delta\psi_2(\dot{x}_2 - u) + \psi_2(\delta\dot{x}_2 - \delta u)\mathrm{d}t = 0. \tag{64}$$

Using integration by parts it follows

$$\int_0^1 (u - \psi_2)\delta u + (\dot{x}_1 - x_2)\delta\psi_1 - \dot{\psi}_1\delta x_1 + (\dot{x}_2 - u)\delta\psi_2 - (\dot{\psi}_2 + \psi_1)\delta x_2 \mathrm{d}t = 0. \tag{65}$$

Finally, from the main theorem of the variational calculus we obtain

$$u = \psi_2$$

and

$$\dot{x}_1 = x_2, \tag{66a}$$
$$\dot{x}_2 = \psi_2, \tag{66b}$$
$$\dot{\psi}_1 = 0, \tag{66c}$$
$$\dot{\psi}_2 = -\psi_1. \tag{66d}$$

Due to their simple structure the equations (66) can be integrated analytically yielding

$$x_1 = -\frac{1}{6}k_1 t^3 + \frac{1}{2}k_2 t^2 + k_3 t + k_4, \tag{67a}$$

$$x_2 = -\frac{1}{2}k_1 t^2 + k_2 t + k_3, \tag{67b}$$

$$\psi_1 = k_1, \tag{67c}$$

$$\psi_2 = -k_1 t + k_2. \tag{67d}$$

Adjusting (67) to the boundary conditions yields

$$k_1 = 12, \qquad k_2 = 6, \qquad k_3 = 0, \qquad k_4 = 1,$$

and the optimal control reads

$$u = -12t + 6. \tag{68}$$

It can be easily verified that the solution actually yields a minimum.

In the general case (61) is often written in the form

$$I = -\int_{t_0}^{t_f} H(\boldsymbol{x}, \boldsymbol{\psi}, \boldsymbol{u}, t) - \boldsymbol{\psi}^T \dot{\boldsymbol{x}} \mathrm{d}t, \tag{69a}$$

defining the Hamilton function

$$H = -L(\boldsymbol{x}, \boldsymbol{u}, t) + \boldsymbol{\psi}^T \boldsymbol{f}(\boldsymbol{x}, \boldsymbol{u}, t). \tag{69b}$$

The variation of (69a) yields

$$\delta I = -\int_{t_0}^{t_f} \left(\frac{\partial H}{\partial \boldsymbol{x}}\right)^T \delta \boldsymbol{x} + \left(\frac{\partial H}{\partial \boldsymbol{\psi}}\right)^T \delta \boldsymbol{\psi} + \left(\frac{\partial H}{\partial \boldsymbol{u}}\right)^T \delta \boldsymbol{u} - \dot{\boldsymbol{x}}^T \delta \boldsymbol{\psi} - \boldsymbol{\psi}^T \delta \dot{\boldsymbol{x}} \, \mathrm{d}t = 0. \tag{70}$$

Using integration by parts on the term with $\delta \dot{\boldsymbol{x}}$ and the main theorem of variational calculus we obtain

$$\dot{\boldsymbol{x}} = \frac{\partial H}{\partial \boldsymbol{\psi}}, \tag{71a}$$

$$\dot{\boldsymbol{\psi}} = -\frac{\partial H}{\partial \boldsymbol{x}}, \tag{71b}$$

$$\frac{\partial H}{\partial \boldsymbol{u}} = \boldsymbol{0}, \tag{71c}$$

which are necessary conditions for an optimum. The equations (71) consist of a boundary value problem of the so called canonical Hamilton equations in $\boldsymbol{x}$ and $\boldsymbol{\psi}$ given by (71a) and (71b) with the boundary conditions $\boldsymbol{x}(t_0) = \boldsymbol{x}_0$ and $\boldsymbol{x}(t_f) = \boldsymbol{x}_f$ and a set of algebraic equations (71c) from which the optimal controls can be calculated. Note that equations (71) imply that $\frac{\mathrm{d}H}{\mathrm{d}t} = \frac{\partial H}{\partial t}$ and that H is therefore constant in the autonomous case. Of course, after solving equations (71) it has to be checked whether a minimum of the cost functional I has been found. In many cases however the equations (71) yield a unique solution with guaranties an optimum provided the problem was well posed.

As an example consider the optimal linear control problem for the system

$$\dot{\boldsymbol{x}} = \boldsymbol{A}(t)\boldsymbol{x} + \boldsymbol{B}(t)\boldsymbol{u}, \tag{72}$$

which is to be brought from $\boldsymbol{x}(t_0) = \boldsymbol{x}_0$ to $\boldsymbol{x}(t_f) = \boldsymbol{x}_f$ by a control $\boldsymbol{u}$ which minimizes the quadratic cost functional

$$\int_{t_0}^{t_f} L(\boldsymbol{x}, \boldsymbol{u}, t)\mathrm{d}t = \frac{1}{2}\int_{t_0}^{t_f} [\boldsymbol{x}^T\boldsymbol{Q}(t)\boldsymbol{x} + \boldsymbol{u}^T\boldsymbol{R}(t)\boldsymbol{u}]\mathrm{d}t. \tag{73}$$

The possibly time dependent matrices $\boldsymbol{Q}(t)$ and $\boldsymbol{R}(t)$ are assumed to be symmetric positive definite and are weighting the minimization of the states and controls of the system. The necessary conditions for an optimum (71) can be calculated as

$$\dot{\boldsymbol{x}} = \frac{\partial H}{\partial \boldsymbol{\psi}} = \boldsymbol{A}(t)\boldsymbol{x} + \boldsymbol{B}(t)\boldsymbol{u}, \tag{74a}$$

$$\dot{\boldsymbol{\psi}} = -\frac{\partial H}{\partial \boldsymbol{x}} = \boldsymbol{Q}(t)\boldsymbol{x} - \boldsymbol{A}^T(t)\boldsymbol{\psi}, \tag{74b}$$

$$\boldsymbol{0} = \frac{\partial H}{\partial \boldsymbol{u}} = \boldsymbol{R}(t)\boldsymbol{u} + \boldsymbol{B}^T(t)\boldsymbol{\psi}. \tag{74c}$$

From (74) we obtain

$$\boldsymbol{u} = \boldsymbol{R}^{-1}(t)\boldsymbol{B}^T(t)\boldsymbol{\psi}(t) \tag{75}$$

and

$$\begin{pmatrix} \dot{\boldsymbol{x}} \\ \dot{\boldsymbol{\psi}} \end{pmatrix} = \begin{bmatrix} \boldsymbol{A}(t) & \boldsymbol{B}(t)\boldsymbol{R}^{-1}\boldsymbol{B}^T(t) \\ \boldsymbol{Q}(t) & -\boldsymbol{A}^T(t) \end{bmatrix} \begin{pmatrix} \boldsymbol{x} \\ \boldsymbol{\psi} \end{pmatrix}. \tag{76}$$

For the solution of (76) we make the ansatz

$$\boldsymbol{\psi} = -\boldsymbol{P}(t)\boldsymbol{x}, \tag{77}$$

where the matrix $\boldsymbol{P}(t)$ is to be determined. From (76) we obtain

$$\dot{\boldsymbol{x}} = [\boldsymbol{A}(t) - \boldsymbol{B}(t)\boldsymbol{R}^{-1}\boldsymbol{B}(t)\boldsymbol{P}(t)]\boldsymbol{x}, \tag{78a}$$

$$\dot{\boldsymbol{\psi}} = -\boldsymbol{P}(t)\dot{\boldsymbol{x}} - \dot{\boldsymbol{P}}(t)\boldsymbol{x} = [\boldsymbol{Q}(t) + \boldsymbol{A}(t)\boldsymbol{P}(t)]\boldsymbol{x}. \tag{78b}$$

Substituting (78a) into (78b) we obtain by comparison of coefficients the determining equation for $\boldsymbol{P}(t)$

$$\dot{\boldsymbol{P}}(t) = -\boldsymbol{Q}(t) - \boldsymbol{P}(t)\boldsymbol{A}(t) - \boldsymbol{A}^T(t)\boldsymbol{P}(t) + \boldsymbol{P}(t)\boldsymbol{B}(t)\boldsymbol{R}^{-1}\boldsymbol{B}^T(t)\boldsymbol{P}(t), \tag{79}$$

which is the well known matrix Ricatti equation which is uniquely solvable. The optimal control therefore reads

$$\boldsymbol{u} = -\boldsymbol{R}^{-1}(t)\boldsymbol{B}^T(t)\boldsymbol{P}(t)\boldsymbol{x}. \tag{80}$$

Note that the result can be interpreted as a feedback control $\boldsymbol{u} = -\boldsymbol{F}\boldsymbol{x}$. For systems with constant coefficients the feedback matrix $\boldsymbol{F}$ can be determined by optimization the cost functional (73) instead of a pole placement.

Returning to the general case we observe that the presented variational approach for optimal control is very powerful and can be extended for systems with variable end time t_f. However, there is the important limitation that for the use of the main theorem of variational calculus the controls have to be unbounded. Thinking of the problem of time optimal control this limitation is not acceptable, since without a limitation of the controls every controllable system can be steered into the desired final state in arbitrary short finite time. The extension of the presented variational approach to systems with limited controls $\boldsymbol{u} \in U$ is given by the maximum principle of Pontryagin. For the formulation of the maximum principle Pontryagin embedded the cost functional into the new state variable x_0 as

$$x_0 = \int_{t_0}^{t} L(\boldsymbol{x}, \boldsymbol{u}, \tau)\mathrm{d}\tau \tag{81}$$

and defined the augmented state vector as

$$\tilde{\boldsymbol{x}} = \begin{bmatrix} x_0 \\ \boldsymbol{x} \end{bmatrix}. \tag{82}$$

Therefore the augmented system reads

$$\dot{\tilde{\boldsymbol{x}}} = \tilde{\boldsymbol{f}}(\tilde{\boldsymbol{x}}, \boldsymbol{u}, t) = \begin{bmatrix} L(\boldsymbol{x}, \boldsymbol{u}, t) \\ \boldsymbol{f}(\boldsymbol{x}, \boldsymbol{u}, t) \end{bmatrix}. \tag{83}$$

If one defines the augmented Hamilton function as

$$\tilde{H}(\tilde{\boldsymbol{x}}, \tilde{\boldsymbol{\psi}}, \boldsymbol{u}, t) = \tilde{\boldsymbol{\psi}}^T \tilde{\boldsymbol{f}}(\boldsymbol{x}, \boldsymbol{u}, t), \qquad \tilde{\boldsymbol{\psi}} = \begin{bmatrix} \psi_0 \\ \boldsymbol{\psi} \end{bmatrix} \tag{84}$$

and requires that for an optimal control the canonical Hamilton equations

$$\dot{\tilde{\boldsymbol{x}}} = \frac{\partial \tilde{H}}{\partial \tilde{\boldsymbol{\psi}}}, \tag{85a}$$

$$\dot{\tilde{\boldsymbol{\psi}}} = -\frac{\partial \tilde{H}}{\partial \tilde{\boldsymbol{x}}}, \tag{85b}$$

are fullfilled, one can state Pontryagin's maximum principle as: A necessary condition for the existence of an optimal control $\boldsymbol{u}^* \in U$ and the corresponding solution $\boldsymbol{x}^*$ is the existence of a solution $\boldsymbol{\psi}^* \neq \boldsymbol{0}$ of (85) for which holds

- for all t in $t_0 \leq t \leq t_f$ the Hamiltonian takes its absolute maximum with respect to $\boldsymbol{u} \in U$ i.e.

$$\tilde{H}(\tilde{\boldsymbol{x}}^*, \tilde{\boldsymbol{\psi}}^*, \boldsymbol{u}^*, t) = \max\nolimits_{\boldsymbol{u} \in U} \tilde{H}(\tilde{\boldsymbol{x}}, \tilde{\boldsymbol{\psi}}, \boldsymbol{u}, t), \tag{86}$$

- ψ_0 does not get positive, i.e.

$$\psi_0(t) \leq 0. \tag{87}$$

Note that equations (85) certify that the optimal control solution actually satisfies the system's equations of motion. Therefore, in the case of unbounded controls they imply that (69a) is made stationary. The calculus of variations is therefore a special case of the Pontryagin maximum principle which also holds for piecewise discontinuous controls. The extension to discontinuous controls is the reason why (86) replaces the condition (71c) that has previously been used.

As an important example where controls need to be assumed bounded we consider a problem of time optimal control. Assume the very simple system (63) from the previous example is to be steered into the origin in minimal time using a control from $u \in \{-1, 1\}$. The Hamilton function in this case reads

$$\tilde{H} = \psi_0 + \psi_1 x_2 + \psi_2 u. \tag{88}$$

From (86) it follows

$$u = \begin{cases} 1 & \psi_2 > 0 \\ -1 & \psi_2 < 0 \end{cases}. \tag{89}$$

Equations (85) yield

$$\dot{x}_1 = \frac{\partial \tilde{H}}{\partial \psi_1} = x_2, \tag{90a}$$

$$\dot{x}_2 = \frac{\partial \tilde{H}}{\partial \psi_2} = u, \tag{90b}$$

$$\dot{\psi}_1 = -\frac{\partial \tilde{H}}{\partial x_1} = 0, \tag{90c}$$

$$\dot{\psi}_2 = -\frac{\partial \tilde{H}}{\partial x_2} = -\psi_1. \tag{90d}$$

The equations (90) can be easily integrated resulting in

$$x_1 = \frac{1}{2}ut^2 + c_3 t + c_4, \tag{91a}$$

$$x_2 = ut + c_3, \tag{91b}$$

$$\psi_1 = c_1, \tag{91c}$$

$$\psi_2 = c_2 - c_1 t, \tag{91d}$$

and the optimal control reads $u = \text{sign}(c_2 - c_1 t)$. Using (91a) and (91a) we can eliminate the time and obtain the system trajectories

$$x_1 = \pm\frac{1}{2}x_2^2 + c_5, \tag{92}$$

which are parabolas depending on the sign of the control u and on the constant c_5 that is a function of c_3, c_4 and u. Using them the optimal trajectories can be obtained by calculating the switching curve on the phase plane Pontryagin and Mishchenko (1962); Ludyk (1995).

In the example we observe that the optimal control always takes its extreme values in the set of admissible controls U. This is a general property of the controls for time optimal problems. Especially for linear systems theorems for existence and uniqueness as well as the number of switching occasions have been derived. For details refer to Pontryagin and Mishchenko (1962); Ludyk (1995) and the references therein.

5.3 Example of passive structural design against self-excited vibrations

In this section we consider an example of a system tending to self-excited vibrations which are unwanted. In order to stabilize the system we will not introduce damping but show that also with an appropriate design of the

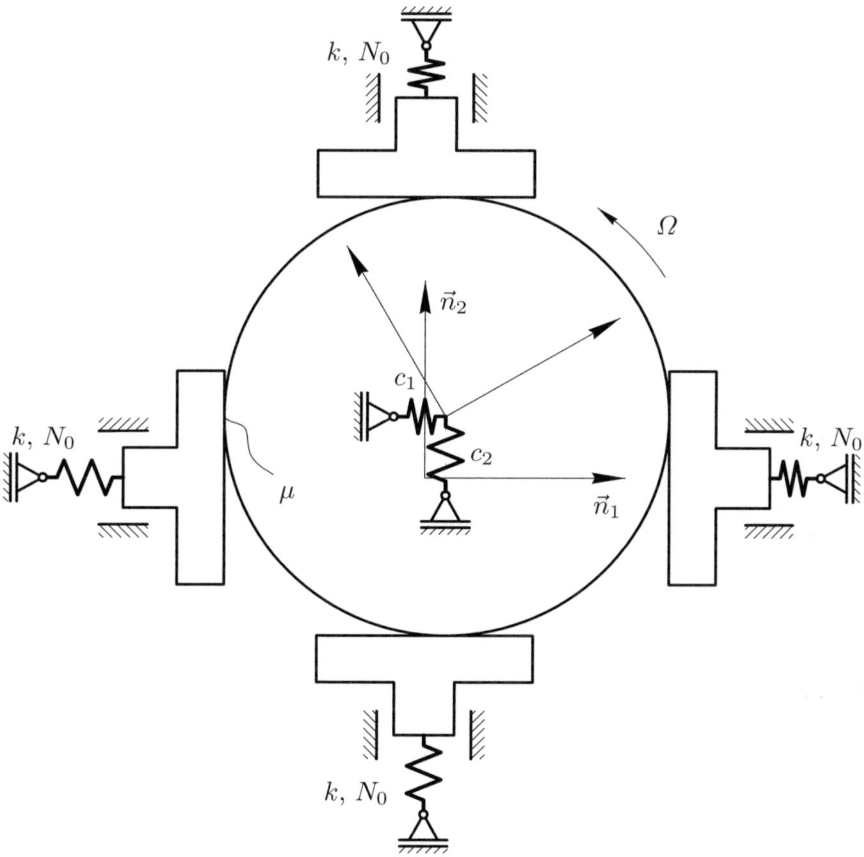

Figure 14. Rotor in frictional contact

stiffness properties the problem can be attacked. For a rather wide class of systems it is helpful to avoid symmetries either in the structure itself or in the bedding. In order to motivate the ideas we consider a simple example from rotor dynamics. Consider the Laval rotor shown in figure 14 which is bedded on two springs with stiffness c_1 and c_2 respectively and assumed to turn with constant angular velocity Ω. The rotor is in contact with friction pads which are pressed onto it's surface by prestressed springs (prestress N_0, stiffness k). Between the pads and the rotor friction occurs which we model by Coulomb's law with friction constant μ. Using

$$q_1 = {}^{N_0}\boldsymbol{p}^P \cdot \vec{n}_1 \tag{93a}$$

$$q_2 = {}^{N_0}\boldsymbol{p}^P \cdot \vec{n}_2 \tag{93b}$$

as degrees of freedom the equations of motion read

$$\begin{bmatrix} 1 & 0 \\ 0 & 1 \end{bmatrix} \begin{pmatrix} \ddot{q}_1 \\ \ddot{q}_2 \end{pmatrix} + \begin{bmatrix} \omega_1^2 & n \\ -n & \omega_2^2 \end{bmatrix} \begin{pmatrix} q_1 \\ q_2 \end{pmatrix} = \begin{pmatrix} 0 \\ 0 \end{pmatrix}, \tag{94}$$

where $\omega_1^2 = \frac{c_1+2k}{M}$, $\omega_2^2 = \frac{c_2+2k}{M}$ and $n = \frac{2\mu k}{M}$. The eigenvalues of the system can readily be calculated as

$$\lambda_{1,2} = \sqrt{-\frac{1}{2}\left(\omega_1^2 + \omega_2^2 \pm \sqrt{(\omega_1^2 - \omega_2^2)^2 - 4n^2}\right)}. \tag{95}$$

By inspection of (95) it can directly be seen that when

$$4n^2 > (\omega_1^2 - \omega_2^2)^2 \tag{96}$$

the inner square root yields a complex number. Since the square root of a complex number is point symmetry with respect to the origin of the complex plane this means that one of the eigenvalue $\lambda_{1,2}$ has a positive real part and the system is unstable. The analysis shows that if we split the eigenvalues ω_1 and ω_2 of the corresponding conservative system, it is stabilized. A splitting of eigenvalues can for example be achieved by changing the geometry of the mountings of the bearings.

The observed effect can be shown to hold for a much more general class of problems given by equations of the type

$$\boldsymbol{M}\ddot{\boldsymbol{q}} + \Delta\boldsymbol{D}(t)\dot{\boldsymbol{q}} + (\boldsymbol{K} + \Delta\boldsymbol{K}(t))\boldsymbol{q} = \boldsymbol{0}, \tag{97a}$$

$$\Delta\boldsymbol{D}(t) = \Delta\boldsymbol{D}(t+T), \qquad \Delta\boldsymbol{K}(t) = \Delta\boldsymbol{K}(t+T), \tag{97b}$$

using analytic perturbation theory for eigenvalues Seyranian and Mailybaev (2003). The matrices $\boldsymbol{M}$, $\boldsymbol{K}$ are constant, symmetric and positive definite and $\Delta\boldsymbol{D}(t)$, $\Delta\boldsymbol{K}(t)$ are time periodic but have arbitrary structure. Note that the constant coefficient case studied in the previous example can be seen as a special case of the periodic setting. In fact if we had attached the pads to the rotor we would have obtained time periodic coefficients. In applications the perturbation matrices $\Delta\boldsymbol{D}(t)$ and $\Delta\boldsymbol{K}(t)$ often arise due to contact forces as we saw in the previous example. In many cases these forces are small compared to elastic restoring terms and can therefore be treated as perturbations.

As motivated by the example system from figure 14 systems of the type (97a) can be stabilized by separating the eigenvalues of the uperturbed conservative $\boldsymbol{M}$, $\boldsymbol{K}$ system. From a constructive point of view this is much easier than to introduce damping since it can be achieved by a change of geometry. Intuitively it is clear that multiple eigenvalues naturally arise due to symmetries of the structure which means that symmetries have to be destroyed in order to make the system robust against instability phenomena. In rotor dynamics this is not so easy since usually one wants the rotor to be balanced. However the approach of designing asymmetric rotors to avoid self-excited vibrations has been successfully been performed in the past. For deeper insights into the underlying mechanisms we refer to Spelsberg-Korspeter (2013) and the references therein.

6 Concluding remarks

In this chapter we discussed the role of variational principles in the derivation of mechanical control systems. The major goal was to show the connection between different methods for deriving the model equations and to emphasize the importance of structural models which significantly reduce the degrees of freedom of control models. We then showed that variational techniques are also useful in the theory of optimal control and discussed strategies for active and passive stabilization of control systems. The chapter does not aim for completeness but tries indicate linkages between different topics which are most often treated more or less independently in university curricula.

Bibliography

W. Becker and D. Gross. *Mechanik elastischer Körper und Strukturen.* Springer, 2002.

O. Föllinger. *Regelungstechnik.* 1994.

P. Hagedorn. *Non-Linear Oscillations.* Calendron Press Oxford, 1988.

P. Hagedorn. *Technische Schwingunslehre*, volume 2. Springer-Verlag, 1989. Lineare Schwingungen kontinuierlicher mechanischer Systeme.

P. Hagedorn and D. Hochlenert. *Technische Schwingunslehre.* Harry Deutsch, 1987.

P. Hagedorn and S. Otterbein. *Technische Schwingunslehre.* Springer-Verlag, 1987.

T. R. Kane and Levinson D. A. *Dynamics: Theory and Applications.* McGraw-Hill, New York, 1985.

G. Ludyk. *Theoretische Regelungstechnik 2: Zustandsrekonstruktion, optimale und nichtlineare Regelsysteme.* Springer Verlag, Berlin, 1995.

L. Meirovitch. *Fundamentals of Vibration.* McGraw-Hill, 2001.

P. C. Müller. *Stabilität und Matrizen.* Springer, 1977.

L. S. Pontryagin and E. F. Mishchenko. *The Mathematical Theory of Optimal Processes.* John Wiley & Sons, 1962.

A. P. Seyranian and A. A. Mailybaev. *Multiparameter Stability Theory with Mechanical Applications.* World Scientific, 2003.

G. Spelsberg-Korspeter. *Robust structural design against self-excited vibrations.* Springer, 2013.

S. Timoshenko and J. N. Goodier. *Theory of Elasticity.* McGraw-Hill, 1951.

H. Unbehauen. *Regelungstechnik.* Vieweg Verlag, 2000.

K. Washizu. *Variational Methods in Elasticity and Plasticity.* Pergamon Press, 1974.

Hybrid Mass Damper: A Tutorial Example

André Preumont, David Alaluf and Renaud Bastaits
ULB, Brussels, Belgium

Abstract This chapter compares three different ways of mitigating the dynamic response of buildings: Dynamic Vibration Absorber (DVA), Active Mass Damper (AMD) and Hybrid Mass Damper (HMD). The methodology is illustrated with a shear frame example subjected to a random seismic input. Two different ways of implementing the HMD are considered, one called *passive* starting from a mistuned DVA, and one called *active* starting from a tuned DVA and using a control system with two feedback loops. It is shown that a well designed HMD may produce performances comparable to that of an AMD while significantly reducing the actuator force and stroke requirements. Besides, the *active* implementation is immune to control system breakdown, because the HMD is reduced to a properly tuned DVA with optimum performances for a passive system.

1 Introduction

1.1 Dynamic Vibration Absorber

The celebrated *Dynamic Vibration Absorber* (DVA) was invented by Frahm in 1909; it is based on simple idea of transferring the kinetic energy of the vibrating structure to a properly tuned and specially designed single d.o.f. oscillator, where it is dissipated. The underlying theory was developed by Ormondroyd and Den Hartog in 1928. Depending on the application, it can also be called *Tuned Mass Damper* (TMD). Because it is simple and robust, the DVA has been used very successfully in many applications, including the mitigation of the wind response of tall building and the vibration alleviation in helicopters. Note, however, that in helicopter applications, the disturbance f is harmonic at a constant frequency ω_0, and the DVA parameters are adjusted to produce a sharp transmission zero at ω_0 in the Frequency Response Function (FRF) between the disturbance and the structural response; this usually implies a lightly damped DVA. On the contrary, when the structure is subjected to a wide band disturbance (as in the case of a tall buildings subjected to wind gusts or an earthquake), the

P. Hagedorn, G. Spelsberg-Korspeter (Eds.), *Active and Passive Vibration Control of Structures*, CISM International Centre for Mechanical Sciences
DOI 10.1007/ 978-3-7091-1821-4_3

DVA parameters are tuned in such a way that the structural response is minimized (in some sense); the DVA parameters appropriate for mitigating the response to wind gusts or to earthquake ground motion are essentially the same, except that the seismic excitation of strong earthquakes usually leads to much larger structural responses than that of wind gusts, and this requests DVAs with larger strokes. Figure 1 shows the DVA mounted at the top of the Taipei 101 building in Taiwan. This building has a height of 509 m and has 101 floors. The absorber consists of a pendulum with a mass of 730 tons suspended with four cables extending over 4 floors.

Figure 1. Dynamic Vibration Absorber of the Taipei 101 building. The DVA consists of a pendulum with a mass of 730 tons suspended with 4 cables extending over 4 floors.

1.2 Active Mass Damper and Hybrid Mass Damper

This study considers only the case of a wide band disturbance and compares various ways of enhancing the performance of a DVA, all based on the

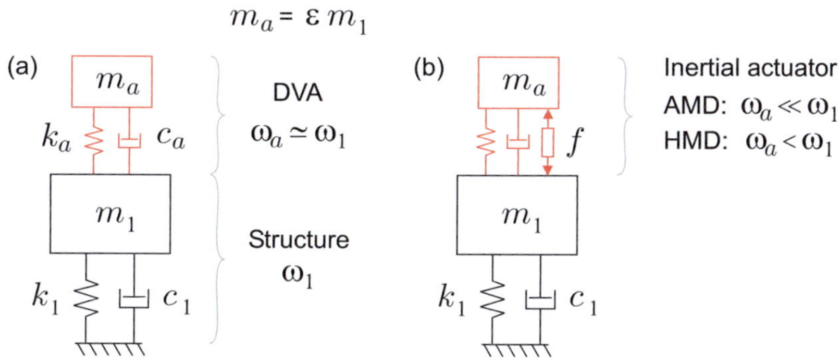

Figure 2. (a) Dynamic Vibration Absorber (DVA): $\omega_a \simeq \omega_1$. (b) Active Mass Damper (AMD) $[\omega_a \ll \omega_1]$ and Hybrid Mass Damper (HMD) $[\omega_a < \omega_1]$.

use of an inertial actuator with different tuning (Fig.2). A DVA is a passive spring and damper device whose natural frequency is close to the targeted mode $\omega_a \simeq \omega_1$. The Active Mass Damper (AMD) includes a force actuator acting in parallel with the spring and damper, between the inertial mass and the supporting structure; the device (called a proof-mass actuator) is usually tuned on a frequency significantly lower than the modes to be damped, so that it behaves as a perfect force actuator for the whole frequency range of interest (from ω_1 and above). The Hybrid Mass Damper (HMD) is based on the same principle as the AMD, except that the frequency ω_a is tuned closer to ω_1, in an attempt to combine the properties of passive and active devices, to achieve one or several of the following features: (1) improved performances; (2) improved robustness with respect to structural changes; (3) improved reliability (e.g. immunity with respect to power failure); (3) reduced construction and maintenance costs. Various examples of HMD used in civil engineering are described in [Seto, Spencer].

In this note, two different ways of implementing the HMD are considered; the first one is based on a passive mistuning of the device before applying active control (velocity feedback); this approach is simple, but has the drawback of reduced reliability in case of control system failure. The second approach starts from a regularly tuned DVA and involves two control layers : the first layer (P+D controller) produces the effect of the passive mistuning of the previous method and the second layer superimposes the active control with velocity feedback. The various approaches are compared on the simple example of shear frame.

This note is organized as follows: Section 2 recalls the equations governing the seismic response of a building, describes the shear frame example used to illustrate the various methods and analyzes its random response to a stationary seismic input. Section 3 recaps the design rules of the DVA and its extension to multiple d.o.f. systems; the shear frame example is used for illustration. Section 4 is devoted to the active control with an AMD; the actuator and sensor are briefly discussed and the system response is formulated in state space. Section 5 discusses the HMD where a DVA is modified to operate as an AMD; the effect on performance and control requirements is analyzed. Finally section 6 discusses how the mistuning of the HMD may be achieved actively with a dual loop controller; it is demonstrated that this offers a better immunity with respect to control system failure.

2 Seismic response

2.1 Basic equations

The seismic excitation is usually specified by the support acceleration, $\ddot{x}_0$. The seismic response of a building excited by a single-axis support excitation is governed by the following equations [e.g. Preumont, 2012, Ch.7]:

The global displacements may be decomposed into the rigid body motion due to the support input motion x_0 and the motion relative to the support; the total displacement is given by

$$\mathbf{x} = x_0 \mathbf{1} + \mathbf{y} \tag{1}$$

where $\mathbf{1}$ is the unit rigid body mode and $\mathbf{y}$ is the motion relative to the support. With these notations, the dynamic equation is

$$M\ddot{\mathbf{y}} + C\dot{\mathbf{y}} + K\mathbf{y} = -M\mathbf{1}\ddot{x}_0 \tag{2}$$

If $\mathbf{y}$ is decomposed into the modes of the structure fixed at its base, $\mathbf{y} = \Phi\mathbf{z}$, where the column of Φ are the modes ϕ_i of the structure clamped at the support, and if a modal damping is assumed, such that $\Phi^T C \Phi = diag(2\xi_i \mu_i \omega_i)$, the modal components z_i satisfy a set of decoupled equations

$$\mu_i \ddot{z}_i + 2\xi_i \mu_i \omega_i \dot{z}_i + \mu_i \omega_i^2 z_i = -\phi_i^T M \mathbf{1} \ddot{x}_0 \tag{3}$$

where $\Gamma_i = -\phi_i^T M \mathbf{1}$ is known as the *modal participation factor* of mode i. The absolute acceleration within the structure is related to the ground acceleration by (in the frequency domain)

$$\ddot{\mathbf{X}}(\omega) = [\mathbf{1} - \sum_{i=1}^{n} \frac{\omega^2 \Gamma_i}{\mu_i(\omega_i^2 - \omega^2 + 2j\xi_i \omega\omega_i)} \phi_i]\ \ddot{X}_0(\omega) \tag{4}$$

The reaction force is related to the ground acceleration by

$$F_0 = \mathbf{1}^T M\ddot{\mathbf{X}} = [m_T + \sum_{i=1}^{n} \frac{\Gamma_i^2}{\mu_i}(\frac{\omega^2}{\omega_i^2 + 2j\xi_i\omega_i\omega - \omega^2})]\ddot{X}_0 \tag{5}$$

where m_T is the total mass of the system. Γ_i^2/μ_i is called the *effective modal mass* of mode i ; it represents the part of the total mass which is associated with mode i for this particular type of excitation (defined by the vector $\mathbf{1}$).

2.2 n-storey shear frame

The structure analyzed in this comparison consists of a n-storey shear frame (Fig.3) made of identical floors of mass m and lateral stiffness k (in the numerical example, $n = 10$).

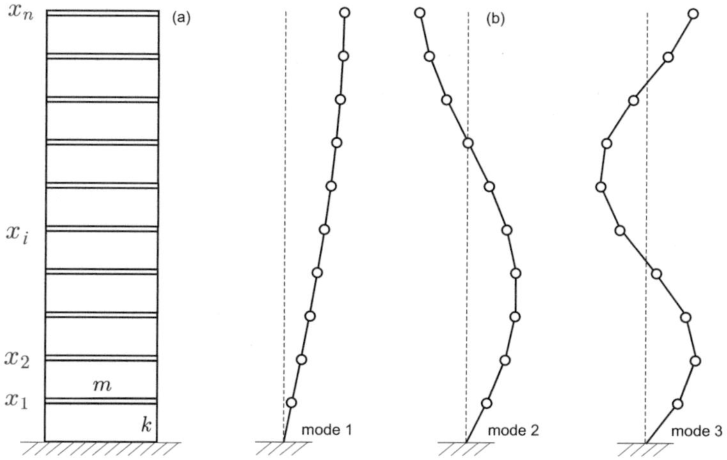

Figure 3. (a) Model of a building with n identical floors; the floor is considered as a rigid slab of mass m and the columns are massless and clamped at both ends; the lateral stiffness of one floor is k. (b) Mode shapes.

Owing to the simple geometry, there are analytical expressions for the natural frequencies:

$$\omega_r = 2\sqrt{\frac{k}{m}} \sin[\frac{(2r-1)}{(2n+1)}\frac{\pi}{2}] \qquad (r = 1, 2, ..., n) \tag{6}$$

where r is the order of the mode and n is the number of floors in the shear frame. The corresponding mode shape is

$$\phi_r(i) = C \sin[i\frac{(2r-1)}{(2n+1)}\pi] \tag{7}$$

(i refers to the floor; r is the order of the mode). The first few mode shapes are illustrated in Fig.3. This structure is assumed to be excited by the horizontal acceleration of the ground, $\ddot{x}_0$. For this particular system, the modal participation factors are

$$\Gamma_r = -\boldsymbol{\phi}_r^T M \mathbf{1} = -m \sum_{i=1}^{n} \phi_r(i) \tag{8}$$

The generalized mass is

$$\mu_r = \boldsymbol{\phi}_r^T M \boldsymbol{\phi}_r = m \sum_{i=1}^{n} \phi_r^2(i) \tag{9}$$

The effective modal mass of mode r is

$$\frac{\Gamma_r^2}{\mu_r} = m \frac{[\sum_{i=1}^{n} \phi_r(i)]^2}{\sum_{i=1}^{n} \phi_r^2(i)} \tag{10}$$

Note that $\sum_{r=1}^{n} \Gamma_r^2/\mu_r = n.m = m_T$, the total mass of the system.

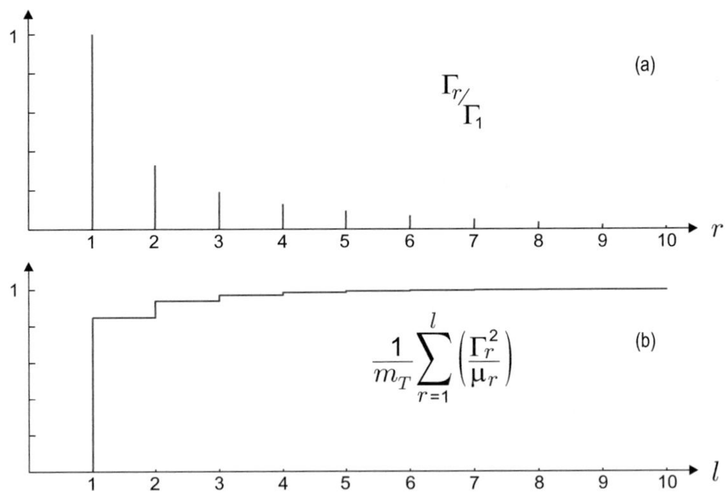

Figure 4. (a) Evolution of the modal participation factor $|\Gamma_r/\Gamma_1|$ with the order r of the mode. (b) Cumulated effective modal mass $\frac{1}{m_T}\sum_{r=1}^{l} \Gamma_r^2/\mu_r$. The amplitude of the steps correspond to the effective modal mass of individual modes.

Figure 4 shows the evolution of the modal participation factor $|\Gamma_r/\Gamma_1|$ and of the cumulated effective modal mass (normalized to the total mass); the modal participation factor and the effective modal mass decrease rapidly with the order of the mode.

2.3 Seismic excitation

In earthquake engineering, it is frequently assumed that the ground acceleration at one point can be modeled by the absolute acceleration response of a linear oscillator excited by a white noise. The natural frequency ω_g and the damping ratio ξ_g are selected to fit the local ground conditions. This leads to the following (two-sided, defined on $-\infty < \omega < \infty$) Power Spectral Density (PSD):

$$\Phi_g(\omega) = \Phi_0 \frac{1 + 4\xi_g^2(\omega^2/\omega_g^2)}{(1 - \omega^2/\omega_g^2)^2 + 4\xi_g^2(\omega^2/\omega_g^2)} \tag{11}$$

It is often referred to as the *Kanai-Tajimi spectrum.* In the numerical example discussed later, the numerical constants are chosen in such a way that the RMS ground acceleration is $\sigma_g = 1$ m/s^2, the central frequency is $\omega_g = 12.56$ rad/sec (2 Hz), and $\xi_g = 0.1$. This leads to the intensity

$$\Phi_0 = \frac{\sigma_g^2}{\pi\omega_g(\frac{1}{2\xi_g} + 2\xi_g)} = 4.87 \times 10^{-3} \quad [\frac{(\mathrm{m/s}^2)^2}{(\mathrm{rad/s})} = \frac{\mathrm{m}^2}{\mathrm{s}^3\mathrm{rad}}] \tag{12}$$

The one-sided[1] PSD of the ground acceleration, $\bar{\Phi}_g(\omega) = 2\Phi_g(\omega)$ is represented in Fig.5.a. This model tends to produce unrealistic displacements and velocities at low frequency, which interfere with the active control based on the measurement of the absolute velocity; this issue may be solved by passing the foregoing excitation in a second order high-pass filter of corner frequency and damping (ω_f, ξ_f); the FRF of the filter is.

$$H_f(\omega) = -\frac{\omega^2}{\omega_f^2 - \omega^2 + 2j\xi_f\omega_f\omega} \tag{13}$$

leading to

$$\Phi_g^*(\omega) = \Phi_g(\omega).|H_f(\omega)|^2 \tag{14}$$

Typical values used in what follows are: $\omega_f = 1.57$ rad/sec (0.25 Hz), $\xi_f = 1$. We will use this form later when necessary, when the active control is added to the system. The modified input PSD $\bar{\Phi}_g^*(\omega)$ is represented in dotted lines in Fig.5.a.

[1] One-sided PSD consider only positive frequencies.

2.4 Seismic response of a n-storey building

Consider a shear frame with 10 identical floors, with $m = 10^5$ kg and $k = 16{\times}10^7$ N/m.[2] The damping is assumed uniform in all modes, $\xi_i = 0.01$. We evaluate the statistics of the random response of this building to the ground acceleration defined in the previous section; in particular, we are interested in the absolute acceleration $\ddot{x}_k$ of the various floors of the building and the support reaction f_0 which is also the shear force in the column of the first floor of the building (excessive shear force in the first floor is a frequent failure mode of civil structures during strong earthquakes).

The transmissibility between the ground acceleration and the absolute acceleration of floor k follows from Equ.(4):

$$\ddot{X}_k(\omega) = [1 - \sum_{i=1}^{n} \frac{\omega^2 \Gamma_i}{\mu_i(\omega_i^2 - \omega^2 + 2j\xi_i\omega\omega_i)} \phi_i(k)]\, \ddot{X}_0(\omega) = T_k^a(\omega)\ddot{X}_0(\omega) \quad (15)$$

where $\phi_i(k)$ is the amplitude of mode i at floor k and Γ_i are the modal participation factors defined earlier. It follows that the one-sided PSD of the absolute acceleration of floor k is given by

$$\bar{\Phi}_{\ddot{x}_k}(\omega) = |T_k^a(\omega)|^2 . \bar{\Phi}_g(\omega) \quad (16)$$

Figure 5.b shows the one-sided PSD of the acceleration of floor 1 and floor 10. Figure 5.c shows the cumulative RMS value of the floor acceleration, again of floor 1 and floor 10, defined by

$$\sigma_{\ddot{x}_k}(\omega) = [\int_{\omega}^{\infty} \bar{\Phi}_{\ddot{x}_k}(\nu)\, d\nu]^{1/2} \quad (17)$$

This diagram is very interesting, because the value for $\omega = 0$ is the RMS acceleration, and the amplitude of the various steps at the natural frequencies indicate how the corresponding mode contributes to the global response. It is clear from this diagram that the top acceleration is dominated by the first mode. Figure 6 shows the evolution of the RMS absolute acceleration within the building; the figure illustrates how the ground acceleration is amplified within the structure; one sees that higher floors experience a larger acceleration than the lower one. This is a general observation, and this is why precision equipments a rarely placed on the upper floors, to minimize their sensitivity to vibrations coming from the environment (traffic, etc...).

[2] these values are consistent with the empirical formula $T_1 = 2\pi/\omega_1 = 0.1n$ relating the first natural period of vibration of buildings with the number n of stories [Newmark and Rosenblueth, p.421]

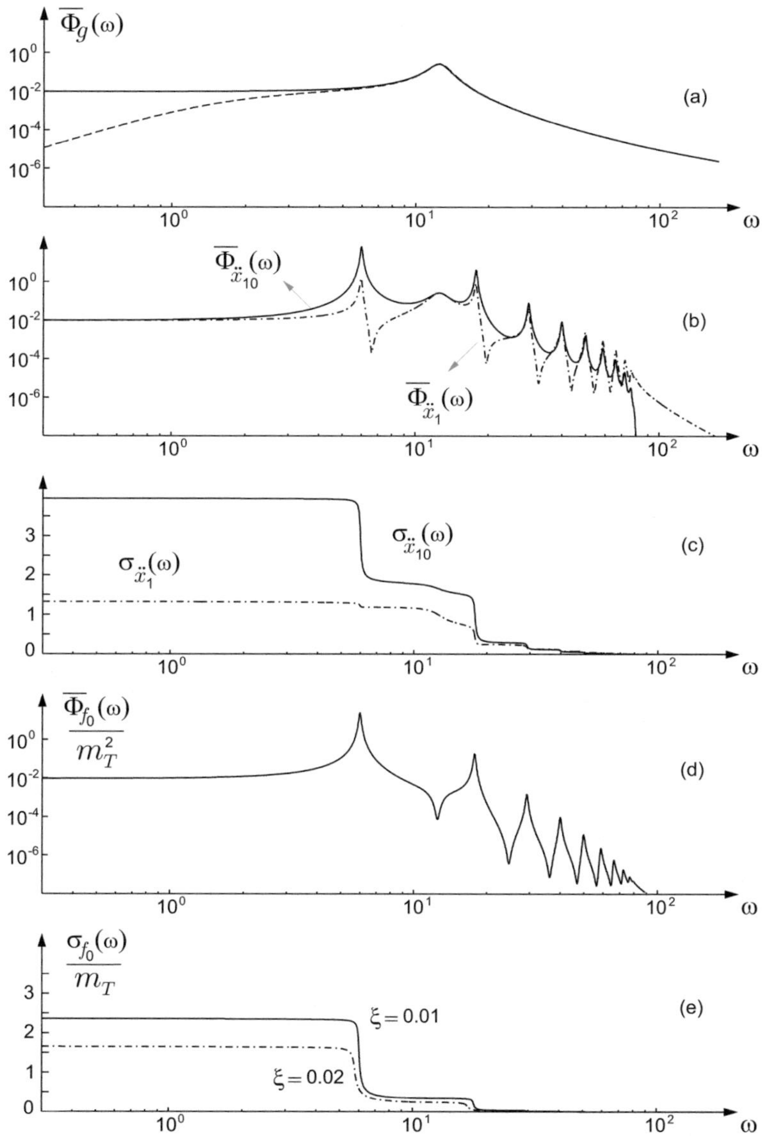

Figure 5. Seismic response of a 10 storey shear frame. (a) One-sided PSD of the ground acceleration; the PSD modified according to Equ.(13) is shown in dotted lines. (b) One-sided PSD of the acceleration of floor 1 and floor 10. (c) Cumulative RMS acceleration of floor 1 and floor 10. (d) One-sided PSD of the non-dimensional reaction force $f_0/\sigma_g m_T$ ($\xi = 0.01$). (e) Cumulative RMS value of the non-dimensional reaction force, for $\xi = 0.01$ and $\xi = 0.02$.

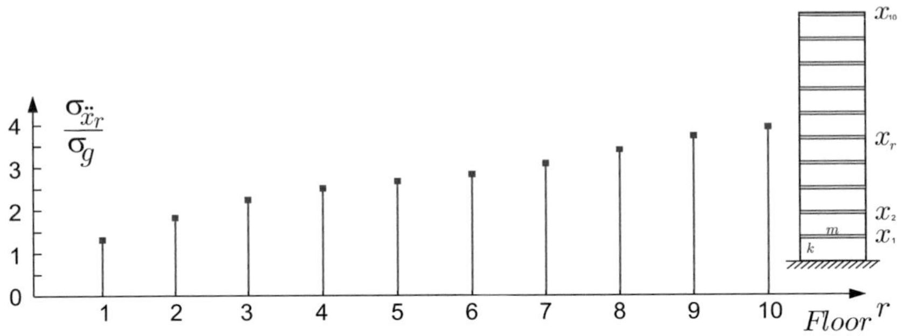

Figure 6. Seismic response of a 10 storey shear frame. Amplification of the RMS floor acceleration $\sigma_{\ddot{x}_k}/\sigma_g$ within the building.

The transmissibility between the ground acceleration and the support reaction f_0 (also called *dynamic mass*) is given by Equ.(5)

$$\frac{F_0(\omega)}{\ddot{X}_0(\omega)} = m_T + \sum_{i=1}^{n} \frac{\Gamma_i^2}{\mu_i}\left(\frac{\omega^2}{\omega_i^2 + 2j\xi_i\omega_i\omega - \omega^2}\right) = T_0(\omega) \tag{18}$$

where m_T is the total mass of the structure and Γ_i^2/μ_i is the effective modal mass of mode i. It follows that

$$\bar{\Phi}_{f_0}(\omega) = |T_0(\omega)|^2.\bar{\Phi}_g(\omega) \tag{19}$$

The one-sided PSD of the non-dimensional reaction force $f_0/\sigma_g m_T$ is represented in Fig.5.d and the cumulative RMS value is represented in Fig.5.e, for two values of the damping ratio ($\xi = 0.01$ and $\xi = 0.02$). Once again, the amplitude of the various steps indicate how the corresponding mode contributes to the global response (the reaction force is dominated by the first mode) and the comparison between the two curves illustrates the effect of the damping (From the random vibration theory, the contribution of mode i to the RMS value is known to vary according to $\xi_i^{-1/2}$). The value at $\omega = 0$,

$$\frac{\sigma_{f_0}}{\sigma_g m_T}$$

ratio between the RMS reaction force and the inertia forces associated with the RMS ground acceleration, is a measure of the dynamic amplification of the shear force in the column of the first floor. One sees that a significant reduction of the shear force may be achieved by increasing the damping of the first mode; various solutions to achieve this, passive and active, are investigated in what follows.

3 Dynamic Vibration Absorber (DVA)

This section recaps the design of a DVA to reduce the structural response to a wide band excitation. The results for a single d.o.f. oscillator are extended to multiple d.o.f. structures. Because of the presence of the DVA, the complete system does not satisfy the assumption of modal damping and the problem is formulated in structural coordinates.

3.1 Equal peak design

Consider the single d.o.f. oscillator of mass m_1, excited by a wide band external force f (Fig.7.a). The DVA consists of a secondary single d.o.f. system, attached to the initial system, with properties such that the response of the primary system to the load f is reduced; the mass of the DVA is usually much smaller than that of the primary system, ($m_2 \ll m_1$). It is usual to introduce the following parameters:

$$\omega_1 = \sqrt{\frac{k_1}{m_1}}, \qquad \omega_2 = \sqrt{\frac{k_2}{m_2}} \tag{20}$$

$$\xi_1 = \frac{c_1}{2m_1\omega_1}, \qquad \xi_2 = \frac{c_2}{2m_2\omega_2} \tag{21}$$

$$\text{Mass ratio:} \quad \mu = \frac{m_2}{m_1}, \quad \text{Frequency ratio:} \quad \nu = \frac{\omega_2}{\omega_1} \tag{22}$$

Typically, the mass ratio is $\mu \simeq 1$ to 3%, to limit the weight penalty associated with the DVA; the frequency ratio ν and the damping ratio ξ_2 of the

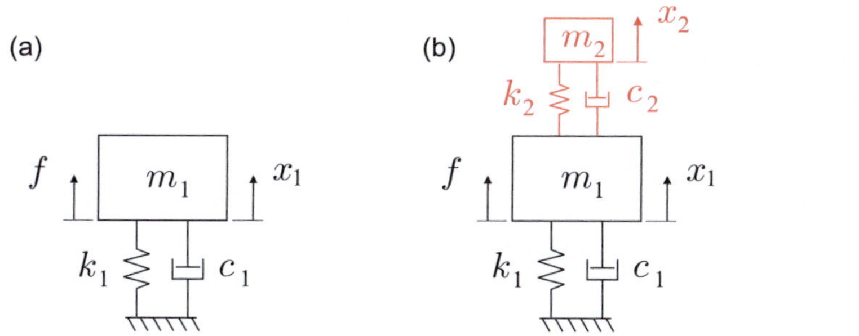

Figure 7. (a) Single d.o.f. oscillator excited by an external force f. (b) Same system equipped with a DVA.

DVA are the design parameters which are adjusted to optimize the performances of the absorber. The optimum values depend on the optimization criteria; in all cases, ν is in general close to 1.

The most popular design method is known as "equal peak design"; it is due to Den Hartog. The optimum is achieved when the frequency ratio is

$$\nu = \frac{\omega_2}{\omega_1} = \frac{1}{1+\mu} \tag{23}$$

and the damping ratio of the DVA is

$$\xi_2 = \sqrt{\frac{3\mu}{8(1+\mu)}} \tag{24}$$

Figure 8 illustrates this design for various values of the mass ratio μ. As μ increases, the frequency difference between the two peaks increases and their amplitude is reduced. The maximum amplification (*quality factor*) of the equal peak design depends only on the mass ratio μ

$$H_{max} = Q = \sqrt{\frac{2+\mu}{\mu}} \simeq \sqrt{\frac{2}{\mu}} \tag{25}$$

Recall that, for a single d.o.f. system, the quality factor and the damping ratio are related by $Q = 1/2\xi$. Therefore, for a mass ratio μ, the maximum amplification will be that of a single d.o.f. with an equivalent damping ratio $\xi_e = \sqrt{\mu/8}$; for $\mu = 0.01$, $\xi_e \simeq 0.032$; for $\mu = 0.02$, $\xi_e \simeq 0.05$, etc...

There are other ways of choosing the absorber parameters: ν and ξ_2 may be selected to minimize the MS response $E[x_1^2]$ of the primary structure to a white noise excitation [Crandall & Mark], or to maximize the stability margin of the poles of the system. Different types of excitations and different optimization criteria will lead to slightly different values of ν and ξ_2; a comprehensive discussion is done in [Warburton] and [Asami et al.] where the "optimal" values are tabulated for a large variety of situations. It turns out, however, that, for small values of μ, the optimum is reasonably flat, and that the DVA operates reasonably well even if the parameters differ from their optimum values; the present study illustrates this fact by using the optimal parameters obtained assuming an external force (f applied on m_1) for a problem involving a seismic excitation. The relatively low sensitivity of the DVA performance to the tuning of its parameters explains why the DVAs are used extensively.

In order to apply DVAs to target one mode of a multiple d.o.f. system, say mode k, one needs to estimate what mass should be considered as m_1 (the mass of the primary system) in the foregoing design procedure. If

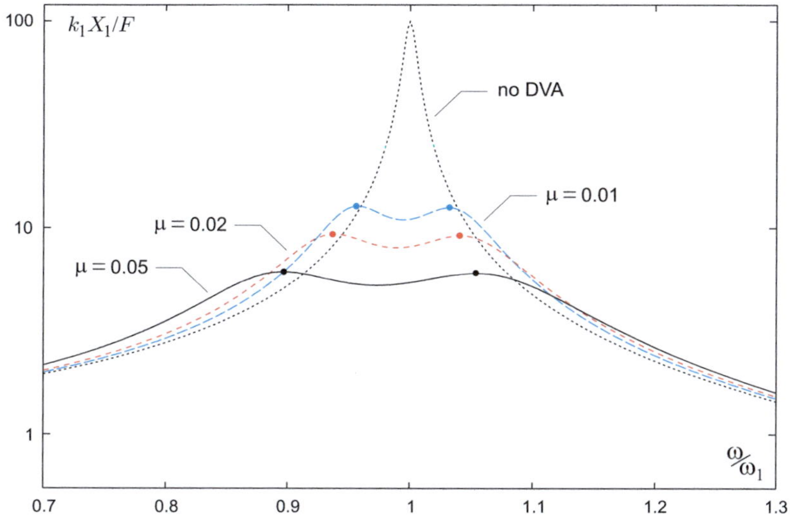

Figure 8. DVA with equal peak design, dynamic amplification near resonance for various values of the mass ratio μ. Larger values of μ lead to more distant peaks, with lower amplitudes. The maximum amplification is $Q \simeq \sqrt{2/\mu}$.

one assumes that the modes are well separated and that, in the vicinity of ω_k, the system response is dominated by mode k, one can show that the equivalent mass to take into account in the design is

$$m_1 = \frac{\mu_k}{\phi_k^2(d)} \tag{26}$$

where μ_k is the generalized mass of mode k and $\phi_k(d)$ is the modal amplitude of mode k projected on the d.o.f. where the DVA is attached. Alternatively, the equivalent mass m_1 is the generalized mass of mode k when the modal amplitude of mode k is normalized according to $\phi_k(d) = 1$ at the d.o.f. where the DVA is attached. Once the mass m_1 has been determined, the design of the DVA proceeds as for a single d.o.f. system. Observe from the previous equation that a larger modal amplitude will tend to decrease m_1, that is to increase the mass ratio for a given mass m_2 of the DVA. Thus, in order to maximize its efficacy, the DVA should be located where the targeted mode has large modal amplitudes.

3.2 n-storey shear frame with a DVA at the top

Consider the shear frame with 10 identical floors (Fig.9); in Fig.5.e, it was observed that the shear force at the base, f_0, is dominated by the response of the first mode. The collapse of buildings during earthquakes is often associated with the maximum shear force exceeding the building resistance; in this section, we examine the use of a DVA (tuned on mode 1) to reduce the amplitude of the shear (reaction) force; all the data used here are identical to those used in the foregoing sections. We proceed in 3 steps: *(i)* construction of a model including the DVA, *(ii)* design of the DVA (determination of the constants k_a and c_a), and *(iii)* calculation of the random response of the structure with DVA to the random seismic input.

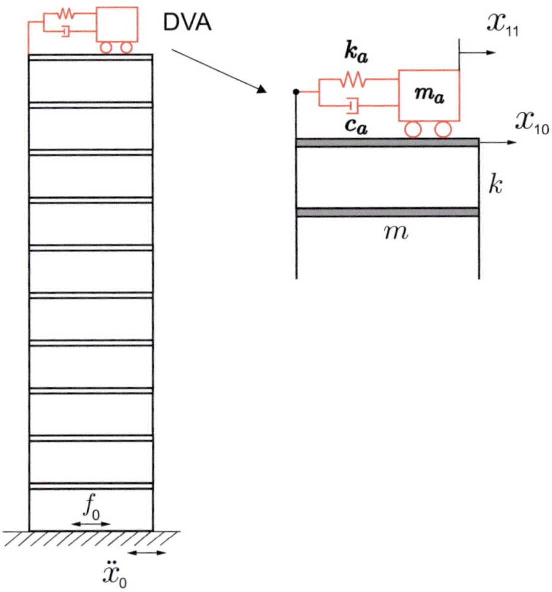

Figure 9. 10-storey shear frame equipped with a DVA at the top. The mass ratio is defined as $\varepsilon = m_a/m_T = m_a/10m$. f_0 is the shear force at the base; the stroke of the DVA is $\Delta = x_{11} - x_{10}$.

Model construction The structure is represented in Fig.9; we assume that the mass of the DVA is only a small fraction of the total mass of the building, $m_a = \varepsilon m_T$ with $\varepsilon = 0.01$. The (11×11) mass and stiffness matrices of the global system can easily be constructed by inspection of Fig.9; m and k being the mass and stiffness of one floor, one finds

$$M = \begin{bmatrix} mI_{10} & 0 \\ 0 & m_a \end{bmatrix} \tag{27}$$

$$K = \begin{bmatrix} 2k & -k & 0 & \dots & 0 & 0 \\ -k & 2k & -k & 0 & 0 & 0 \\ 0 & \dots & & \dots & & 0 \\ 0 & \dots & -k & 2k & -k & 0 \\ 0 & \dots & 0 & -k & k+k_a & -k_a \\ 0 & \dots & 0 & 0 & -k_a & k_a \end{bmatrix} \tag{28}$$

(the geometric stiffness due to the gravity load is ignored). The construction of the damping matrix is more difficult, because the system consists of a shear frame where a uniform modal damping ξ is assumed and a discrete damper (the DVA) connecting x_{10} and x_{11}; the global system does not satisfy the assumption of modal damping. The (10×10) damping matrix C^* of the shear frame alone can be reconstructed as follows: because of the uniform mass distribution within the frame, $M^* = mI_{10}$ and the orthogonality condition reads

$$\Phi^T M^* \Phi = m\, \Phi^T \Phi = \mathrm{diag}(\mu_i)$$

Thus, the mode shapes are orthogonal:

$$\Phi^T \Phi = \mathrm{diag}(\frac{\mu_i}{m})$$

It follows that

$$C^* = \Phi\ \mathrm{diag}(\frac{2m^2 \xi_i \omega_i}{\mu_i})\ \Phi^T \tag{29}$$

satisfies the assumption of modal damping, because $\Phi^T C^* \Phi = \mathrm{diag}(2\xi_i \omega_i \mu_i)$. The global (11×11) damping matrix of the system reads

$$C = \begin{bmatrix} C^* & & 0 \\ & C^*_{10,10} + c_a & -c_a \\ 0 & -c_a & c_a \end{bmatrix} \tag{30}$$

where $C^*_{10,10}$ is the component (10,10) of the matrix C^* and c_a the damper constant of the DVA. In a more general case where the mode shapes are not orthogonal, a full damping matrix may be constructed by making use of the Rayleigh damping assumption.

Design of the DVA The DVA is designed to operate on mode 1; it is placed at the top, where the amplitude of mode 1 is maximum. Once the mass m_a of the DVA has been chosen, $m_a = \varepsilon\, m_T$ (with $\varepsilon = 0.01$ in this case), the equivalent mass of mode 1 is calculated from Equ.(26); combining with (9), one finds

$$m_1 = \frac{\mu_1}{\phi_1^2(n)} = \frac{m \sum_{i=1}^{n} \phi_1^2(i)}{\phi_1^2(n)} \tag{31}$$

In this formula, n is the index of the upper floor ($n = 10$ in this case).[3] Next, the mass ratio is calculated, $\mu = \frac{m_a}{m_1}$, and the optimum parameters of the absorber, k_a and c_a, are obtained from Equ.(23) and (24)

$$\omega_a = \sqrt{\frac{k_a}{m_a}} = \frac{\omega_1}{1+\mu}; \qquad \xi_a = \frac{c_a}{2m_a\omega_a} = \sqrt{\frac{3\mu}{8(1+\mu)}} \tag{32}$$

The parameters of the DVA, m_a, k_a and c_a have all been determined; however, for practical applications, an important design parameter remains to be determined: the stroke, which depends on the seismic input; it will result from the calculation of the random response of the structure with DVA to the random seismic input.

Random response of the structure with DVA The random response of the structure cannot be analyzed in modal coordinates, because the complete structure does not satisfy the assumption of modal damping, and owing to the small size of the model, it is convenient to perform the analysis in global coordinates. The relative displacement response satisfies Equ.(2) where M, K and C are given above and $\mathbf{1} = (1, 1, \ldots, 1)^T$ is the unit rigid body mode (all translational d.o.f. along the axis of excitation are equal to 1 and the rotational d.o.f. are equal to 0). Thus, the FRF between the relative displacement vector and the ground acceleration is

$$\mathbf{Y} = -(K + j\omega C - \omega^2 M)^{-1} M\mathbf{1}\ddot{X}_0 \tag{33}$$

The absolute accelerations of the various floors are

$$\ddot{\mathbf{x}} = \mathbf{1}\ddot{x}_0 + \ddot{\mathbf{y}} \tag{34}$$

leading to the FRF

$$\ddot{\mathbf{X}} = [\mathbf{1} + \omega^2(K + j\omega C - \omega^2 M)^{-1} M\mathbf{1}]\ddot{X}_0 = \mathbf{H}_{\ddot{x}}(\omega)\,\ddot{X}_0 \tag{35}$$

Similarly, the reaction force is given by $f_0 = \mathbf{1}^T M\ddot{\mathbf{x}}$, leading to the FRF

$$F_0 = [m_T + \omega^2 \mathbf{1}^T M(K + j\omega C - \omega^2 M)^{-1} M\mathbf{1}]\ddot{X}_0 = H_{f_0}(\omega)\,\ddot{X}_0 \tag{36}$$

after using $m_T = \mathbf{1}^T M\mathbf{1}$.

[3]Since the DVA is placed on the upper floor where the amplitude $\phi_1(n)$ is maximum, m_1 is significantly smaller than m_T, which makes the mass ratio μ significantly larger than ε; $\mu = 1.89\varepsilon$ in this case. According to (25), the equivalent damping is $\xi_e \simeq \sqrt{\mu/8} = 0.048$, which is quite substantial.

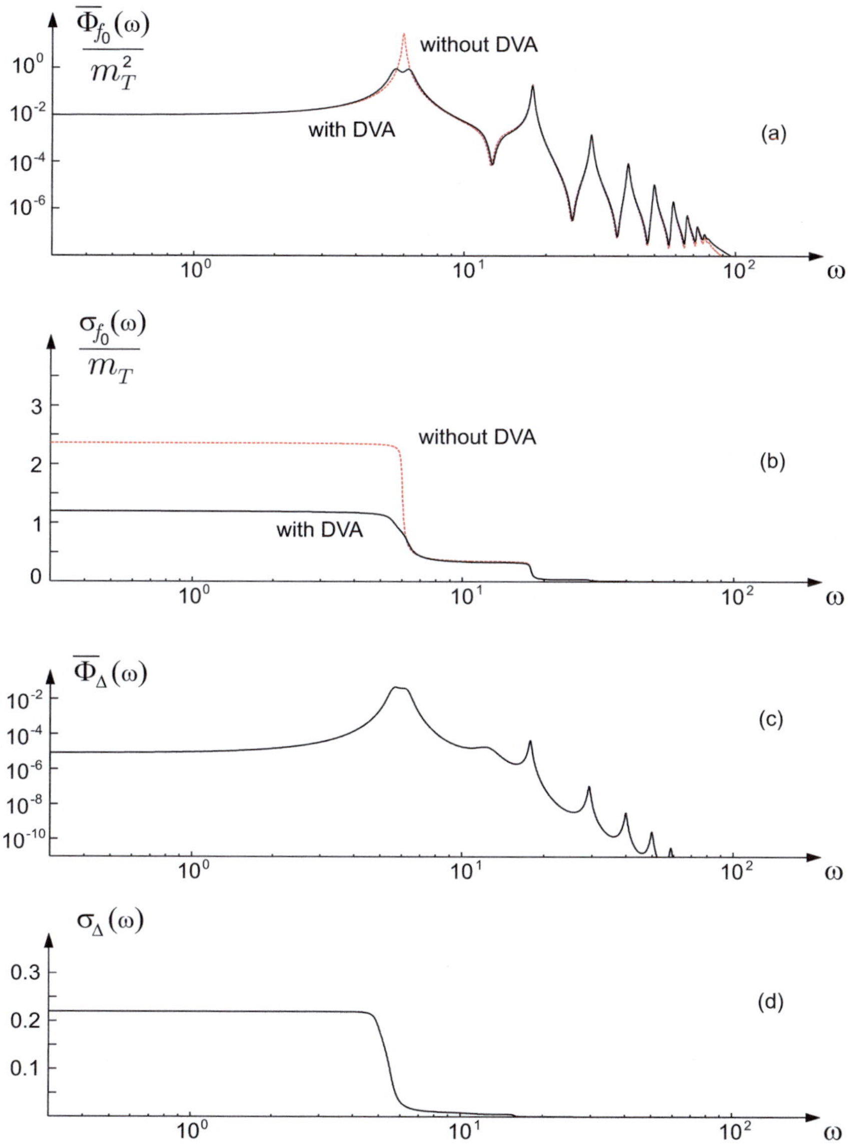

Figure 10. Seismic response of a 10-storey shear frame equipped with a DVA at the top ($\varepsilon = 0.01$). (a) One-sided response PSD of the non-dimensional reaction force $f_0/\sigma_g m_T$ with and without DVA. (b) Cumulative RMS value of the non-dimensional reaction force, with and without DVA. (c) One-sided PSD of the relative displacement of the DVA, $\Delta = x_{11} - x_{10}$. (d) Cumulative RMS value of the relative displacement, $\sigma_\Delta(\omega)$ [m].

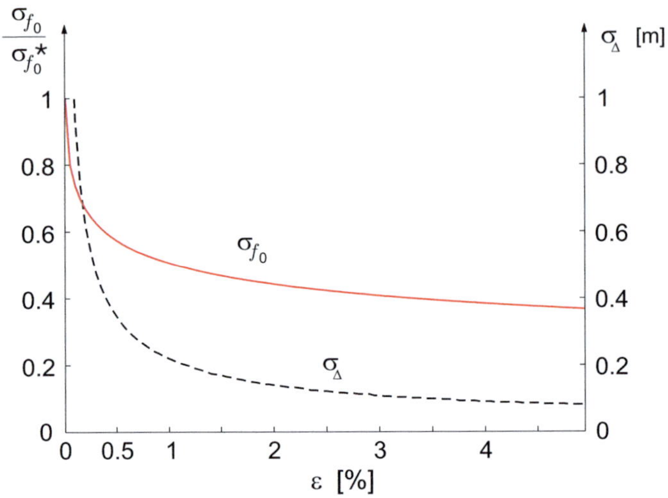

Figure 11. 10-storey shear frame equipped with a DVA at the top. Influence of the fraction of mass $\varepsilon = m_a/m_T$ on the reduction of the reaction force ($\sigma^*_{f_0}$ refers to the response without DVA) and on the RMS relative displacement σ_Δ.

The displacement of the DVA relative to the upper floor, $\Delta = y_{11} - y_{10}$ is an important design parameter; it may be obtained by defining a vector $\mathbf{b}^T = (0, \ldots, 0, -1, 1)$ such that $\Delta = \mathbf{b}^T \mathbf{y}$; the FRF between the complex amplitudes of Δ and $\ddot{x}_0$ is given by

$$\Delta = -\mathbf{b}^T (K + j\omega C - \omega^2 M)^{-1} M \mathbf{1} \ddot{X}_0 = H_\Delta(\omega)\, \ddot{X}_0 \tag{37}$$

The power spectral density of the relative displacement Δ is

$$\Phi_\Delta(\omega) = |H_\Delta(\omega)|^2\, \Phi_{\ddot{x}_0}(\omega) \tag{38}$$

and similarly for all the response quantities. Figure 10 shows the influence of the DVA on the structural response. Figure 10.a shows the impact of the DVA on the resonance peak of the first mode in the reaction force f_0. The reduction of the RMS of the shear force is illustrated in Fig.10.b. The relative response of the DVA with respect to the top floor is illustrated in Fig.10.c and d. Finally, Fig.11 illustrates the influence of the mass of the DVA on the attenuation of the RMS reaction force σ_{f_0} and on the stroke σ_Δ. For every value of the mass fraction ε, the parameters are those of the equal peak design.

4 Active control with an AMD

In this section, the DVA is replaced by an Active Mass Damper (AMD) and a feedback control is considered, based on the absolute velocity of the top floor.

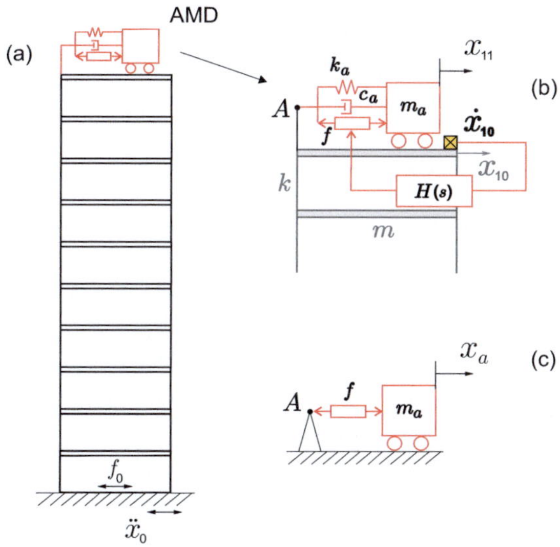

Figure 12. (a) 10-storey building with an Active Mass Damper (AMD) at the top. (b) Active Mass Damper: a force actuator f operates in parallel with the spring k_a and the damper c_a. The parameters are selected according to $m_a = \varepsilon m_T$, $\omega_a = \sqrt{k_a/m_a} \ll \omega_1$, $0.5 \leq \xi_a = c_a/2m_a\omega_a \leq 0.7$. (c) Perfect inertial actuator.

4.1 Control hardware

AMD actuator The AMD consists of a spring mass system similar to a DVA, with the addition of a force actuator f acting in parallel with the spring k_a and damper c_a (Fig.12). However, there are two notable differences:
(i) the stiffness k_a is significantly lower than that of a DVA, so that the natural frequency of the AMD satisfies:

$$\omega_a = \sqrt{\frac{k_a}{m_a}} \ll \omega_1 \tag{39}$$

(ii) the damper c_a is selected to introduce a significant damping in the actuator system:

$$0.5 \leq \xi_a = \frac{c_a}{2m_a\omega_a} \leq 0.7 \tag{40}$$

With these characteristics, the inertial actuator behaves as a near-perfect force actuator in the frequency band containing the flexible modes of the structure. To understand this, consider first a perfect inertial actuator consisting of a force actuator acting on a mass m_a; if it is attached to a fixed point A (Fig.12.c), expressing the equilibrium of the massless actuator, the force applied to A is $f = -m_a\ddot{x}_a$ (positive in traction). Next, if an AMD is connected to a fixed point A, it is governed by the equation

$$m_a\ddot{x}_a + c_a\dot{x}_a + k_a x_a = -f$$

(with again f being positive in traction). Thus, the transfer function between f and x_a is (s is the Laplace variable)

$$x_a = \frac{-f}{m_a s^2 + c_a s + k_a};$$

the force F applied to point A is $F = -m_a\ddot{x}_a$ and the transfer function between the actuator force f and the force F transmitted to the structure is that of a second order high-pass filter:

$$\frac{F}{f} = \frac{s^2}{s^2 + 2\xi_a\omega_a s + \omega_a^2} \tag{41}$$

where ξ_a and ω_a are defined according to Equ.(39) and (40). Thus, when A is fixed, the AMD behaves as a perfect force generator for frequencies $\omega \gg \omega_a$.

Geophone In addition to an actuator, the feedback loop needs a sensor; in this study, we assume that the structure is equipped with a sensor measuring the horizontal velocity of the top floor, $\dot{x}_{10}$; this can be obtained either with a *geophone*, or by integrating the output signal of an accelerometer (although accelerometers do not work well at low frequency).

A geophone (Fig.13) is a spring mass system connected with a voice coil transducer; the relationship between the output voltage and the floor velocity is that of a high-pass filter

$$\frac{e}{\dot{x}_0} = \frac{-s^2 T}{s^2 + 2\xi_g\omega_g s + \omega_g^2} \tag{42}$$

where e is the output voltage, $\dot{x}_0$ is the input velocity, T is the voice coil constant (in volt.sec/m) and ω_g and ξ_g are the frequency and the damping constant of the device. In the subsequent development, we will assume that the corner frequency of the geophone, ω_g is smaller than the corner frequency of the actuator, ω_a, so that the velocity sensor may be regarded as perfect.

Note that, although both the velocity sensor and the AMD are attached to the same floor, the control system is not *collocated*, because the system output $\dot{x}_{10}$ is the top floor absolute velocity while the system input f consists of a pair of opposing forces acting respectively on the top floor and on the inertial mass m_a. However, if the inertial actuator parameters are chosen as explained before, it behaves closely to a perfect force actuator f applied to the top floor (collocated), and the open-loop FRF, $G = \dot{x}_{10}/f$, exhibits alternating poles and zeros (Fig.14).

4.2 System modeling

As compared to the system considered in the foregoing section on DVA, the only difference is the presence of a force f acting between the top floor x_{10} and the inertial mass x_{11}. The governing equation is

$$M\ddot{\mathbf{y}} + C\dot{\mathbf{y}} + K\mathbf{y} = -M\mathbf{1}\ddot{x}_0 + \mathbf{b}_a f \tag{43}$$

where $\mathbf{b}_a = (0, \ldots, 0, 1, -1)^T$ is the influence vector of the control force (f assumed positive in traction). The matrices M, C and K have been

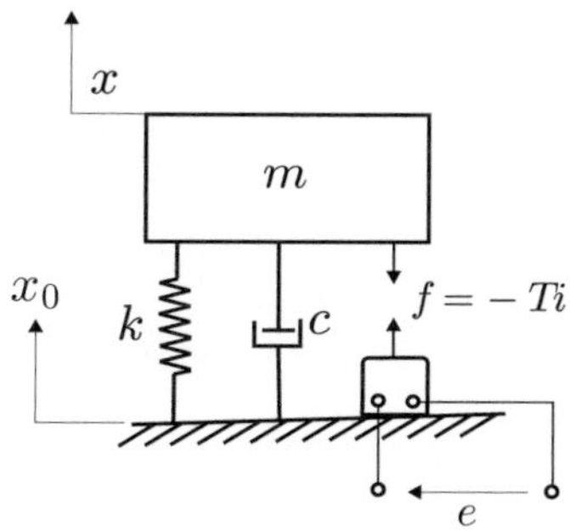

Figure 13. Geophone based on a voice coil transducer of constitutive equations $f = -Ti$ and $e = T(\dot{x} - \dot{x}_0)$ (T is the transducer constant). The input is the motion x_0 of the support and the output is the voltage e of the voice coil. Since the electrical circuit is open, $i = 0$.

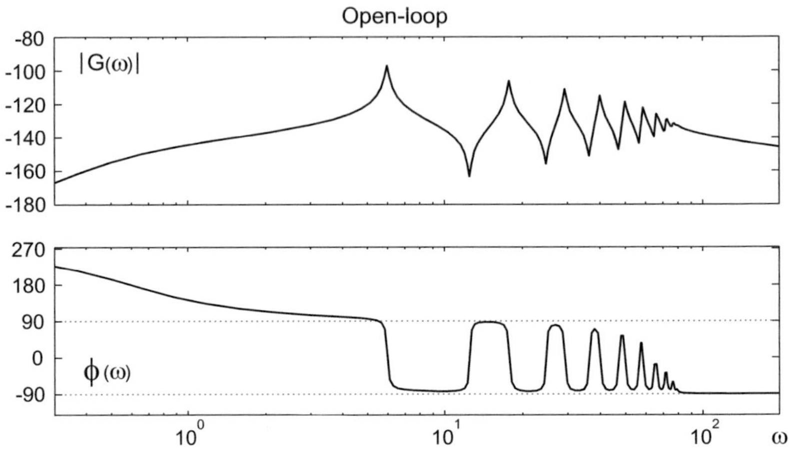

Figure 14. 10-storey building with an Active Mass Damper (AMD) at the top. Open-loop FRF $G = v/f$ between the force f in the actuator and the velocity of the top floor (amplitude in dB and phase). The FRF exhibits alternating poles and zeros above ω_1.

constructed before. Introducing the state vector

$$\mathbf{z} = (\mathbf{y}^T, \dot{\mathbf{y}}^T)^T \tag{44}$$

it can be rewritten in state variable form (as a system of first order differential equations)

$$\dot{\mathbf{z}} = A\mathbf{z} + \mathbf{b}f + \mathbf{e}\ddot{x}_0 \tag{45}$$

where A is the system matrix

$$A = \begin{bmatrix} 0 & I \\ -M^{-1}K & -M^{-1}C \end{bmatrix}, \quad \mathbf{b} = \left\{ \begin{matrix} \mathbf{0} \\ M^{-1}\mathbf{b}_a \end{matrix} \right\}, \quad \mathbf{e} = \left\{ \begin{matrix} \mathbf{0} \\ -\mathbf{1} \end{matrix} \right\} \tag{46}$$

(the size of the control input vector $\mathbf{b}$ and disturbance input vector $\mathbf{e}$ is twice the number of d.o.f., 22 in this case).

The sensor output is the velocity of the top floor,

$$v = \dot{x}_{10} = \dot{x}_0 + \mathbf{b}_s^T \dot{\mathbf{y}} \tag{47}$$

where the vector $\mathbf{b}_s = (0, \ldots, 1, 0)^T$ defines the sensor location (note that $\mathbf{b}_s \neq \mathbf{b}_a$, because the control force is acting also on m_a). This equation is rewritten in state space form

$$v = \mathbf{c}^T\mathbf{z} + \dot{x}_0 \tag{48}$$

with the output vector $\mathbf{c}^T = (\mathbf{0}^T, \mathbf{b}_s^T)$. In addition to the system equation (45) and the output equation (48), the feedback control law relates the control input to the sensor output; in this case a *direct velocity feedback* is used:

$$f = -gv \tag{49}$$

g is the control gain.

4.3 System response

The open-loop transfer function of the system is the relationship between the control input f and the sensor output v without feedback control and without disturbance; it is readily obtained from Equ.(45) and (48):

$$G(s) = \mathbf{c}^T(sI - A)^{-1}\mathbf{b} \tag{50}$$

The open-loop FRF $G(j\omega)$ is obtained by substituting $s = j\omega$. It is represented in Fig.14. The FRF exhibits alternating poles and zeros for all flexible modes (ω_1 and above), typical of collocated control systems. The

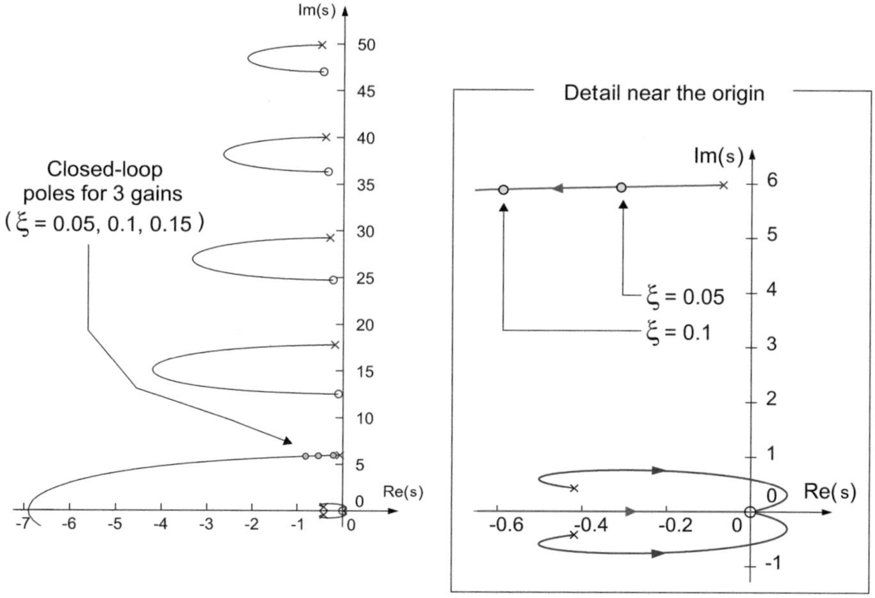

Figure 15. 10-storey building with an Active Mass Damper (AMD) at the top and velocity feedback. Evolution of the closed-loop poles for increasing values of the control gain g (root locus).

closed-loop system equation is obtained by combining Equ.(45), (48) and (49); one finds easily

$$\dot{\mathbf{z}} = (A - g\mathbf{b}\mathbf{c}^T)\mathbf{z} - g\mathbf{b}\dot{x}_0 + \mathbf{e}\ddot{x}_0 \tag{51}$$

The eigenvalues of the closed-loop system matrix $A - g\mathbf{b}\mathbf{c}^T$ are the closed-loop poles. Their evolution for increasing values of the gain g is represented in the root locus of Fig.15. The transmissibility (in the frequency domain) between the the ground acceleration $\ddot{x}_0$ and the state $\mathbf{z}$ is

$$\mathbf{Z}(j\omega) = [j\omega I - (A - g\mathbf{b}\mathbf{c}^T)]^{-1}(-\frac{g\mathbf{b}}{j\omega} + \mathbf{e})\ddot{X}_0(j\omega) \tag{52}$$

The shear force at the base due to a seismic excitation can be expressed either as a function of the absolute accelerations:

$$f_0 = \mathbf{1}^T M\ddot{\mathbf{x}} \tag{53}$$

or as a function of the relative displacements and velocities:

$$f_0 = -\mathbf{1}^T(K\mathbf{y} + C\dot{\mathbf{y}}) = -\mathbf{1}^T(K, C)\mathbf{z} \tag{54}$$

(which is simpler in this case). The closed-loop transmissibility between the ground acceleration $\ddot{x}_0$ and f_0 is readily obtained by combining with the previous equation. Figure 16.a shows the PSD of the shear force f_0 resulting from the steady state response to a seismic acceleration with a modified Kanai-Tajimi profile $\Phi_g^*(\omega)$, for various control gains, respectively g_1 leading to a damping ratio of $\xi = 0.05$ in the first mode, g_2 leading to $\xi = 0.1$ and g_3 to $\xi = 0.15$. The corresponding values of the closed-loop poles are indicated in Fig.15. An important observation is that the AMD damps all the modes, unlike the DVA. The cumulative RMS value of the non-dimensional shear force is shown in Fig.16.b, for the same values of the gain, and the cumulative RMS control force f is represented in Fig.16.c; this figure shows that the control effort increases rapidly with the gain, and so does the stroke of the actuator which may become unacceptable for large values of the gain. Note that the control effort and the stroke will eventually fix the size of the actuator. We now examine how the control effort may be reduced with a Hybrid Mass Damper.

5 Hybrid Mass Damper

The DVA is a purely passive device, tuned on the targeted mode (in this case mode 1) and it leaves the other modes unchanged. The AMD is fully

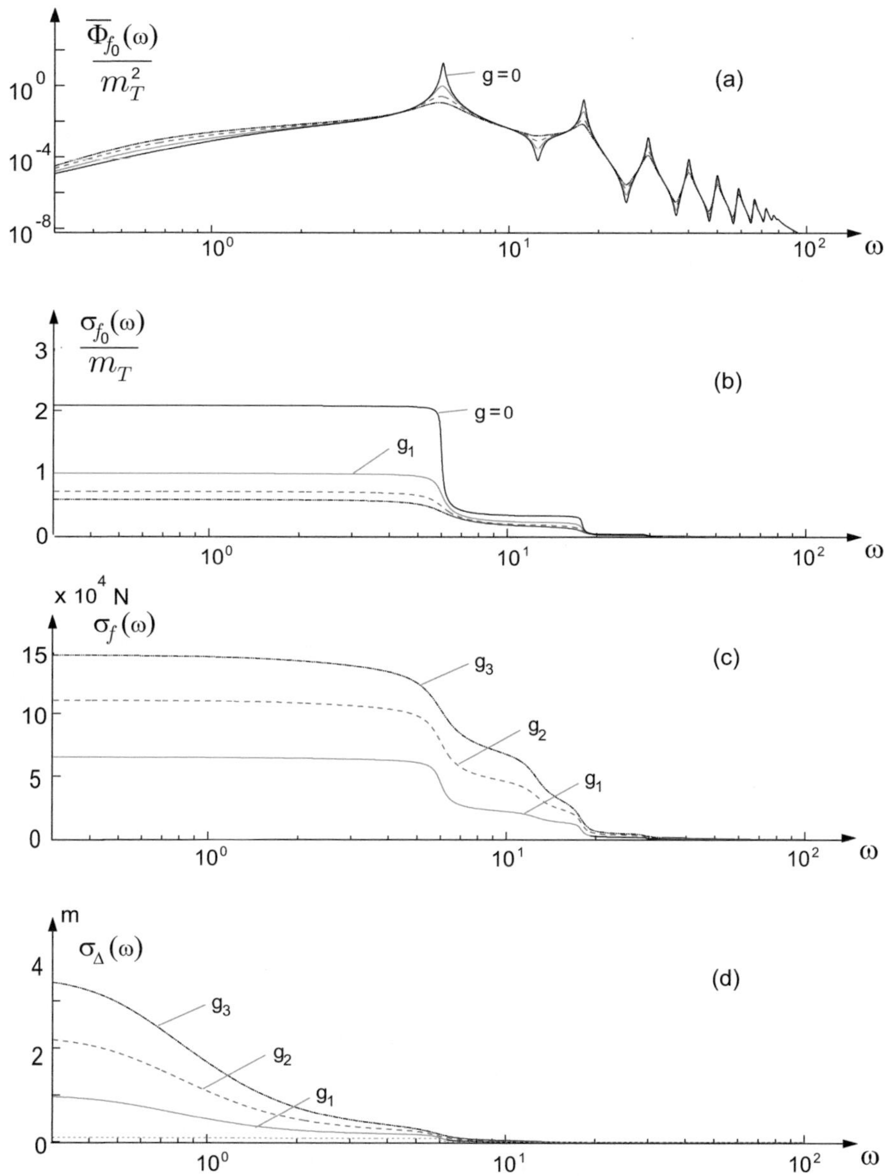

Figure 16. 10-storey building with an Active Mass Damper (AMD) at the top ($\varepsilon = 0.01$) and velocity feedback. (a) PSD of the shear force f_0 due to the seismic acceleration with a modified Kanai-Tajimi profile $\Phi_g^*(\omega)$, for various values of the control gain, g_1, g_2, g_3, leading respectively to a closed-loop damping ξ_1 of 5%, 10% and 15 % in the first mode. (b) Cumulative RMS value of the non-dimensional reaction force. (c) Cumulative RMS value of the control force. (d) Cumulative RMS value of the stroke Δ of the actuator.

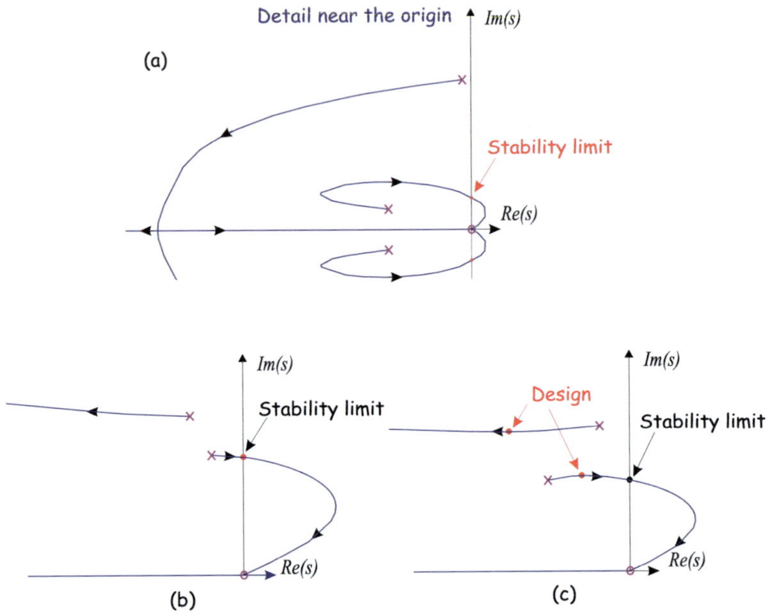

Figure 17. (a) Detail of the root locus of the AMD near the origin ($\omega_a \ll \omega_1$). (b) Same root locus when the actuator is tuned as a DVA [$\omega_a = \omega_1/(1+\mu)$]; the branch of the root locus starting from the pole of lower frequency points towards the imaginary axis, reducing the damping ratio. (c) Root locus after reducing the stiffness and increasing the damping of the actuator, so that all the closed-loop poles have appropriate locations.

active and operates on all the controllable modes; it requires an absolute velocity (or acceleration) sensor and an actuator of appropriate size and stroke, with a corner frequency satisfying the condition (39). The hybrid controller is attempting to get the best of the two worlds by modifying a DVA to use it as an AMD (with the same absolute velocity feedback), with the objectives of increasing the performance and/or decreasing the control effort of the actuator. The idea is explained in Fig.17.[4] Figure 17.a shows the detail of the root locus of an AMD near the origin, and in particular the trajectory of the actuator poles; Fig.17.b shows the root locus for an absolute velocity feedback when the actuator is tuned as a DVA with the equal peak design; one observes that the control increases the damping of the structure pole (with a larger frequency) by moving it to the left, while the actuator pole is moved to the right (it can even become unstable if the

[4] All the root locus plots are symmetrical with respect to the real axis; only the upper half is shown.

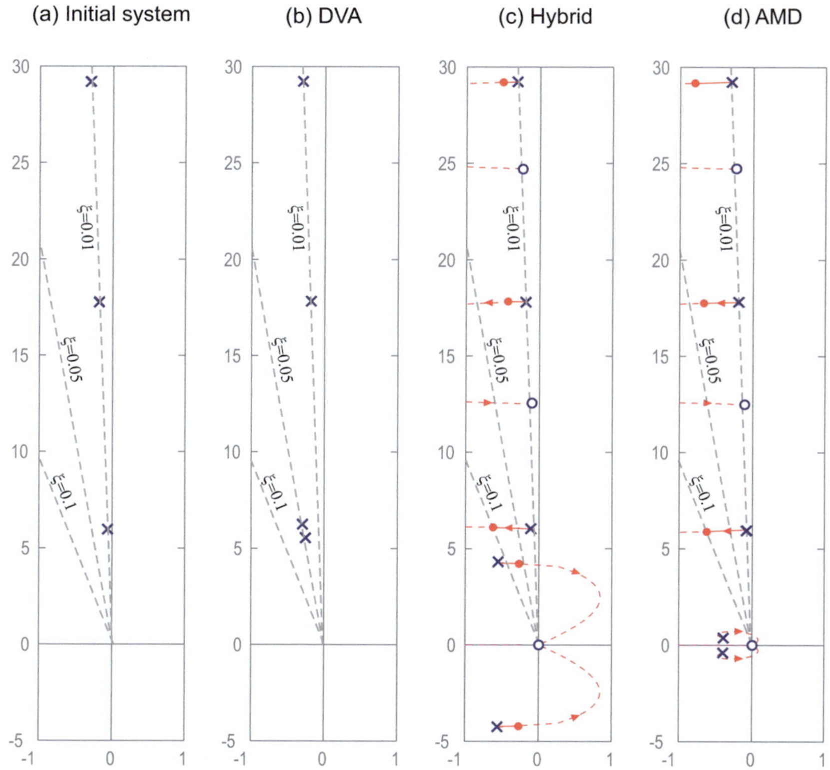

Figure 18. Position of the poles (×) and zeros (○) in the complex plane. (a) Initial system with uniform damping of $\xi_i = 0.01$. (b) DVA ($\varepsilon = m_a/m_T = 0.01$) with ω_a and c_a corresponding to the equal peak design; the pair of poles have a damping ratio close to $\xi \simeq 0.05$. (c) Root locus of the HMD starting from the location obtained for $\omega_a' = 0.74\omega_a$ and $c_a' = 1.2c_a$. The red • indicates the position of the closed loop poles for $g = 2.7\ 10^5$, corresponding to the double peaked curve of Fig.19.a. (d) Root locus of the AMD as a function of the control gain g. For $g_2 = 5.5\ 10^5$, the pole of the first structural mode (indicated by the red •) has a damping ratio $\xi_1 \simeq 0.1$.

gain is large enough). The actuator of the HMD is tuned slightly differently from the DVA, by reducing the stiffness k_a and increasing the damping constant c_a of the actuator, so that the position of the poles after applying the feedback gain is where one wants them to be (Fig.17.c). This part of the design requires some trial and error, but may be done very quickly once the effect of the parameters k_a and c_a has been understood. In this study, the original system was assumed to have a structural damping of $\xi_i = 0.01$ uniformly in all modes (Fig.18.a); the DVA with equal peak design led to

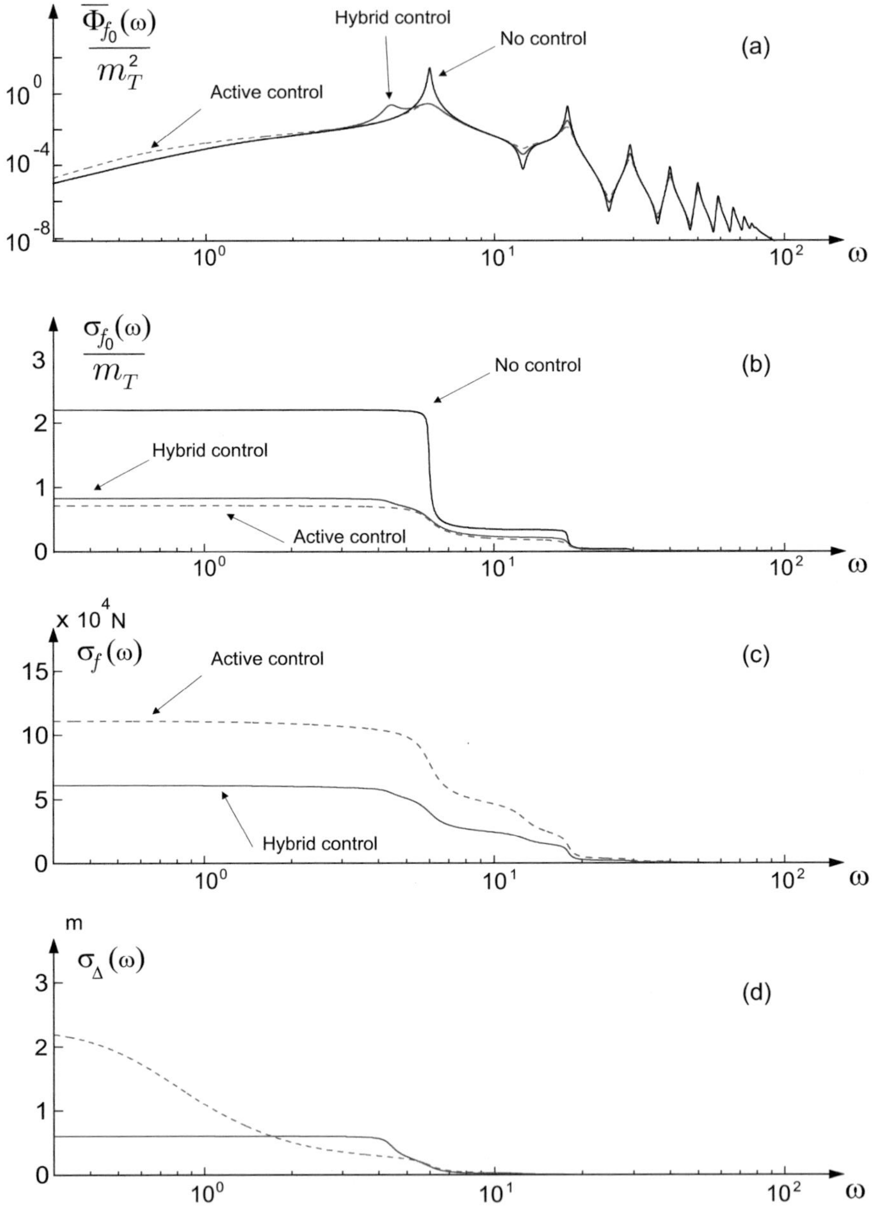

Figure 19. Comparison of the HMD with the AMD. (a) PSD of the shear force f_0 due to the seismic acceleration with modified Kanai-Tajimi profile. (b) Cumulative RMS value of the non-dimensional reaction force. (c) Cumulative RMS value of the control force. (d) Cumulative RMS value of the stroke Δ of the actuator. The position of the poles are indicated in Fig.18.c and d.

a pair of poles with about $\xi_i = 0.05$ (Fig.18.b); the HMD was designed *(i)* to bring the peak in the FRF corresponding to the first mode to the same level as that obtained with an AMD when $\xi_1 = 0.1$ (this corresponds to the gain g_2 in Fig.16) and *(ii)* to achieve equal peak design for the two low frequency peaks in the FRF (Fig.19.a); this is achieved for $\omega_a' = 0.74\omega_a$ and $c_a' = 1.2c_a$ where ω_a and c_a are given by Equ.(32).

Figure 19.a compares the PSD of the shear force f_0 due to the seismic acceleration with modified Kanai-Tajimi profile for the AMD and HMD with the poles being represented in Fig.18.c and d. Figure 19.b compares the cumulative RMS value of the non-dimensional reaction force; Figure 19.c compares the cumulative RMS value of the control force and Fig.19.d that of the actuator stroke. One sees that the performance of the HMD (non-dimensional reaction force) is only slightly worse than that of the AMD, but the actuator force is significantly reduced, by more than 50%, and the stroke even more, by 75%. Note also that the authority of the HMD on the higher modes is slightly reduced as compared to the AMD, because the control gain is reduced, but this is without consequence on this problem which is dominated by the first mode.

6 Dual loop Hybrid Mass Damper

The HMD performs better than the DVA, and has substantially less control force and stroke requirements than the AMD. However, in case of control system failure, the HMD degenerates into a mistuned DVA with degraded performances; this motivates to investigate the possibility to tune actively the HMD from the DVA with equal peak parameters ω_a and c_a to the

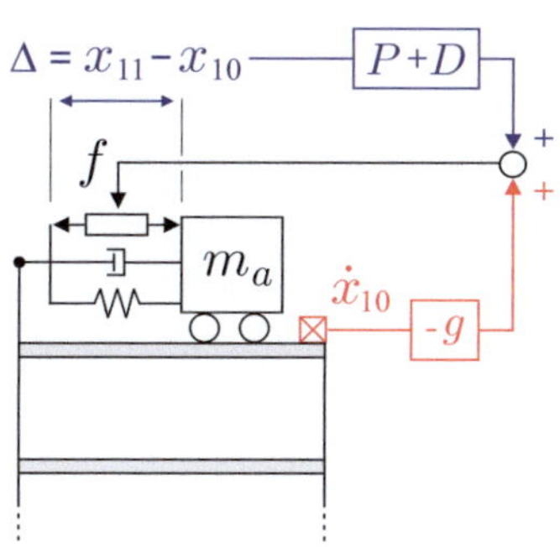

Figure 20. Conceptual design of the dual loop HMD. The control system consists of two loops, one P+D controller acting on a sensor measuring the relative displacement of the actuator, $\Delta = x_{11} - x_{10}$, and a proportional controller on the absolute velocity $\dot{x}_{10}$.

modified parameters $\omega_a' = 0.74\omega_a$ and $c_a' = 1.2c_a$, by means of an additional sensor (e.g. linear encoder) measuring the relative displacement between the inertial mass and the upper floor, $\Delta = x_{11} - x_{10}$ and a proportional plus derivative (P+D) compensator acting on the same actuator. The complete control system is represented in Fig.20. The purpose of the P+D loop is to bring the pair of poles of the DVA from the position in Fig.18.b to the poles $\times$ for $g = 0$ in Fig.18.c. This is achieved with an appropriate negative stiffness and a positive damping.

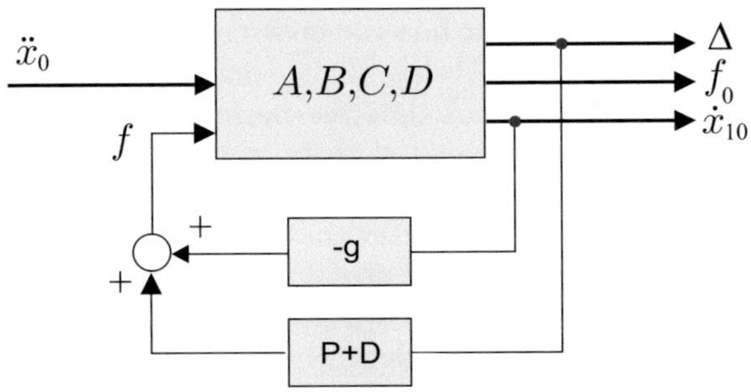

Figure 21. Block diagram for the analysis of the dual loop HMD in MATLAB.

6.1 System modeling

In order to analyze this more complex control configuration in MATLAB, it is simpler to cast the equation in state space form as represented in Fig.21. Because the absolute ground velocity $\dot{x}_0$ appears explicitly in output equation (48), it is necessary to include it in the state vector. The extended state vector is now

$$\mathbf{z} = (\mathbf{y}^T, \dot{\mathbf{y}}^T, \dot{x}_0)^T \tag{55}$$

The input vector includes the control force f and the disturbance applied to the system

$$\mathbf{u} = (f,\ \ddot{x}_0)^T \tag{56}$$

With these notations, the system equation in state variable form reads

$$\dot{\mathbf{z}} = A\mathbf{z} + B\mathbf{u} \tag{57}$$

where

$$A = \begin{bmatrix} 0 & I & 0 \\ -M^{-1}K & -M^{-1}C & 0 \\ 0 & 0 & 0 \end{bmatrix}, \quad B = \begin{bmatrix} 0 & 0 \\ M^{-1}\mathbf{b}_a & -\mathbf{1} \\ 0 & 1 \end{bmatrix} \tag{58}$$

The output equation reads

$$\mathbf{y} = (\dot{x}_{10}, \Delta)^T = C\mathbf{z} \tag{59}$$

with

$$C = \begin{bmatrix} \mathbf{0}^T & \mathbf{b}_s^T & 1 \\ -\mathbf{b}_a^T & \mathbf{0}^T & 0 \end{bmatrix} \tag{60}$$

where $\mathbf{b}_s$ has been defined in Equ.(47) and $\mathbf{b}_a$ in Equ.(43) (the relative displacement sensor is collocated with the force actuator).

6.2 System response

In the dual loop implementation of the HMD, the initial system is a DVA with equal peak design parameters, ω_a, c_a; the poles location of the system is that of Fig.18.b. The P+D loop modifies the system in such a way that the poles are moved to the initial position $\times$ of Fig.18.c, from where they are moved further to the $\bullet$ by the second loop (with gain g). Since the closed loop poles are placed at the same location as in the previous section, the overall performance is identical; the PSD of the shear force is that displayed in Fig.19.a and the cumulative RMS of the shear force is that of Fig.19.b. However, the control force f is different, because there is an additional contribution coming from the P+D loop. This is shown in Fig.22; one sees that the P+D loop is responsible for an increment Δf in the RMS control force; this is the price to pay for changing actively the tuning of the system. Note that the actuator stroke is identical in both cases and is that shown in Fig.19.d.

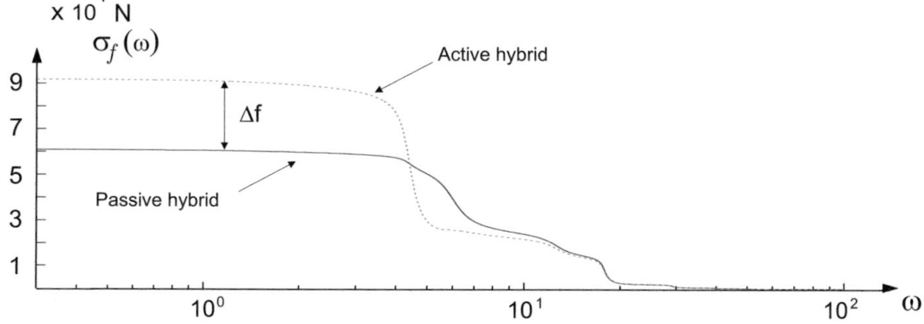

Figure 22. Cumulative RMS value of the control force f of the HMD; comparison of the passive hybrid (initial system tuned on ω'_a, c'_a) and the active hybrid (initial system tuned on ω_a, c_a and P+D loop). The P+D loop is responsible for an increment Δf in the RMS control force.

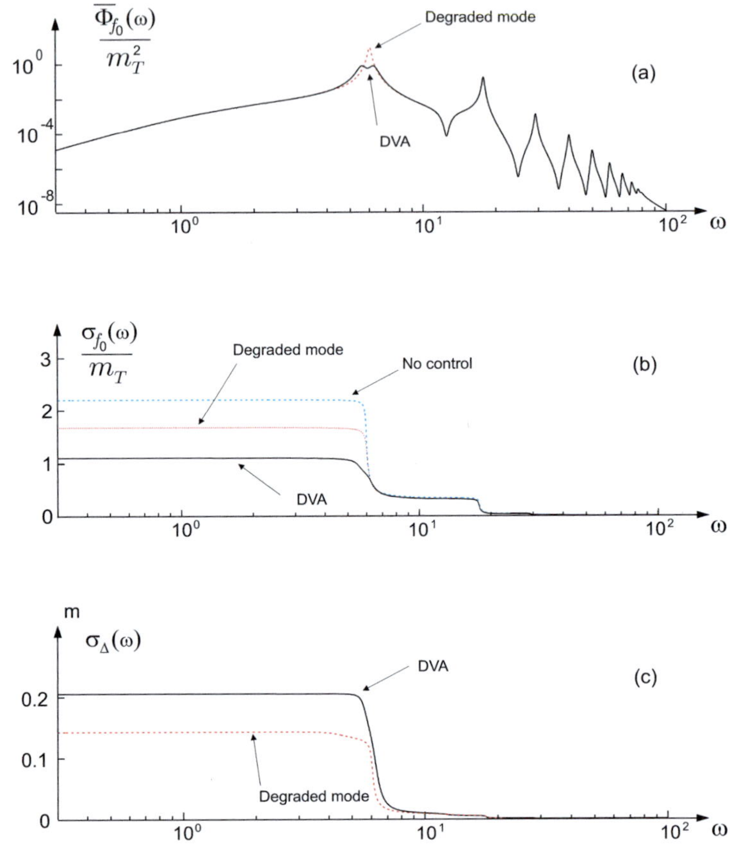

Figure 23. HMD when the control is disabled; comparison of the *passive* HMD (the curve is referred to as "degraded mode") with the *active* HMD (which degenerates into an equal peak DVA). (a) PSD of the shear force f_0 due to the seismic acceleration with modified Kanai-Tajimi profile. (b) Cumulative RMS value of the non-dimensional reaction force; the cumulative RMS response of the initial system is also included for comparison. (c) Cumulative RMS value of the stroke Δ of the actuator.

6.3 HMD in degraded mode

Active systems are more sensitive than passive ones to reliability issues; their normal operation may be prevented by sensor failure, or cut off of external power. It is therefore interesting to investigate the behavior of the system in case of control system breakdown. When the control system is switched off, the single loop HMD (called here *passive* HMD) is reduced to a DVA with imperfect tuning (ω_a', c_a'), while the dual loop HMD (*ac-*

tive HMD) is reduced to a DVA with equal peak design (ω_a, c_a). The degraded responses of the two systems are compared in Fig.23; one notices that the *passive* HMD is still better than the initial system, but the *active* one degrades into the DVA with equal peak design, leading to much better performances (at the expense, however, of a larger control effort when in active operation).

7 Conclusions

Using a multistory shear frame subjected to a seismic excitation, this study has compared three different ways of mitigating the dynamic response of buildings: Dynamic Vibration Absorber (DVA), Active Mass Damper (AMD) and Hybrid Mass Damper (HMD). It has been shown that a well designed HMD may produce performances comparable to that of an AMD while significantly reducing the actuator force and stroke requirements. Besides, if a dual loop design is used, the system is immune to control system breakdown, because the HMD is reduced to a properly tuned DVA with optimum performances for a passive system.

Bibliography

[1] T. Asami, O. Nishihara and A. M. Baz. Analytical solutions to H_∞ and H_2 optimization of dynamic vibration absorbers attached to damped linear systems, *ASME Journal of Vibration and Acoustics*, Vol.124, 284-295, April 2002.
[2] S. H. Crandall and W. D. Mark. *Random Vibration in Mechanical Systems*, Academic Press, 1963.
[3] J. P. Den Hartog *Mechanical Vibrations*, 4th Edition, Dover, 1985.
[4] N. M. Newmark and E. Rosenblueth. *Fundamental of Earthquake Engineering*, Prentice Hall, 1971.
[5] J. Ormondroyd and J. P. Den Hartog. The theory of the damped vibration absorber, *Trans. ASME, J. of Applied Mechanics*, 50:7, 1928.
[6] A. Preumont. *Twelve Lectures on Structural Dynamics*, Springer, 2013.
[7] A. Preumont. *Vibration Control of Active Structures, An Introduction*, 3rd Edition, Springer, 2011.
[8] A. Preumont and K. Seto. *Active Control of Structures*, Wiley, 2008.
[9] B. F. Jr. Spencer and M. k. Sain. Controlling Buildings: A new Frontier in Feedback, Special Issue of *IEEE Control Systems Magazine* on emerging Technology, Vol .17, No 6, pp.19-35, Dec. 1997.
[10] G. B. Warburton. Optimum absorber parameters for various combinations of response and excitation parameters, *Earthquake Engineering and Structural Dynamics*, Vol.10, 381-401, 1982.

Electromagnetic and Piezoelectric Transducers

André Preumont and Bilal Mokrani
ULB, Brussels, Belgium

Abstract This chapter analyzes the two most popular classes of transducers used in active vibration control: the electromagnetic transducer known as *voice coil*, and the piezoelectric transducer. The first part of the chapter discusses the theory of the transducers and the second part discusses some applications in structural control.

1 Introduction

Transducers are critical in active structures technology; they can play the role of actuator, sensor, or simply energy converter, depending on the application and the electrical connections. In many applications, the actuators are the most critical part of the system; however, the sensors become very important in precision engineering where sub-micron amplitudes must be detected. This chapter begins with a description of the voice coil transducer and its application to the proof-mass actuator and the geophone (absolute velocity sensor). The remaining of the chapter is devoted to the piezoelectric materials and the constitutive equations of a discrete piezoelectric transducer.

2 Voice coil transducer

A *voice coil* transducer is an energy transformer which converts electrical power into mechanical power and vice versa. The system consists of a permanent magnet (Fig.1) which produces a uniform magnetic flux density B normal to the gap, and a coil which is free to move axially within the gap. Let v be the velocity of the coil, f the external force acting to maintain the coil in equilibrium against the electromagnetic forces, e the voltage difference across the coil and i the current into the coil. In this ideal transducer, we neglect the electrical resistance and the self inductance of the coil, as well as its mass and damping (if necessary, these can be handled by adding

P. Hagedorn, G. Spelsberg-Korspeter (Eds.), *Active and Passive Vibration Control of Structures*, CISM International Centre for Mechanical Sciences
DOI 10.1007/ 978-3-7091-1821-4_4

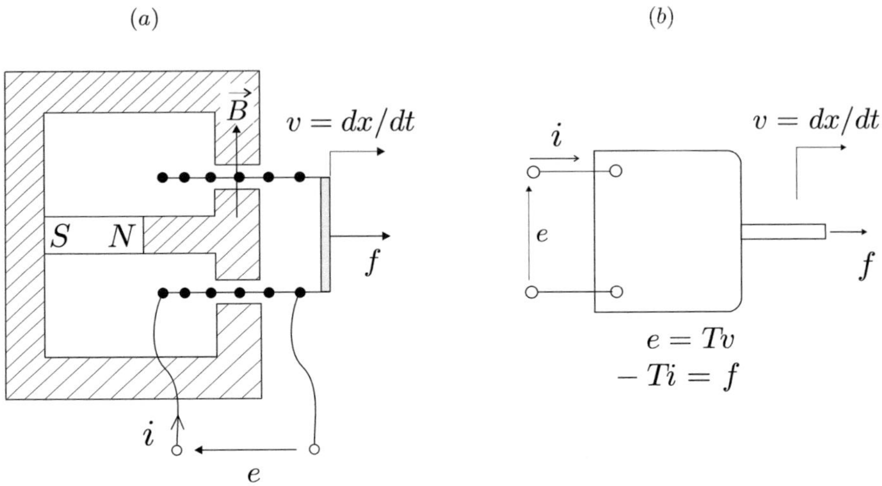

Figure 1. Voice-coil transducer: (a) Physical principle. (b) Symbolic representation.

R and L to the electrical circuit of the coil, or a mass and damper to its mechanical model). The voice coil actuator is one of the most popular actuators in mechatronics (e.g. it is used in electromagnetic loudspeakers), but it is also used as sensor in geophones.

The first constitutive equation of the voice coil transducer follows from *Faraday's law*: A coil of n turns moving at the velocity v with respect to the magnetic flux density B generates an electromotive force (voltage) e given by

$$e = 2\pi n r B v = T v \tag{1}$$

where

$$T = 2\pi n r B \tag{2}$$

is the *transducer constant*, equal to the product of the length of the coil exposed to the magnetic flux, $2\pi nr$, and the magnetic flux density B. The second equation follows from the L*orentz force* law: The external force f required to *balance* the total force of the magnetic field on n turns of the conductor is

$$f = -i\, 2\pi n r B = -T i \tag{3}$$

where i is the current intensity in the coil and T is again the transducer constant (2). Equation (1) and (3) are the constitutive equations of the voice coil transducer. Notice that the transducer constant T appearing in Faraday's law (1), expressed in $volt.sec/m$, is the same as that appearing in the Lorentz force (3), expressed in N/Amp.

The total power delivered to the moving-coil transducer is equal to the sum of the electric power, ei, and the mechanical power, fv. Combining with (1) and (3), one gets

$$ei + fv = Tvi - Tiv = 0 \tag{4}$$

Thus, at any time, there is an equilibrium between the electrical power absorbed by the device and the mechanical power delivered (and vice versa). The moving-coil transducer cannot store energy, and behaves as a perfect electromechanical converter. In practice, however, the transfer is never perfect due to eddy currents, flux leakage and magnetic hysteresis, leading to slightly different values of T in (1) and (3).

Let us now examine various applications of the voice coil transducer.

2.1 Proof-mass actuator

A proof-mass actuator (Fig.2) is an inertial actuator which is used in various applications of vibration control. A reaction mass m is connected to the support structure by a spring k, a damper c and a force actuator f which can be either magnetic or hydraulic. In the electromagnetic actuator discussed here, the force actuator consists of a voice coil transducer of constant T excited by a current generator i; the spring is achieved with membranes which also guide the linear motion of the moving mass. The system is readily modelled as in Fig.2.a. Combining the equation of a single d.o.f. oscillator with the Lorentz force law (3), one finds

$$m\ddot{x} + c\dot{x} + kx = Ti \tag{5}$$

or, in the Laplace domain,

$$x = \frac{Ti}{ms^2 + cs + k} \tag{6}$$

(s is the Laplace variable). The total force applied to the support is equal and opposite to the force applied to the proof-mass, $-m\ddot{x}$, or in Laplace form:

$$F = -ms^2x = \frac{-ms^2Ti}{ms^2 + cs + k} \tag{7}$$

It follows that the transfer function between the total force F and the current i applied to the coil is

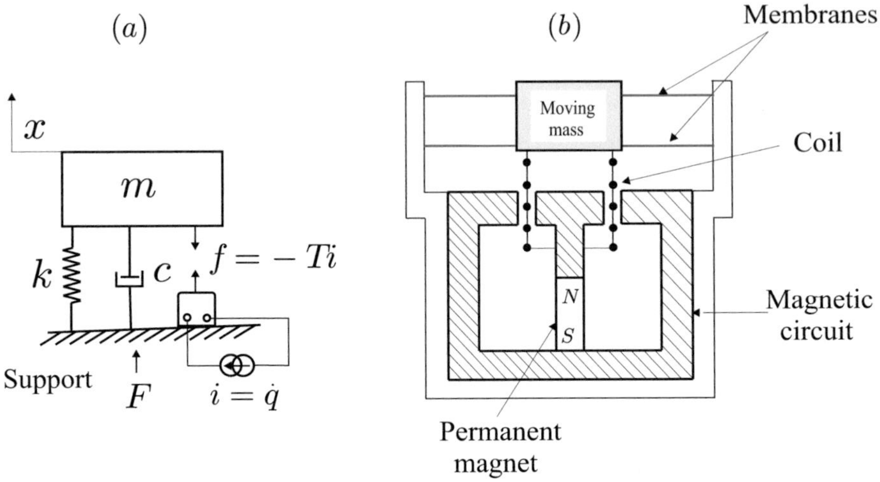

Figure 2. Proof-mass actuator (a) model assuming a current generator; (b) conceptual design of an electrodynamic actuator based on a voice coil transducer. The mass is guided by the membranes.

$$\frac{F}{i} = \frac{-s^2 T}{s^2 + 2\xi_p \omega_p s + \omega_p^2} \tag{8}$$

where T is the transducer constant (in N/Amp), $\omega_p = (k/m)^{1/2}$ is the natural frequency of the spring-mass system and ξ_p is the damping ratio, which in practice is fairly high, typically 20 % or more [the negative sign in (8) is irrelevant.] The Bode plots of (8) are shown in Fig.3; one sees that the system behaves like a high-pass filter with a high frequency asymptote equal to the transducer constant T; above some critical frequency $\omega_c \simeq 2\omega_p$, the proof-mass actuator can be regarded as an *ideal force generator*. It has no authority over the rigid body modes (at zero frequency) and the operation at low frequency requires a large stroke, which is technically difficult. Medium to high frequency actuators (40 Hz and more) are relatively easy to obtain with low cost components (loudspeaker technology).

If the current source is replaced by a voltage source (Fig.4), the modeling is slightly more complicated and combines the mechanical equation (5) and an electrical equation which is readily derived from Faraday's law:

$$T\dot{x} + L\frac{di}{dt} + Ri = E(t) \tag{9}$$

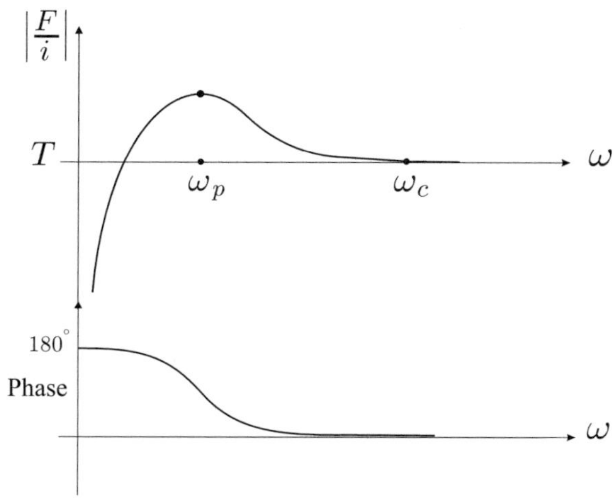

Figure 3. Bode plot F/i of an electrodynamic proof-mass actuator (amplitude and phase).

where L is the inductance R is the resistance of the electrical circuit and $E(t)$ is the external voltage source applied to the transducer.

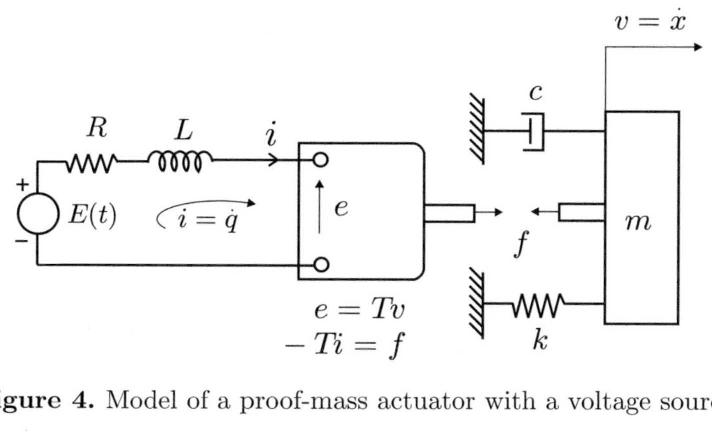

Figure 4. Model of a proof-mass actuator with a voltage source.

2.2 Geophone

The geophone is a transducer which behaves like an *absolute velocity sensor* above some cut-off frequency which depends on its mechanical construction. The system of Fig.2.a is readily transformed into a geophone by using the voltage e as the sensor output (Fig.5). If x_0 is the displacement of the support and if the voice coil is open ($i = 0$), the governing equations are

$$m\ddot{x} + c(\dot{x} - \dot{x}_0) + k(x - x_0) = 0$$

$$T(\dot{x} - \dot{x}_0) = e$$

combining these equations, one readily finds that

$$x - x_0 = \frac{-ms^2 x_0}{ms^2 + cs + k}$$

$$e = Ts(x - x_0) = \frac{-s^2 T}{s^2 + (c/m)s + k/m} sx_0$$

$$\frac{e}{\dot{x}_0} = \frac{-s^2 T}{s^2 + 2\xi_p \omega_p s + \omega_p^2} \tag{10}$$

Thus, there is a perfect *duality* between a proof-mass actuator and the geophone. The same device may be used either as actuator or sensor, depending on the electrical boundary conditions. The proof-mass actuator uses a current source while the geophone is connected to an infinite resistor. Above the corner frequency, the gain of the geophone is equal to the transducer constant T. Designing geophones with very low corner frequency is in general difficult, especially if their orientation with respect to the gravity

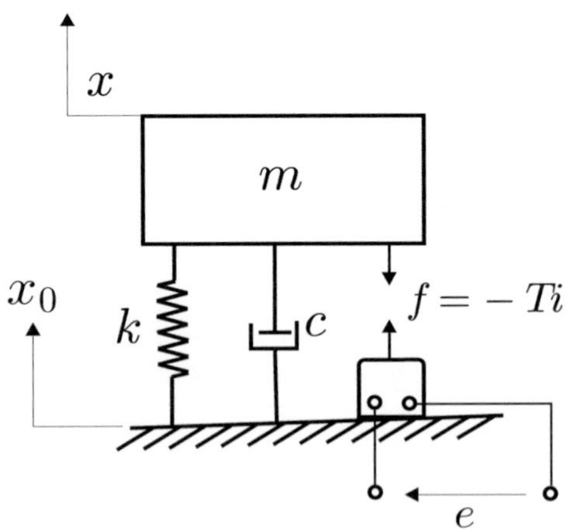

Figure 5. Model of a geophone based on a voice coil transducer.

vector is variable; active geophones where the corner frequency is lowered electronically may constitute a good alternative option.

3 General electromechanical transducer

3.1 Constitutive equations

The constitutive behavior of a wide class of electromechanical transducers can be modeled as in Fig.6, where the central box represents the conversion mechanism between electrical energy and mechanical energy, and vice versa. In Laplace form, the constitutive equations read

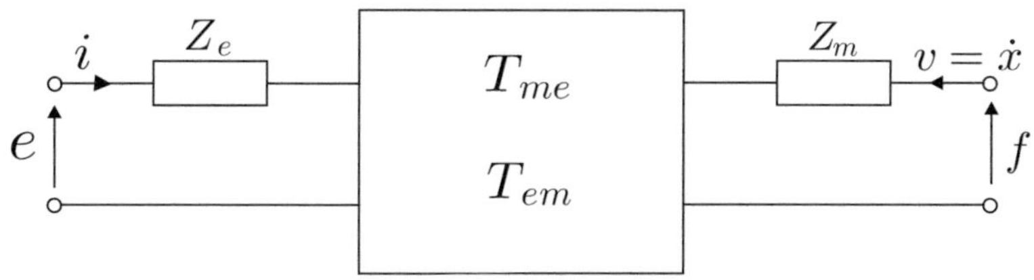

Figure 6. Electrical analog representation of an electromechanical transducer.

$$e = Z_e i + T_{em} v \tag{11}$$

$$f = T_{me} i + Z_m v \tag{12}$$

where e is the Laplace transform of the input voltage across the electrical terminals, i the input current, f the force applied to the mechanical terminals, and v the velocity of the mechanical part. Z_e is the blocked electrical impedance, measured for $v = 0$; T_{em} is the transduction coefficient representing the electromotive force (voltage) appearing in the electrical circuit per unit velocity in the mechanical part (in *volt.sec/m*). T_{me} is the transduction coefficient representing the force acting on the mechanical terminals to balance the electromagnetic force induced per unit current input on the electrical side (in *N/Amp*), and Z_m is the mechanical impedance, measured when the electrical side is open ($i = 0$).

To illustrate this representation, consider the proof-mass actuator with the voltage source of Fig.4; the electrical equation reads

$$E(t) = Ri + L\frac{di}{dt} + Tv$$

or in Laplace form

$$E = (Ls + R)i + Tv$$

If F is the external force applied to the mechanical terminal (positive in the positive direction of v), the mechanical equation reads

$$F(t) = m\ddot{x} + c\dot{x} + kx - Ti$$

or in Laplace form (using the velocity as mechanical variable)

$$F = (ms + c + k/s)v - Ti$$

Thus, the constitutive equations may be written in the form (11) and (12) with

$$Z_e = Ls + R, \qquad Z_m = ms + c + k/s, \qquad T_{em} = T, \qquad T_{me} = -T$$

In absence of external force ($f = 0$), v can be resolved from Equ.(12) and substituted into Equ.(11), leading to

$$e = (Z_e - \frac{T_{em}T_{me}}{Z_m})i$$

$-T_{em}T_{me}/Z_m$ is called the *motional impedance.* The total driving point electrical impedance is the sum of the blocked and the motional impedances.

3.2 Self-sensing

Equation (11) shows that the voltage drop across the electrical terminals of any electromechanical transducer is the sum of a contribution proportional to the current applied and a contribution proportional to the velocity of the mechanical terminals. Thus, if $Z_e i$ can be measured and subtracted from e, a signal proportional to the velocity is obtained. This suggests the bridge structure of Fig.7. The bridge equations are as follows: for the branch containing the transducer,

$$e = Z_e I + T_{em}v + Z_b I$$

$$I = \frac{1}{Z_e + Z_b}(e - T_{em}v)$$

$$V_4 = Z_b I = \frac{Z_b}{Z_e + Z_b}(e - T_{em}v)$$

For the other branch,

$$e = kZ_e i + kZ_b i$$

$$V_2 = kZ_b i = \frac{Z_b}{Z_e + Z_b}e$$

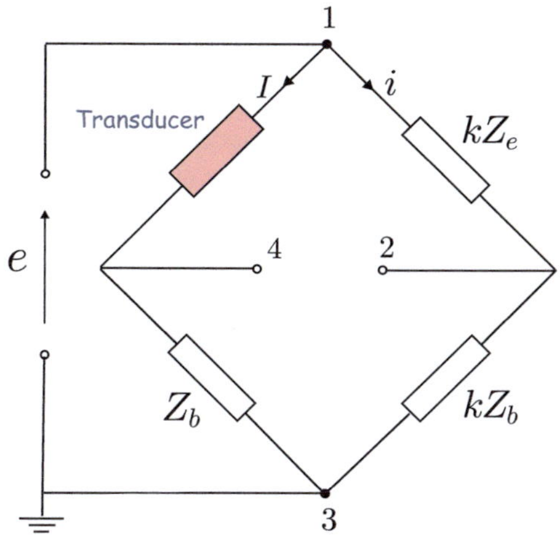

Figure 7. Bridge circuit for self-sensing actuation.

and the bridge output

$$V_4 - V_2 = (\frac{-Z_b T_{em}}{Z_e + Z_b})\, v \tag{13}$$

is indeed a linear function of the velocity v of the mechanical terminals. Note, however, that $-Z_b T_{em}/(Z_e+Z_b)$ acts as a filter; the bridge impedance Z_b must be adapted to the transducer impedance Z_e to avoid amplitude distortion and phase shift between the output voltage $V_4 - V_2$ and the transducer velocity in the frequency band of interest.

4 Smart materials

Piezoelectric materials belong to the so-called *smart materials*, or *multi-functional materials*, which have the ability to respond significantly to stimuli of different physical natures. Figure 8 lists various effects that are observed in materials in response to various inputs: mechanical, electrical, magnetic, thermal, light. The coupling between the physical fields of different types is expressed by the non-diagonal cells in the figure; if its magnitude is sufficient, the coupling can be used to build discrete or distributed transducers of various types, which can be used as sensors, actuators, or even integrated in structures with various degrees of tailoring and complexity (e.g. as fibers), to make them controllable or responsive to their environment (e.g. for shape morphing, precision shape control, damage detection,

Input \ Output	Strain	Electric charge	Magnetic flux	Temperature	Light
Stress	Elasticity	Piezo-electricity	Magneto-striction		Photo-elasticity
Electric field	Piezo-electricity	Permittivity			Electro-optic effect
Magnetic field	Magneto-striction	Magneto-electric effect	Permeability		Magneto-optic
Heat	Thermal expansion	Pyro-electricity		Specific heat	
Light	Photostriction	Photo-voltaic effect			Refractive index

Figure 8. Stimulus-response relations indicating various effects in materials. The smart materials correspond to the non-diagonal cells.

dynamic response alleviation,...).

Figure 9 summarizes the mechanical properties of a few smart materials which are considered for actuation in structural control applications. Figure 9.a shows the maximum (blocked) stress versus the maximum (free) strain; the diagonal lines in the diagram indicate a constant energy density. Figure 9.b shows the specific energy density (i.e. energy density by unit mass) versus the maximum frequency; the diagonal lines indicate a constant specific power density. Note that all the material characteristics vary by several orders of magnitude. Among them all, the piezoelectric materials are undoubtedly the most mature and those with the most applications.

5 Piezoelectric transducer

The piezoelectric effect was discovered by Pierre and Jacques Curie in 1880. The direct piezoelectric effect consists in the ability of certain crystalline materials to generate an electrical charge in proportion to an externally applied force; the direct effect is used in force transducers. According to the inverse piezoelectric effect, an electric field parallel to the direction of polarization induces an expansion of the material. The piezoelectric effect

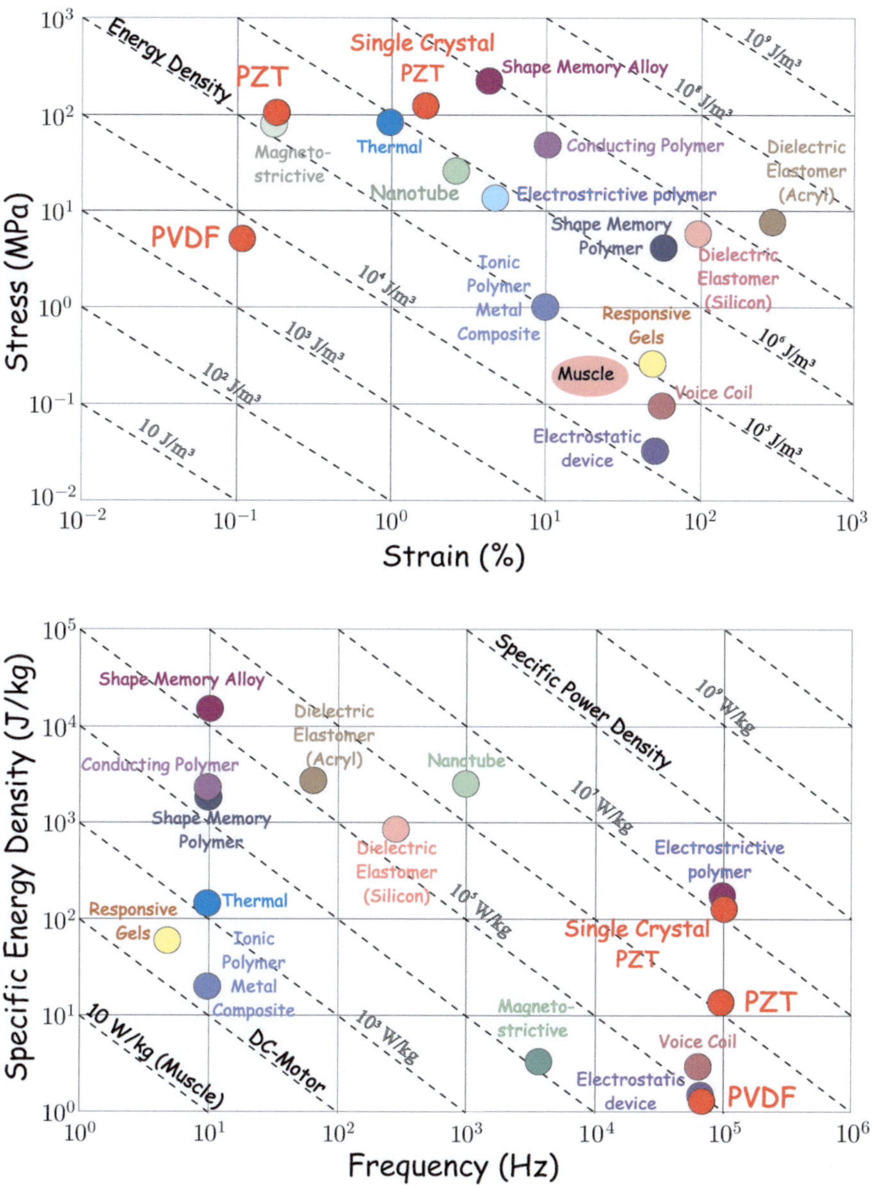

Figure 9. (a) Maximum stress vs. maximum strain of various smart material actuators. (b) Specific energy density vs. maximum frequency (by courtesy of R.Petricevic, *Neue Materialen Würzburg*).

is anisotropic; it can be exhibited only by materials whose crystal structure has no center of symmetry; this is the case for some ceramics below a certain temperature called the *Curie temperature*; in this phase, the crystal has built-in electric dipoles, but the dipoles are randomly orientated and the net electric dipole on a macroscopic scale is zero. During the poling process, when the crystal is cooled in the presence of a high electric field, the dipoles tend to align, leading to an electric dipole on a macroscopic scale. After cooling and removing of the poling field, the dipoles cannot return to their original position; they remain aligned along the poling direction and the material body becomes permanently piezoelectric, with the ability to convert mechanical energy to electrical energy and vice versa; this property will be lost if the temperature exceeds the Curie temperature or if the transducer is subjected to an excessive electric field in the direction opposed to the poling field.

The most popular piezoelectric materials are *Lead-Zirconate-Titanate (PZT)* which is a ceramic, and *Polyvinylidene fluoride (PVDF)* which is a polymer. In addition to the piezoelectric effect, piezoelectric materials exhibit a *pyroelectric* effect, according to which electric charges are generated when the material is subjected to temperature; this effect is used to produce heat detectors; it will not be discussed here.

In this section, we consider a transducer made of a one-dimensional piezoelectric material of constitutive equations (we use the notations of the IEEE Standard on Piezoelectricity)

$$D = \varepsilon^T E + d_{33} T \tag{14}$$
$$S = d_{33} E + s^E T \tag{15}$$

where D is the electric displacement (charge per unit area, expressed in $Coulomb/m^2$), E the electric field (V/m), T the stress (N/m^2) and S the strain. ε^T is the dielectric constant (permittivity) under constant stress, s^E is the compliance when the electric field is constant (inverse of the Young's modulus) and d_{33} is the piezoelectric constant, expressed in m/V or $Coulomb/Newton$; the reason for the subscript 33 is that, by convention, index 3 is always aligned to the poling direction of the material, and we assume that the electric field is parallel to the poling direction. Note that the same constant d_{33} appears in (14) and (15).

In the absence of an external force, a transducer subjected to a voltage with the same polarity as that during poling produces an elongation, and a voltage opposed to that during poling makes it shrink (inverse piezoelectric effect). In (15), this amounts to a positive d_{33}. Conversely (direct piezoelectric effect), if we consider a transducer with open electrodes ($D = 0$), according to (14), $E = -(d_{33}/\varepsilon^T)T$, which means that a traction stress will produce a voltage with polarity opposed to that during poling, and a compressive stress will produce a voltage with the same polarity as that during poling.

5.1 Constitutive relations of a discrete transducer

Equations (14) and (15) can be written in a matrix form

$$\left\{ \begin{array}{c} D \\ S \end{array} \right\} = \left[\begin{array}{cc} \varepsilon^T & d_{33} \\ d_{33} & s^E \end{array} \right] \left\{ \begin{array}{c} E \\ T \end{array} \right\} \tag{16}$$

where (E, T) are the independent variables and (D, S) are the dependent variables. If (E, S) are taken as the independent variables, they can be rewritten

$$D = \frac{d_{33}}{s^E} S + \varepsilon^T \left(1 - \frac{{d_{33}}^2}{s^E \varepsilon^T}\right) E$$

$$T = \frac{1}{s^E} S - \frac{d_{33}}{s^E} E$$

or

$$\left\{ \begin{array}{c} D \\ T \end{array} \right\} = \left[\begin{array}{cc} \varepsilon^T(1-k^2) & e_{33} \\ -e_{33} & c^E \end{array} \right] \left\{ \begin{array}{c} E \\ S \end{array} \right\} \tag{17}$$

where $c^E = 1/s^E$ is the Young's modulus under $E = 0$ (short circuited electrodes), in N/m^2 (Pa); $e_{33} = d_{33}/s^E$, the product of d_{33} by the Young modulus, is the constant relating the electric displacement to the strain for short-circuited electrodes (in $Coulomb/m^2$), and also that relating the compressive stress to the electric field when the transducer is blocked ($S = 0$).

$$k^2 = \frac{{d_{33}}^2}{s^E \varepsilon^T} = \frac{{e_{33}}^2}{c^E \varepsilon^T} \tag{18}$$

k is called the *Electromechanical coupling factor* of the material; it measures the efficiency of the conversion of mechanical energy into electrical energy, and vice versa, as discussed below. From (17), we note that $\varepsilon^T(1-k^2)$ is the dielectric constant under zero strain.

If one assumes that all the electrical and mechanical quantities are uniformly distributed in a linear transducer formed by a stack of n disks of thickness t and cross section A (Fig.9), the global constitutive equations of

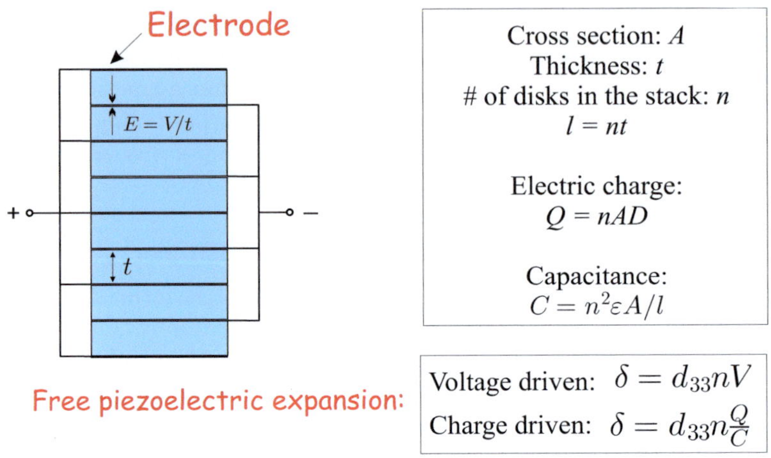

Figure 10. Piezoelectric linear transducer.

the transducer are obtained by integrating Equ.(16) or (17) over the volume of the transducer; one finds

$$\left\{ \begin{array}{c} Q \\ \Delta \end{array} \right\} = \left[\begin{array}{cc} C & nd_{33} \\ nd_{33} & 1/K_a \end{array} \right] \left\{ \begin{array}{c} V \\ f \end{array} \right\} \tag{19}$$

or

$$\left\{ \begin{array}{c} Q \\ f \end{array} \right\} = \left[\begin{array}{cc} C(1-k^2) & nd_{33}K_a \\ -nd_{33}K_a & K_a \end{array} \right] \left\{ \begin{array}{c} V \\ \Delta \end{array} \right\} \tag{20}$$

where $Q = nAD$ is the total electric charge on the electrodes of the transducer, $\Delta = Sl$ is the total extension ($l = nt$ is the length of the transducer), $f = AT$ is the total force and V the voltage applied between the electrodes of the transducer, resulting in an electric field $E = V/t = nV/l$. $C = \varepsilon^T A n^2/l$ is the capacitance of the transducer with no external load ($f = 0$), $K_a = A/s^E l$ is the stiffness with short-circuited electrodes ($V = 0$). Note that the electromechanical coupling factor can be written alternatively

$$k^2 = \frac{{d_{33}}^2}{s^E \varepsilon^T} = \frac{n^2 {d_{33}}^2 K_a}{C} \tag{21}$$

Equation (19) can be inverted

$$\left\{ \begin{array}{c} V \\ f \end{array} \right\} = \frac{K_a}{C(1-k^2)} \left[\begin{array}{cc} 1/K_a & -nd_{33} \\ -nd_{33} & C \end{array} \right] \left\{ \begin{array}{c} Q \\ \Delta \end{array} \right\} \tag{22}$$

from which we can see that the stiffness with open electrodes ($Q = 0$) is $K_a/(1-k^2)$ and the capacitance for a fixed geometry ($\Delta = 0$) is $C(1-k^2)$. Note that typical values of k are in the range $0.3 - 0.7$; for large k, the stiffness changes significantly with the electrical boundary conditions, and similarly the capacitance depends on the mechanical boundary conditions.

Next, let us write the total stored electromechanical energy and coenergy functions.[1] Consider the discrete piezoelectric transducer of Fig.11; the

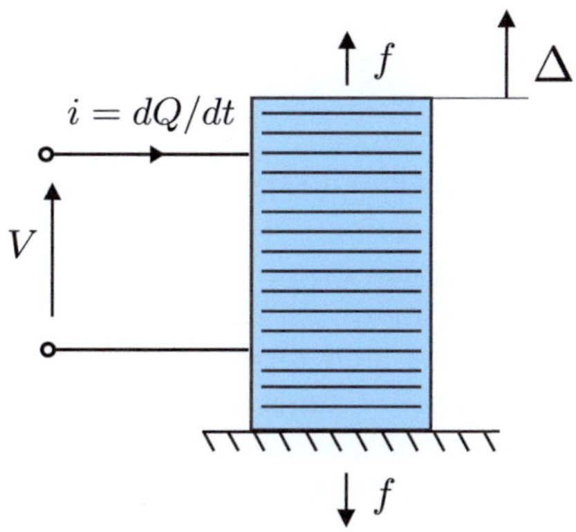

Figure 11. Discrete Piezoelectric transducer.

total power delivered to the transducer is the sum of the electric power, Vi and the mechanical power, $f\dot{\Delta}$. The net work on the transducer is

$$dW = Vidt + f\dot{\Delta}dt = VdQ + fd\Delta \tag{23}$$

For a conservative element, this work is converted into stored energy, dW_e, and the total stored energy, $W_e(\Delta, Q)$ can be obtained by integrating (23) from the reference state to the state (Δ, Q).[2] Upon differentiating $W_e(\Delta, Q)$,

$$dW_e(\Delta, Q) = \frac{\partial W_e}{\partial \Delta} d\Delta + \frac{\partial W_e}{\partial Q} dQ \tag{24}$$

and, comparing with (23), we recover the constitutive equations

$$f = \frac{\partial W_e}{\partial \Delta} \qquad\qquad V = \frac{\partial W_e}{\partial Q} \tag{25}$$

[1]Energy and coenergy functions are needed in connection with energy formulations such as Hamilton principle, Lagrange equations or finite elements.

[2]Since the system is conservative, the integration can be done along any path leading from $(0, 0)$ to (Δ, Q).

Substituting f and V from (22) into (23), one gets

$$dW_e = V dQ + f d\Delta$$

$$= \frac{Q\,dQ}{C(1-k^2)} - \frac{nd_{33}K_a}{C(1-k^2)}(\Delta\,dQ + Q\,d\Delta) + \frac{K_a}{1-k^2}\Delta\,d\Delta$$

which is the total differential of

$$W_e(\Delta, Q) = \frac{Q^2}{2C(1-k^2)} - \frac{nd_{33}K_a}{C(1-k^2)}Q\Delta + \frac{K_a}{1-k^2}\frac{\Delta^2}{2} \tag{26}$$

This is the analytical expression of the stored electromechanical energy for the discrete piezoelectric transducer. The partial derivatives (25) allow to recover the constitutive equations (22). The first term on the right hand side of (26) is the electrical energy stored in the capacitance $C(1-k^2)$ (corresponding to a fixed geometry, $\triangle = 0$); the third term is the elastic strain energy stored in a spring of stiffness $K_a/(1-k^2)$ (corresponding to open electrodes, $Q = 0$); the second term is the piezoelectric energy.

The electromechanical energy function uses Δ and Q as independent state variables. A *coenergy* function using Δ and V as independent variables can be defined by the *Legendre transformation*

$$W_e^*(\Delta, V) = VQ - W_e(\Delta, Q) \tag{27}$$

The total differential of the coenergy is

$$dW_e^* = Q\,dV + V\,dQ - \frac{\partial W_e}{\partial \Delta}d\Delta - \frac{\partial W_e}{\partial Q}dQ$$

$$dW_e^* = Q\,dV - f\,d\Delta \tag{28}$$

where Equ.(25) have been used. It follows that

$$Q = \frac{\partial W_e^*}{\partial V} \qquad \text{and} \qquad f = -\frac{\partial W_e^*}{\partial \Delta} \tag{29}$$

Introducing the constitutive equations (20) into (28),

$$dW_e^* = \left[C(1-k^2)V + nd_{33}K_a\Delta\right] dV + (nd_{33}K_aV - K_a\Delta)\,d\Delta$$

$$= C(1-k^2)VdV + nd_{33}K_a\,(\Delta dV + Vd\Delta) - K_a\Delta\,d\Delta$$

which is the total differential of

$$W_e^*(\Delta, V) = C(1-k^2)\frac{V^2}{2} + nd_{33}K_aV\Delta - K_a\frac{\Delta^2}{2} \tag{30}$$

This is the analytical form of the coenergy function for the discrete piezoelectric transducer. The first term on the right hand side of (30) is recognized as the electrical coenergy in the capacitance $C(1-k^2)$ (corresponding to a fixed geometry, $\Delta = 0$); the third is the strain energy stored in a spring of stiffness K_a (corresponding to short-circuited electrodes, $V = 0$). The second term of (30) is the piezoelectric coenergy; using the fact that the uniform electric field is $E = nV/l$ and the uniform strain is $S = \Delta/l$, it can be rewritten

$$\int_{\Omega} S e_{33} E \, d\Omega \tag{31}$$

where the integral extends to the volume Ω of the transducer.

The analytical form (26) of the electromechanical energy, together with the constitutive equations (25) can be regarded as an alternative definition of a discrete piezoelectric transducer, and similarly for the analytical expression of the coenergy (30) and the constitutive equations (29).

5.2 Interpretation of k^2

Consider a piezoelectric transducer subjected to the following mechanical cycle: first, it is loaded with a force F with short-circuited electrodes; the resulting extension is

$$\Delta_1 = \frac{F}{K_a}$$

where $K_a = A/(s^E l)$ is the stiffness with short-circuited electrodes. The energy stored in the system is

$$W_1 = \int_0^{\Delta_1} f \, dx = \frac{F\Delta_1}{2} = \frac{F^2}{2K_a}$$

At this point, the electrodes are open and the transducer is unloaded according to a path of slope $K_a/(1-k^2)$, corresponding to the new electrical boundary conditions,

$$\Delta_2 = \frac{F(1-k^2)}{K_a}$$

The energy recovered in this way is

$$W_2 = \int_0^{\Delta_2} f \, dx = \frac{F\Delta_2}{2} = \frac{F^2(1-k^2)}{2K_a}$$

leaving $W_1 - W_2$ stored in the transducer. The ratio between the remaining stored energy and the initial stored energy is

$$\frac{W_1 - W_2}{W_1} = k^2$$

Similarly, consider the following electrical cycle: first, a voltage V is applied to the transducer which is mechanically unconstrained ($f = 0$). The electric charges appearing on the electrodes are

$$Q_1 = CV$$

where $C = \varepsilon^T A n^2 / l$ is the unconstrained capacitance, and the energy stored in the transducer is

$$W_1 = \int_0^{Q_1} v\,dq = \frac{VQ_1}{2} = \frac{CV^2}{2}$$

At this point, the transducer is blocked mechanically [changing its capacitance from C to $C(1-k^2)$] and electrically unloaded from V to 0. The electrical charges are removed according to

$$Q_2 = C(1-k^2)V$$

The energy recovered in this way is

$$W_2 = \int_0^{Q_2} v\,dq = \frac{C(1-k^2)V^2}{2}$$

leaving $W_1 - W_2$ stored in the transducer. Here again, the ratio between the remaining stored energy and the initial stored energy is

$$\frac{W_1 - W_2}{W_1} = k^2$$

Although the foregoing relationships provide a clear physical interpretation of the electromechanical coupling factor, they do not bring a practical way of measuring k^2; the experimental determination of k^2 is often based on impedance (or admittance) measurements.

5.3 Admittance of the piezoelectric transducer

Consider the system of Fig.12, where the piezoelectric transducer is assumed massless and is connected to a mass M. The force acting on the mass is the negative of that acting on the transducer, $f = -M\ddot{x}$; using (20),

$$\left\{ \begin{array}{c} Q \\ -M\ddot{x} \end{array} \right\} = \left[\begin{array}{cc} C(1-k^2) & nd_{33}K_a \\ -nd_{33}K_a & K_a \end{array} \right] \left\{ \begin{array}{c} V \\ x \end{array} \right\} \tag{32}$$

From the second equation, one gets (in Laplace form)

$$x = \frac{nd_{33}K_a}{Ms^2 + K_a} V$$

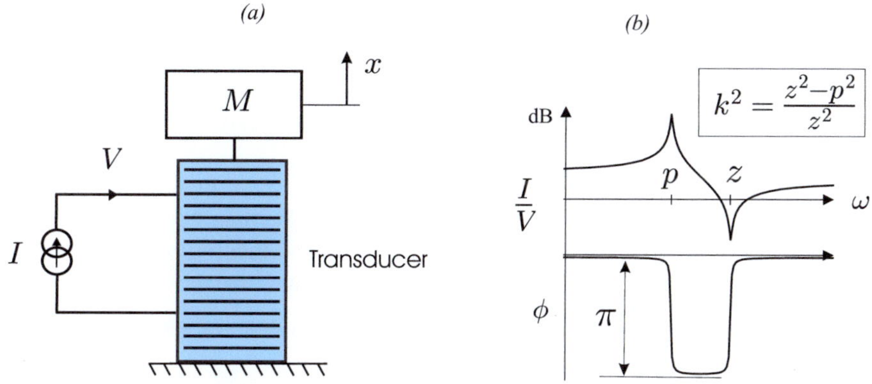

Figure 12. (a) Elementary dynamical model of the piezoelectric transducer. (b) Typical admittance FRF of the transducer, in the vicinity of its natural frequency.

and, substituting in the first one and using (21), one finds

$$\frac{Q}{V} = C(1-k^2)\left[\frac{Ms^2 + K_a/(1-k^2)}{Ms^2 + K_a}\right] \tag{33}$$

It follows that the admittance reads:

$$\frac{I}{V} = \frac{sQ}{V} = sC(1-k^2)\frac{s^2+z^2}{s^2+p^2} \tag{34}$$

where the poles and zeros are respectively

$$p^2 = \frac{K_a}{M} \qquad \text{and} \qquad z^2 = \frac{K_a/(1-k^2)}{M} \tag{35}$$

p is the natural frequency with short-circuited electrodes ($V = 0$) and z is the natural frequency with open electrodes ($I = 0$). From the previous equation one sees that

$$\frac{z^2 - p^2}{z^2} = k^2 \tag{36}$$

This relationship constitutes a practical way of determining the electromechanical coupling factor: An impedance meter is used to measure the Frequency Response Function (FRF) of the admittance (or the impedance), on which the position of the poles p and zeros z are identified and introduced in Equ.(36).

6 Vibration isolation with voice coil transducers

6.1 Viscous damping isolator

Consider the spring mass system of Fig.13. A voice coil connects the mass M to the moving support and a resistor R is connected to the electrical terminals of the voice coil. The governing equations are

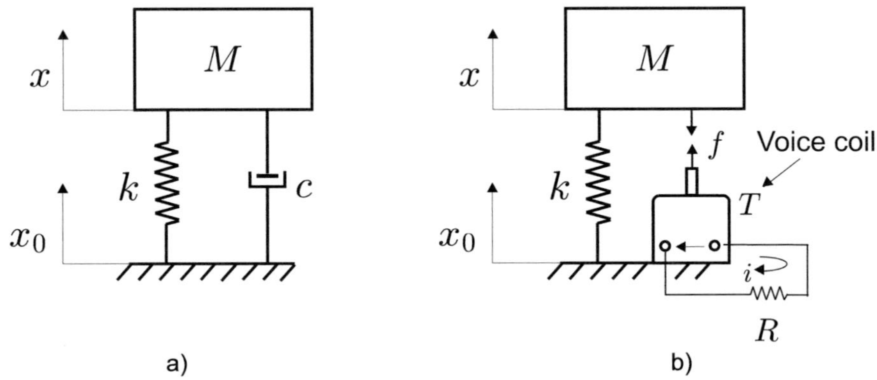

Figure 13. Voice coil used as viscous damper.

$$M\ddot{x} + k(x - x_0) = Ti$$

$$e = -Ri = Tv = T(\dot{x} - \dot{x}_0)$$

where the constitutive equations of the voice coil [Equ.(1) and (3)] have been used. Upon eliminating i between these equations, one finds

$$M\ddot{x} + \frac{T^2}{R}(\dot{x} - \dot{x}_0) + k(x - x_0) = 0 \tag{37}$$

Thus, when a resistor connects the electrical terminals of the voice coil, it behaves as a viscous damper of damping coefficient $c = T^2/R$; a lower resistance R will increase the damping (the minimum value of R is that of the coil itself). From the foregoing equation, the transmissibility of the isolator is readily obtained:

$$\frac{X}{X_0} = \frac{1 + 2j\xi\omega/\omega_n}{1 + 2j\xi\omega/\omega_n - \omega^2/\omega_n^2} \tag{38}$$

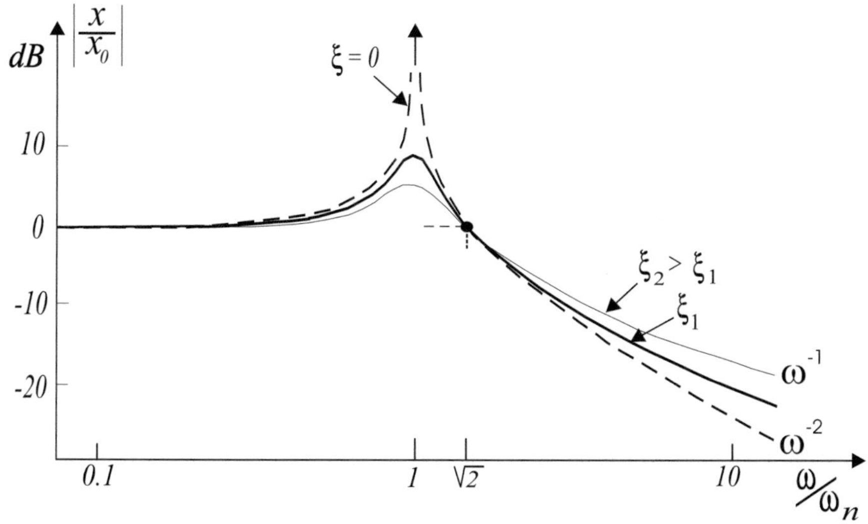

Figure 14. Transmissibility of the passive isolator for various values of the damping ratio ξ. The high frequency decay rate is ω^{-1}.

with the usual notations $\omega_n^2 = k/M$ and $2\xi\omega_n = c/M$.
It is displayed in Fig.14 for various values of the damping ratio ξ: *(i)* All the curves are larger than 1 for $\omega < \sqrt{2}\,\omega_n$ and become smaller than 1 for $\omega > \sqrt{2}\,\omega_n$. Thus the critical frequency $\sqrt{2}\,\omega_n$ separates the domains of amplification and attenuation of the isolator. *(ii)* When $\xi = 0$, the high frequency decay rate is ω^{-2}, that is -40 *dB/decade*, while very large amplitudes occur near the corner frequency ω_n (the natural frequency of the spring-mass system).

Figure 14 illustrates the trade-off in passive isolator design: large damping is desirable at low frequency to reduce the resonant peak while low damping is needed at high frequency to maximize the isolation. One observes that if the disturbance is generated by a rotating unbalance of a motor with variable speed, there is an obvious benefit to use a damper with variable damping characteristics which can be adjusted according to the rotation velocity: high damping when $\omega < \sqrt{2}\omega_n$ and low damping when $\omega > \sqrt{2}\omega_n$. Such (*adaptive*) devices can be readily obtained with a variable resistor R. The following section discusses another electrical circuit which improves the high frequency decay rate of the isolator.

6.2 Relaxation isolator

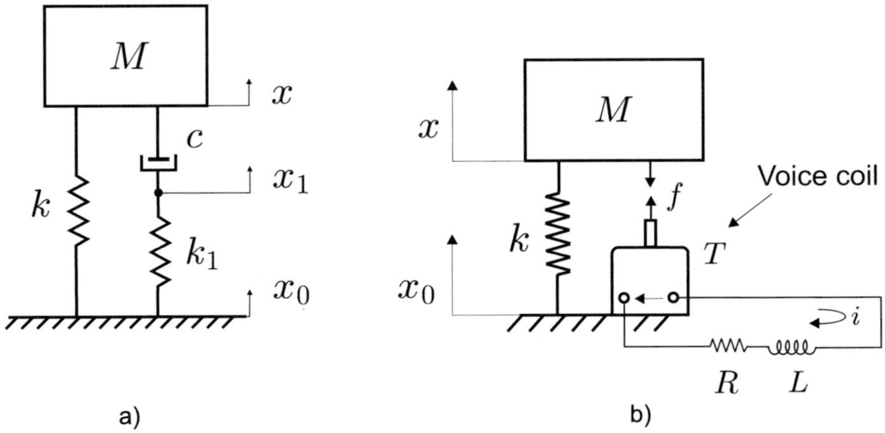

Figure 15. (a) Relaxation isolator. (b) Electromagnetic realization.

In the *relaxation* isolator, the viscous damper c is replaced by a Maxwell unit consisting of a damper c and a spring k_1 in series (Fig.15.a). The governing equations are

$$M\ddot{x} + k(x - x_0) + c(\dot{x} - \dot{x}_1) = 0 \tag{39}$$

$$c(\dot{x} - \dot{x}_1) = k_1(x_1 - x_0) \tag{40}$$

or, in matrix form using the Laplace variable s,

$$\begin{bmatrix} Ms^2 + cs + k & -cs \\ -cs & k_1 + cs \end{bmatrix} \begin{Bmatrix} x \\ x_1 \end{Bmatrix} = \begin{Bmatrix} k \\ k_1 \end{Bmatrix} x_0 \tag{41}$$

Upon inverting this system of equations, the transmissibility is obtained in Laplace form:

$$\frac{x}{x_0} = \frac{(k_1 + cs)k + k_1 cs}{(Ms^2 + cs + k)(k_1 + cs) - c^2 s^2} = \frac{(k_1 + cs)k + k_1 cs}{(Ms^2 + k)(k_1 + cs) + k_1 cs} \tag{42}$$

One sees that the asymptotic decay[3] rate for large frequencies is in s^{-2}, that is -40 *dB/decade*. Physically, this corresponds to the fact that, at high

[3]the asymptotic decay rate is governed by the largest power of s of the numerator and the denominator.

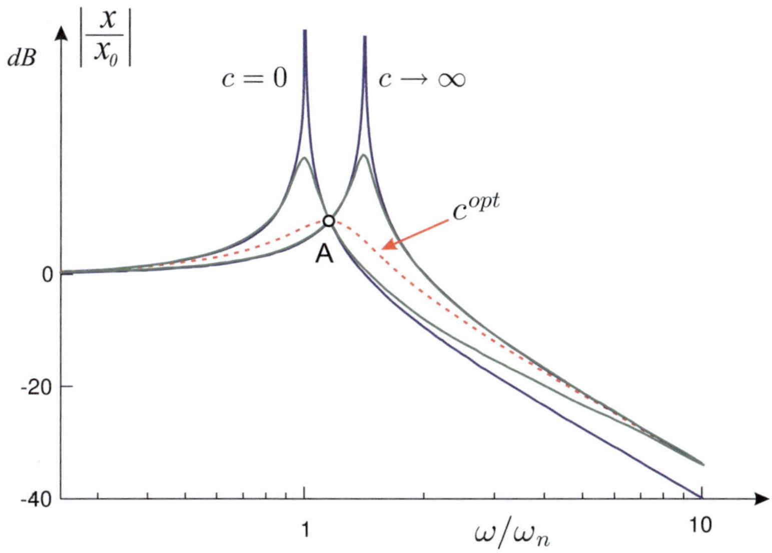

Figure 16. Transmissibility of the relaxation oscillator for fixed values of k and k_1 and various values of c. The first peak corresponds to $\omega = \omega_n = (k/M)^{1/2}$; the second one corresponds to $\omega = \Omega_n = [(k+k_1)/M]^{1/2}$. All the curves cross each other at A and have an asymptotic decay rate of -40 *dB/decade*. The curve corresponding to c_{opt} is nearly maximum at A.

frequency, the viscous damper tends to be blocked, and the system behaves like an undamped isolator with two springs acting in parallel. Figure 16 compares the transmissibility curves for given values of k and k_1 and various values of c. For $c = 0$, the relaxation isolator behaves like an undamped isolator of natural frequency $\omega_n = (k/M)^{1/2}$. Likewise, for $c \to \infty$, it behaves like an undamped isolator of frequency $\Omega_n = [(k+k_1)/M]^{1/2}$. In between, the poles of the system are solution of the characteristic equation

$$(Ms^2 + k)(k_1 + cs) + k_1 cs = (Ms^2 + k)k_1 + cs(Ms^2 + k + k_1) = 0$$

which can be rewritten in root locus form

$$1 + \frac{k_1}{c} \, \frac{s^2 + \omega_n^2}{s(s^2 + \Omega_n^2)} = 0 \qquad (43)$$

It is represented in Fig.17 when c varies from 0 to ∞; it can be shown that

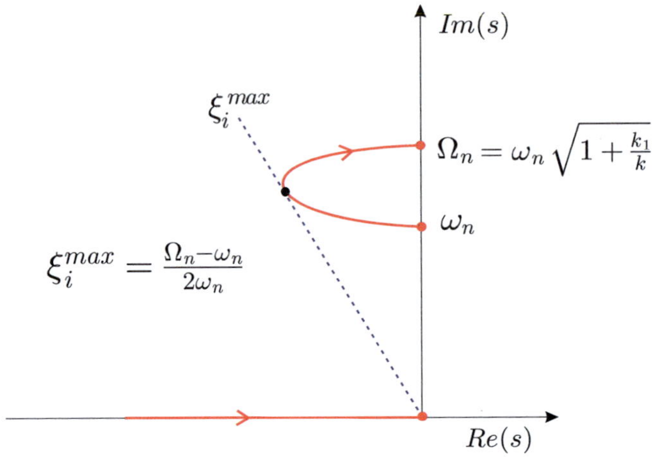

Figure 17. Root locus of the solutions of Equ.(43) as c goes from zero to infinity. The maximum damping is achieved for $k_1/c = \Omega_n^{3/2}\omega_n^{-1/2}$.

the maximum damping ratio is achieved for

$$\frac{k_1}{c} = \frac{\Omega_n^{3/2}}{\omega_n^{1/2}} \tag{44}$$

and the corresponding damper constant is

$$c_{opt} = \frac{k_1}{\Omega_n}(\frac{\omega_n}{\Omega_n})^{1/2} = \frac{k_1}{\Omega_n}(1+\frac{k_1}{k})^{-1/4} = \frac{k_1}{\omega_n}(1+\frac{k_1}{k})^{-3/4} \tag{45}$$

The transmissibility corresponding to c_{opt} is also represented in Fig.16; it is nearly maximum at A.

Electromagnetic realization The principle of the relaxation isolator is simple and it can be realized with viscoelastic materials. However, it may be difficult to integrate in the system, and also viscoelastic materials are notorious for their thermal sensitivity. In some circumstances, especially when thermal stability is critical, it may be more convenient to achieve the same effect through a voice coil transducer whose electrical terminals are connected to an inductor L and a resistor R (Fig.15.b). The governing equations of the system are in this case

$$M\ddot{x} + k(x - x_0) - Ti = 0 \tag{46}$$

$$L\frac{di}{dt} + T(\dot{x} - \dot{x}_0) + Ri = 0 \tag{47}$$

where T is the transducer constant. In matrix form, using the Laplace variable,

$$\begin{bmatrix} Ms^2 + k & -T \\ Ts & Ls + R \end{bmatrix} \begin{Bmatrix} x \\ i \end{Bmatrix} = \begin{Bmatrix} k \\ Ts \end{Bmatrix} x_0 \tag{48}$$

It follows that the transmissibility reads

$$\frac{x}{x_0} = \frac{(Ls + R)k + T^2 s}{(Ms^2 + k)(Ls + R) + T^2 s} \tag{49}$$

Comparing with Equ.(42), one sees that the electromechanical isolator behaves exactly like a relaxation isolator provided that

$$\frac{Ls + R}{T^2} = \frac{cs + k_1}{k_1 c} \tag{50}$$

or

$$k_1 = \frac{T^2}{L} \qquad c = \frac{T^2}{R} \tag{51}$$

These are the two relationships between the three parameters T, L and R so that the transmissibility of the electromechanical system of Fig.15.b is the same as that of Fig.15.a.

7 Controlling structures with piezo transducers

Consider a structure with a single discrete piezoelectric transducer (Fig.18); the transducer is governed by Equ.(20):

$$\begin{Bmatrix} Q \\ f \end{Bmatrix} = \begin{bmatrix} C(1 - k^2) & nd_{33}K_a \\ -nd_{33}K_a & K_a \end{bmatrix} \begin{Bmatrix} V \\ b^T x \end{Bmatrix} \tag{52}$$

where $\Delta = b^T x$ is the relative displacement at the extremities of the transducer. The dynamics of the structure is governed by

$$M\ddot{x} + K^* x = -bf \tag{53}$$

where K^* is the stiffness matrix of the structure without the transducer and b is the influence vector of the transducer in the global coordinate system of the structure. The non-zero components of b are the direction cosines of the active bar. The minus sign on the right hand side of the previous equation comes from the fact that the force acting on the structure is opposed to that acting on the transducer. Note that the same vector b appears in both

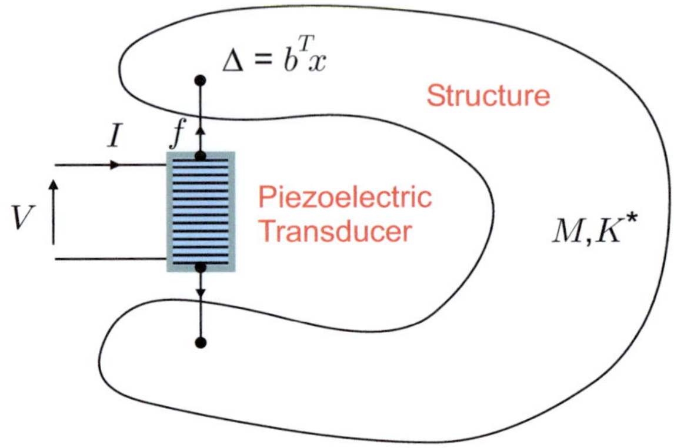

Figure 18. Structure with a piezoelectric transducer. b is the influence vector of the transducer in the global coordinate system of the structure.

equations because the relative displacement is measured along the direction of f. Substituting f from the constitutive equation into the second equation, one finds

$$M\ddot{x} + (K^* + bb^T K_a)x = b\,K_a n d_{33} V$$

or

$$M\ddot{x} + Kx = bK_a\delta \tag{54}$$

where $K = K^* + bb^T K_a$ is the global stiffness matrix of the structure including the piezoelectric transducer in short-circuited conditions (which contributes for $bb^T K_a$); $\delta = nd_{33}V$ is the *free expansion* of the transducer induced by a voltage V; $K_a\delta$ is the equivalent piezoelectric loading: the effect of the piezoelectric transducer on the structure consists of a pair of *self-equilibrating* forces applied axially to the ends of the transducer; as for thermal loads, their magnitude is equal to the product of the stiffness of the transducer (in short-circuited conditions) by the unconstrained piezoelectric expansion; this is known as the *thermal analogy.*

Let ϕ_i be the normal modes, solutions of the eigenvalue problem

$$(K - \omega_i^2 M)\phi_i = 0 \tag{55}$$

They satisfy the usual orthogonality conditions

$$\phi_i^T M\phi_j = \mu_i\delta_{ij} \tag{56}$$

$$\phi_i^T K\phi_j = \mu_i\omega_i^2\delta_{ij} \tag{57}$$

where ω_i is the natural frequency when the transducer is short-circuited. If the global displacements are expanded into modal coordinates,

$$x = \sum_i z_i \phi_i \tag{58}$$

where z_i are the modal amplitudes, Equ.(54) is easily transformed into

$$\mu_i(\ddot{z}_i + \omega_i^2 z_i) = \phi_i^T b K_a \delta \tag{59}$$

Upon taking the Laplace transform, one easily gets

$$x = \sum_{i=1}^{n} \frac{\phi_i \phi_i^T}{\mu_i(\omega_i^2 + s^2)} b K_a \delta \tag{60}$$

and the transducer extension

$$\Delta = b^T x = \sum_{i=1}^{n} \frac{K_a (b^T \phi_i)^2}{\mu_i \omega_i^2 (1 + s^2/\omega_i^2)} \delta \tag{61}$$

From Equ.(57), $\mu_i \omega_i^2/2$ is clearly the strain energy in the structure when it vibrates according to mode i, and $K_a(b^T\phi_i)^2/2$ represents the strain energy in the transducer when the structure vibrates according to mode i. Thus, the ratio

$$\nu_i = \frac{K_a (b^T \phi_i)^2}{\mu_i \omega_i^2} \tag{62}$$

is readily interpreted as the *fraction of modal strain energy* in the transducer for mode i. With this notation, the previous equation is rewritten

$$\Delta = b^T x = \sum_{i=1}^{n} \frac{\nu_i}{(1 + s^2/\omega_i^2)} \delta \tag{63}$$

which relates the actual displacement of the transducer with the free expansion due to the voltage V.

7.1 Force feedback open-loop transfer function

A frequent control configuration is that of an active strut where the piezoelectric actuator is coupled with a collocated force sensor. From the second constitutive equation (52), the open-loop transfer function between the free expansion $\delta = n d_{33} V$ of the transducer (proportional to the applied voltage) and the output force f in the active strut is readily obtained:

$$f = -K_a \delta + K_a \Delta$$

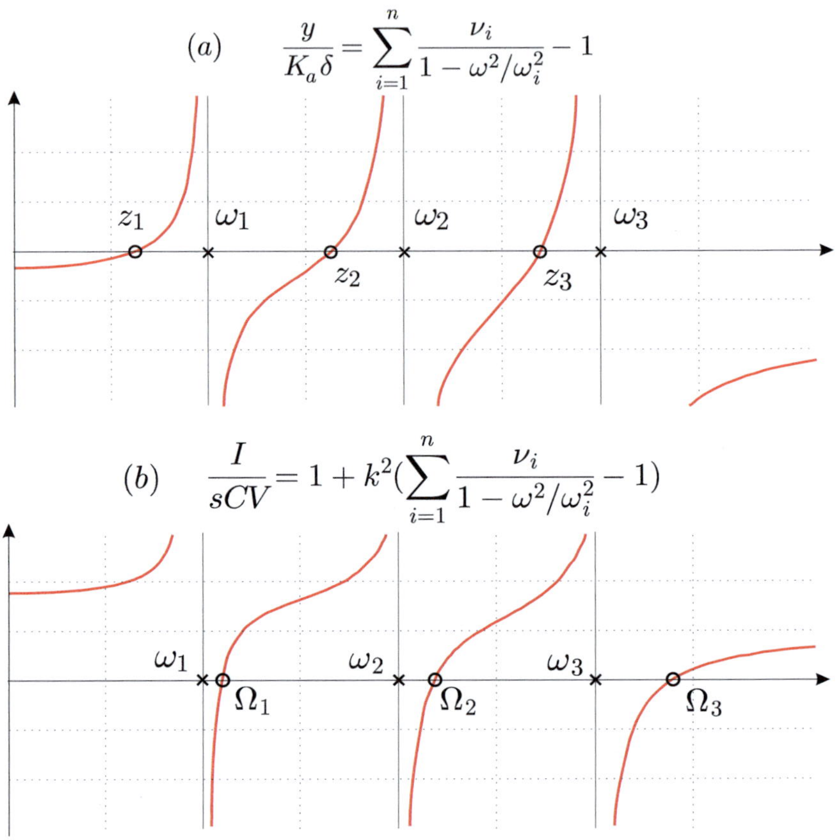

Figure 19. (a) Open-loop FRF of the active strut mounted in the structure (undamped). (b) Admittance of the transducer mounted in the structure; the poles are the natural frequencies with short-circuited electrodes ω_i and the zeros are the natural frequencies with open electrodes Ω_i.

or

$$\frac{f}{\delta} = K_a[\sum_{i=1}^{n} \frac{\nu_i}{(1+s^2/\omega_i^2)} - 1] \tag{64}$$

All the residues being positive, there will be alternating poles and zeros along the imaginary axis. Note the presence of a feedthrough in the transfer function. Figure 19.a shows the open-loop FRF in the undamped case; as expected the poles at $\pm j\omega_i$ are interlaced with the zeros at $\pm z_i$. The transfer function can be truncated after m modes, assuming that the modes above a certain order m have no dynamic amplification:

$$\frac{f}{\delta} = K_a[\sum_{i=1}^{m} \frac{\nu_i}{(1+s^2/\omega_i^2)} + \sum_{i=m+1}^{n} \nu_i - 1] \tag{65}$$

Collocated force feedback can be used very efficiently for active damping of structures, using Integral Force Feedback (IFF) and its variants; this topic is discussed extensively in [Preumont, 2011].

7.2 Admittance function

According to the first constitutive equation (52),

$$Q = C(1-k^2)V + nd_{33}K_a b^T x$$

Using (63),

$$Q = C(1-k^2)V + n^2 d_{33}^2 K_a \sum_{i=1}^{n} \frac{\nu_i}{(1+s^2/\omega_i^2)} V \tag{66}$$

and, taking into account the definition (21) of the electromechanical coupling factor, one finds the dynamic capacitance

$$\frac{Q}{V} = C(1-k^2)[1 + \frac{k^2}{1-k^2} \sum_{i=1}^{n} \frac{\nu_i}{(1+s^2/\omega_i^2)}] \tag{67}$$

The admittance is related to the dynamic capacitance by $I/V = sQ/V$:

$$\frac{I}{V} = \frac{sQ}{V} = sC(1-k^2)[1 + \sum_{i=1}^{n} \frac{K_i^2}{(1+s^2/\omega_i^2)}] \tag{68}$$

where

$$K_i^2 = \frac{k^2 \nu_i}{1-k^2} \tag{69}$$

is the effective electromechanical coupling factor for mode i.[4] The corresponding FRF is represented in Fig.19.b. The zeros of the admittance (or the dynamic capacitance) function correspond to the natural frequencies Ω_i with open electrodes (ω_i is the natural frequency with short-circuited electrodes) and

$$K_i^2 \simeq \frac{\Omega_i^2 - \omega_i^2}{\omega_i^2} \tag{70}$$

[4]Note that k^2 is a material property while ν_i depends on the mode shape, the size and the location of the transducer inside the structure.

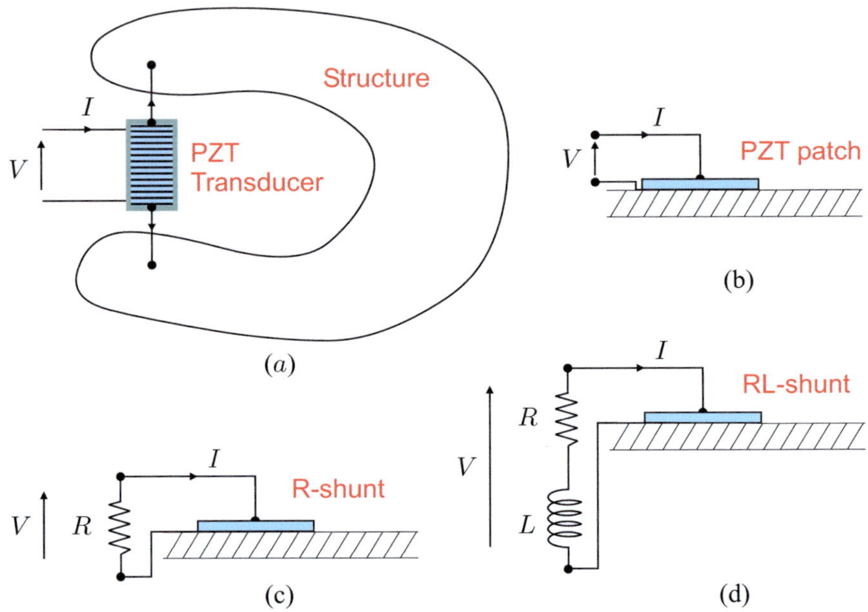

Figure 20. Structure with a piezoelectric transducer (a) in d_{33} mode (b) in d_{31} mode (c) R shunt (d) RL shunt.

The admittance of the transducer integrated in the structure may be written

$$\frac{I}{V} = sC_{\text{stat}} \cdot \frac{\prod_{i=1}^{n}(1 + s^2/\Omega_i^2)}{\prod_{j=1}^{n}(1 + s^2/\omega_j^2)} \tag{71}$$

where C_{stat} is the static capacitance of the transducer when integrated in the structure; it lies between C and $C(1-k^2)$ depending on the restraint offered by the structure.

7.3 Passive damping with a piezoelectric transducer

It is possible to achieve passive damping by integrating piezoelectric transducers at proper locations in a structure and shunting them on passive electrical networks. The theory is explained here with the simple case of a discrete transducer, but more complicated configurations are possible (Fig.20).

Resistive shunting Using the same positive signs for V and I as for the structure (Fig.20.c), the voltage drop in the resistor is $V = -RI$; therefore, the admittance of the shunt is $-1/R$. The characteristic equation of the system is obtained by expressing the equality between the admittance of the structure and that of the passive shunt:

$$-\frac{1}{R} = sC(1-k^2)[1+\sum_{i=1}^{n} \frac{K_i^2}{1+s^2/\omega_i^2}] \tag{72}$$

or

$$-\frac{1}{sRC(1-k^2)} = 1+\sum_{i=1}^{n} \frac{K_i^2\omega_i^2}{s^2+\omega_i^2} \tag{73}$$

In the vicinity of $\pm j\omega_i$, the sum is dominated by the contribution of mode i and the other terms can be neglected; defining $\gamma = [RC(1-k^2)]^{-1}$, the equation may be simplified as

$$-\frac{\gamma}{s} = 1+\frac{K_i^2\omega_i^2}{s^2+\omega_i^2}$$

which, using Equ.(70), can be rewritten

$$1+\gamma\frac{s^2+\omega_i^2}{s(s^2+\Omega_i^2)} = 0 \tag{74}$$

This form of the characteristic equation is identical to Equ.(43) that we met earlier in this chapter. The root locus is represented in Fig.21; the parameter γ acts as the feedback gain in classical root locus plots. For $\gamma = 0$ ($R = \infty$), the poles are purely imaginary, $\pm j\Omega_i$, corresponding to the natural frequency of the system with open electrodes; the system is undamped. As the resistance decreases (γ increases), the poles move to the left and some damping appears in the system; it can be shown that the maximum damping is achieved for $\gamma = \Omega_i\sqrt{\Omega_i/\omega_i} \simeq \Omega_i$ and is

$$\xi_i^{\max} = \frac{\Omega_i-\omega_i}{2\omega_i} \simeq \frac{\Omega_i^2-\omega_i^2}{4\omega_i^2} = \frac{K_i^2}{4} \tag{75}$$

Inductive shunting Since the electrical behavior of a piezoelectric transducer is essentially that of a capacitor, the idea with the RL shunt is to produce a RLC circuit which will be tuned on the natural frequency of the targeted mode and will act as a dynamic vibration absorber. We proceed in the same way as in the previous section, but with a RL-shunt (Fig.20.d); the admittance of the shunt is now $I/V = -1/(R+Ls)$. The characteristic

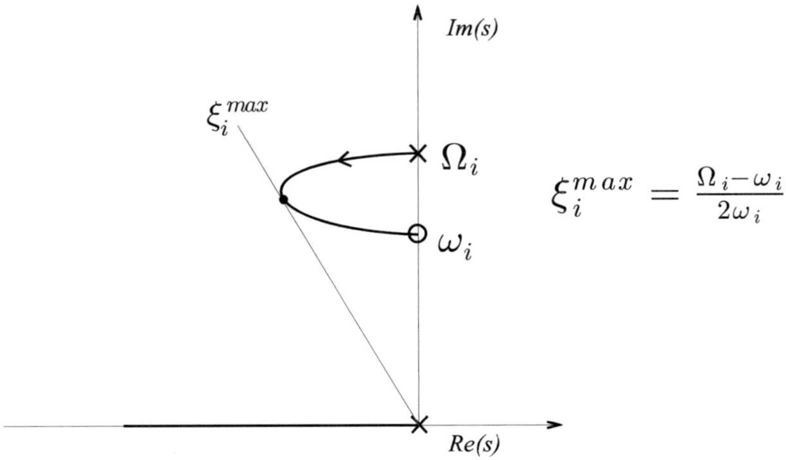

Figure 21. Resistive shunt. Evolution of the poles of the system as $\gamma = [RC(1-k^2)]^{-1}$ goes from 0 to ∞ (the diagram is symmetrical with respect to the real axis, only the upper half is shown).

equation is obtained by expressing the equality between the admittance of the structure and that of the passive shunt:

$$-\frac{1}{R+Ls} = sC(1-k^2)[1+\sum_{i=1}^{n}\frac{K_i^2}{1+s^2/\omega_i^2}] \tag{76}$$

or

$$-\frac{1}{(R+Ls)sC(1-k^2)} = 1+\sum_{i=1}^{n}\frac{K_i^2\omega_i^2}{s^2+\omega_i^2} \tag{77}$$

Once again, in the vicinity of $\pm j\omega_i$, the sum is dominated by the contribution of mode i and the equation is simplified as

$$-\frac{1}{(R+Ls)sC(1-k^2)} = 1+\frac{K_i^2\omega_i^2}{s^2+\omega_i^2} \tag{78}$$

Defining the electrical frequency

$$\omega_e^2 = \frac{1}{LC(1-k^2)} \tag{79}$$

and the electrical damping

$$2\xi_e\omega_e = \frac{R}{L} \tag{80}$$

Equ.(78) is rewritten

$$-\frac{\omega_e^2}{2\xi_e\omega_e s + s^2} = 1 + \frac{K_i^2\omega_i^2}{s^2+\omega_i^2} = \frac{s^2+\Omega_i^2}{s^2+\omega_i^2} \tag{81}$$

or

$$s^4 + 2\xi_e\omega_e s^3 + (\Omega_i^2+\omega_e^2)s^2 + 2\Omega_i^2\xi_e\omega_e s + \omega_i^2\omega_e^2 = 0 \tag{82}$$

This can be rewritten in a root locus form

$$1 + 2\xi_e\omega_e \frac{s(s^2+\Omega_i^2)}{s^4+(\Omega_i^2+\omega_e^2)s^2+\omega_i^2\omega_e^2} = 0 \tag{83}$$

In this formulation, $2\xi_e\omega_e = R/L$ plays the role of the gain in a classical root locus. Note that, for large R, the poles tend to $\pm j\Omega_i$, as expected. For $R = 0$ (i.e. $\xi_e = 0$), they are the solutions p_1 and p_2 of the characteristic equation $s^4+(\Omega_i^2+\omega_e^2)s^2+\omega_i^2\omega_e^2 = 0$ which accounts for the classical double peak of resonant dampers, with p_1 above $j\Omega_i$ and p_2 below $j\omega_i$. Figure 22 shows the root locus for a fixed value of ω_i/Ω_i and various values of the electrical tuning, expressed by the ratio

$$\alpha_e = \frac{\omega_e\omega_i}{\Omega_i^2} \tag{84}$$

The locus consists of two loops, starting respectively from p_1 and p_2; one of them goes to $j\Omega_i$ and the other goes to the real axis, near $-\Omega_i$. If $\alpha_e > 1$ (Fig.22.a), the upper loop starting from p_1 goes to the real axis, and that starting from p_2 goes to $j\Omega_i$, and the upper pole is always more heavily damped than the lower one (note that, if $\omega_e \to \infty$, $p_1 \to \infty$ and $p_2 \to j\omega_i$; the lower branch of the root locus becomes that of the resistive shunting). The opposite situation occurs if $\alpha_e < 1$ (Fig.22.b): the upper loop goes from p_1 to $j\Omega_i$ and the lower one goes from p_2 to the real axis; the lower pole is always more heavily damped. If $\alpha_e = 1$ (Fig.22.c), the two poles are always equally damped until the two branches touch each other in Q. This double root is achieved for

$$\alpha_e = \frac{\omega_e\omega_i}{\Omega_i^2} = 1 \quad , \qquad \xi_e^2 = 1 - \frac{\omega_i^2}{\Omega_i^2} \simeq K_i^2 \tag{85}$$

This can be regarded as the optimum tuning of the inductive shunting. The corresponding eigenvalues satisfy

$$s^2 + \Omega_i^2 + \Omega_i(\frac{\Omega_i^2}{\omega_i^2} - 1)^{1/2} s = 0 \tag{86}$$

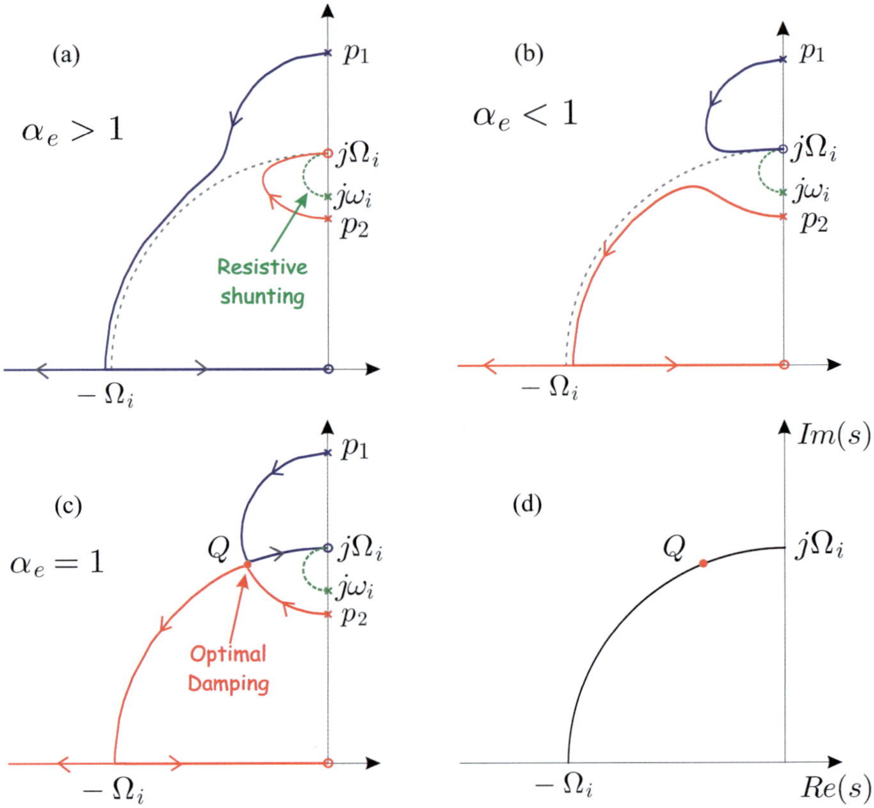

Figure 22. Root locus plot for inductive shunting (only the upper half is shown). The optimum damping at Q is achieved for $\alpha_e = 1$ and $\xi_e = K_i$; the maximum modal damping is $\xi_i \simeq K_i/2$.

For various values of ω_i/Ω_i (or K_i), the optimum poles at Q move along a circle of radius Ω_i (Fig.22.d). The corresponding damping ratio can be obtained easily by identifying the previous equation with the classical form of the damped oscillator, $s^2 + 2\xi_i\Omega_i s + \Omega_i^2 = 0$, leading to

$$\xi_i = \frac{1}{2}(\frac{\Omega_i^2}{\omega_i^2} - 1)^{1/2} = \frac{K_i}{2} = \frac{1}{2}(\frac{k^2\nu_i}{1-k^2})^{1/2} \tag{87}$$

This value is significantly higher than that achieved with purely resistive shunting [it is exactly the square-root of (75)]. Note, however, that it is much more sensitive to the tuning of the electrical parameters on the tar-

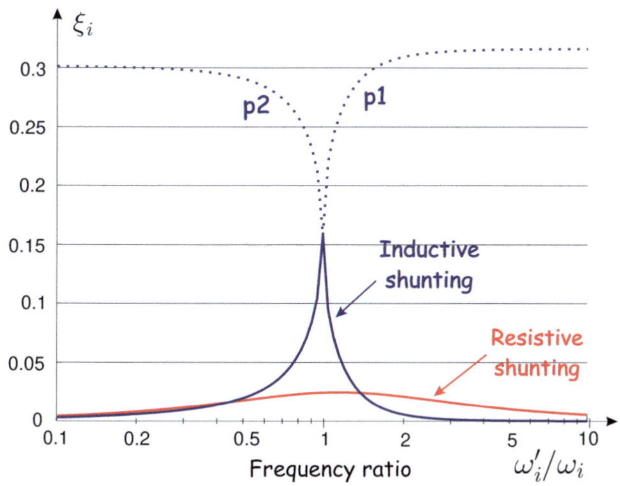

Figure 23. Evolution of the damping ratio of the inductive and resistive shunting with the de-tuning of the structural mode. ω_i is the natural frequency for which the shunt has been optimized, ω_i' is the actual value ($k = 0.5$, $\nu_i = 0.3$).

geted modes. This is illustrated in Fig.23, which displays the evolution of the damping ratio ξ_i when the actual natural frequency ω_i' moves away from the nominal frequency ω_i for which the shunt has been optimized (the damping ratio associated with p_1 and p_2 is plotted in dotted lines; the ratio ω_i'/Ω_i' is kept constant in all cases). One sees that the performance of the inductive shunting drops rapidly below that of the resistive shunting when the de-tuning increases. Note that, for low frequency modes, the optimum inductance value can be very large; such large inductors can be synthesized electronically. The multimodal passive damping via resonant shunt has been investigated by [Hollkamp, 1994].

Bibliography

[1] W. G. Cady. *Piezoelectricity: an Introduction to the Theory and Applications of Electromechanical Phenomena in Crystals*, , McGrawHill, 1946.

[2] S. H. Crandall. D. C. Karnopp, E. F. Jr. Kurtz, D. C. Pridmore-Brown *Dynamics of Mechanical and Electromechanical Systems*, McGraw-Hill, N-Y, 1968.

[3] E. De Boer. Theory of Motional Feedback, *IRE Transactions on Audio*, 15-21, Jan.-Feb., 1961.

[4] J. Holterman and P. Groen. *An Introduction to Piezoelectric Materials and Components*, Stichting Applied Piezo, 2012.

[5] J. J. Hollkamp. Multimodal passive vibration suppression with piezoelectric materials and resonant shunts, *J. Intell. Material Syst. Structures*, Vol.5, Jan. 1994.

[6] F. v. Hunt. *Electroacoustics: The Analysis of Transduction, and its Historical Background*, Harvard Monographs in Applied Science, No 5, 1954. Reprinted, Acoustical Society of America, 1982.

[7] *IEEE Standard on Piezoelectricity.* (ANSI/IEEE Std 176-1987).

[8] J. van Randeraat. & SETTERINGTON, R.E. (Edts) *Philips Application Book on Piezoelectric Ceramics*, Mullard Limited, London, 1974.

[9] *Physik Intrumente* catalogue, Products for Micropositioning (*PI* GmbH).

[10] J. Pratt and A. Flatau. Development and analysis of self-sensing magnetostrictive actuator design, SPIE Smart Materials and Structures Conference, Vol.1917, 1993.

[11] A. Preumont. *Vibration Control of Active Structures, An Introduction*, Third Edition, Springer, 2011.

[12] A. Preumont. *Mechatronics, Dynamics of Electromechanical and Piezoelectric Systems*, Springer, 2006.

[13] C. A. Rosen. Ceramic transformers and filters, Proc. Electronic Component Symposium, p.205-211, 1956.

[14] K. Uchino. *Ferroelectric Devices*, Marcel Dekker, 2000.

[15] H. H. Woodson and J. R. Melcher. *Electromechanical Dynamics, Part I: Discrete Systems*, Wiley, 1968.

LMIs in Control Optimization

Robert E. Skelton
University of California, San Diego, CA, U.S.

1 Early Optimization History

Hamilton invented state space models of nonlinear dynamic systems with his generalized momenta work in the 1800s, but, at that time, the lack of computational tools prevented broad acceptance of the first order form of dynamic equations. With the rapid development of computers in the 1960s, State Space models evoked a formal control theory for minimizing a scalar function of control and state, propelled by the calculus of variations and Pontryagin's maximal principle.

In the case of linear dynamic models, these methods led to the popularization of Linear Quadratic Gaussian (LQG) optimal control, which had globally optimal solutions, Skelton (1988). Since guarantees about the first two moments of the stochastic process (the mean and the covariance) can be made, regardless of the distribution of the random variables involved, LQG became just an acronym for the solution of quadratic functionals of control and state variables, even when the stochastic processes were not Gaussian, The label LQG was often used even for deterministic problems, where an integral operator, rather than an expectation operator, was minimized, with given initial conditions or impulse excitations. These were formally called LQR (Linear Quadratic Regulator) problems. Later the book Skelton (1988) gave the formal conditions under which the LQG and the LQR answers were numerically identical, and this version was called the *Deterministic LQG*.

It was always recognized that the quadratic form of the state and control in the LQG problem was an artificial goal. The real control goals usually involved pre-specified performance bounds on *each* of the errors and bounds on *each* channel of control. This leads to Matrix inequalities rather than scalar minimizations. While it was known early that *any* stabilizing linear controller could be obtained by some choice of weights in an LQG optimization problem, It was not known until the1980s what particular choice of weights in an LQG problem would yield a solution to the matrix inequality problem. A globally convergent algorithm by Zhu and Skelton (1992);

P. Hagedorn, G. Spelsberg-Korspeter (Eds.), *Active and Passive Vibration Control of Structures*, CISM International Centre for Mechanical Sciences
DOI 10.1007/ 978-3-7091-1821-4_5

Zhu et al. (1997) shows how to find such LQG weights when the matrix inequality problem has a solution. Since then, linear control problems can be stated simply in terms of norm bounds on *each* input or *each* output of the system (L_2 bounds , L-infinity bounds, or variance bounds and covariance bounds). These problems are convex for state feedback or full order controllers (the focus of this elementary introduction), and will be solved using Linear Matrix Inequalities in this paper. However the earliest approach to these problems were iterative LQG solutions (to find the correct weights to use in the quadratic penalty of the state), Zhu and Skelton (1992); Zhu et al. (1997).

2 Matrix Inequalities

Let Q be any square matrix. The Matrix Inequality "$Q > 0$" is just a short-hand notation to represent a certain scalar inequality. That is, the matrix notation "$Q > 0$" means "the scalar $x^T Qx$ is positive for all values of x, except $x = 0$". Obviously this is a property of Q, not x, hence the abbreviated matrix notation $Q > 0$. This is called a Linear Matrix Inequality (LMI), since the matrix unknown Q appears linearly in the matrix inequality $Q > 0$. Note also that any square matrix, Q, can be written as the sum of a symmetric matrix $Q_y = \frac{1}{2}(Q + Q^T)$, and a skew-symmetric matrix $Q_k = \frac{1}{2}(Q - Q^T)$, but $x^T Q_k x = 0$, so only the symmetric part of the matrix Q affects the scalar $x^T Qx$. We assume hereafter without loss of generality that Q is symmetric. The notation "$Q \geq 0$" means "the scalar $x^T Qx$ cannot be negative for any x".

Liapunov proved that $x(t)$ converges to zero if, along the nonzero trajectory of a dynamic system, (e.g. the system $\dot{x} = Ax$), two scalars have the property, $x(t)^T Qx(t) > 0$ and $d/dt((x^T(t)Qx(t)) < 0$. This proves that the following statements are all equivalent:

1. For any initial condition $x(0)$ of the system $\dot{x} = Ax$, $x(t)$ will converge to zero as time approaches infinity
2. All eigenvalues of A lie in the open left-half plane.
3. There exists a matrix Q with these two properties: $Q > 0$, $QA + A^T Q < 0$.
4. The set of all quadratic Liapunov functions that can be used to prove the stability or instability the null solution of $\dot{x} = Ax$ is $x^T Qx$, where Q is any square matrix with the two properties of item 3, above.

For the linear system example $\dot{x} = Ax + Bu$, $y = Cx$, the "Observability Gramian" is the infinite integral $Q = \int e^{A^T t} C^T C e^{At} dt$ which is affected only by the observable modes of the system. Furthermore $Q > 0$ if and only if

(A, C) is an observable pair, and Q is bounded only if the observable modes are asymptotically stable. If Q exists then the solution of $QA + A^T Q + C^T C = 0$ is $Q > 0$ if and only if the matrix pair (A, C) is observable.

Likewise the "Controllability Gramian" $X = \int e^{At} BB^T e^{A^T t} dt > 0$ if and only if the pair (A, B) is controllable. If X exists it satisfies $XA^T + AX + BB^T = 0$, and $X > 0$ if and only if (A, B) is a controllable pair. Note also that (A, B) is controllable for any A if $BB^T > 0$, and (A, C) is observable for any A if $C^T C > 0$. Hence the existence of $Q > 0$ or $X > 0$ satisfying either $(QA + A^T Q < 0)$ or $(AX + XA^T < 0)$ is equivalent to the statement that "all eigenvalues of A lie in the open left half plane".

It should be clear now that the set of all stabilizing state feedback controllers, $u = Gx$, is parametrized by the inequalities $Q > 0$, $Q(A + BG) + (A + BG)^T Q < 0$. The difficulty here is the appearance of the product of the two unknowns Q and G.

Here we borrow some tricks from linear algebraic equations, where any Linear Matrix Equality (LME), $\Gamma G \Lambda = \Theta$ has a solution G if and only if $\Gamma \Gamma^+ \Theta \Lambda^+ \Lambda = \Theta$. Notation M^+ denotes the Moore-Penrose inverse of matrix M. If these existence conditions are satisfied, then all solutions are parametrized by $G = \Gamma^+ \Theta \Lambda^+ + Z - \Gamma^+ \Gamma Z \Lambda \Lambda^+$. The point is that in LMEs there are two separate questions and answers. The first question that is answered is "Does there exist a solution?". The second question is "What is the set of all solutions?". LMI approaches will employ the same two steps by formulating the necessary and sufficient conditions for the existence of the LMI solution, and then parametrize all solutions. The earliest book on LMI control methods was Boyd et al. (1994), but the notation used herein is taken from Skelton et al. (1998).

3 Relation to Linear Algebraic Equations

Linear Matrix Equalities have been completely solved many years ago, and the theory gives i) the necessary and sufficient conditions for the existence of a solution, and then ii) the set of all solutions. The LME (Linear Matrix Equality) is supported by the following facts;

There exists a solution G to the LME

$$AGB = C \tag{1}$$

if and only if, matrices A, B, C have the following property:

$$AA^+ CB^+ B = C. \tag{2}$$

If the existence condition (2) is satisfied, then all solutions of (1) are given by

$$G = A^+CB^+ + Z - A^+AZBB^+, \tag{3}$$

where Z is an arbitrary matrix, and the matrix A^+ is the Moore Penrose inverse, which is the unique matrix which satisfies $AA^+A = A$.

Much of the mystery of the above equations is removed by writing them in terms of the Singular Value Decomposition (SVD) of the matrices A and B. For any real matrix M the SVD is defined

$$M = U\Sigma V^T = \begin{bmatrix} U_1 & U_2 \end{bmatrix} \begin{bmatrix} \Sigma & 0 \\ 0 & 0 \end{bmatrix} \begin{bmatrix} V_1^T \\ V_2^T \end{bmatrix}. \tag{4}$$

with these special properties of the SVD components,

$$\Sigma > 0, \quad \Sigma = diagonal \tag{5}$$
$$U^T U = I = UU^T \tag{6}$$
$$V^T V = I = VV^T \tag{7}$$
$$U_2^T M = 0 \tag{8}$$
$$MV_2 = 0 \tag{9}$$
$$M^+ = U_1 \Sigma^{-1} V_1^T, \tag{10}$$

we call U_2^T the basis for the left nullspace of M (the entire nullspace is KU_2^T for arbitrary K). Likewise, we call V_2 the basis for the right nullspace of M. Using these properties of the SVD of matrices A, B, one can write (2) and (3) as

$$U_{A2}^T C = 0, \quad CV_{B2} = 0 \tag{11}$$
$$G = V_{A1}(\Sigma_A^{-1} U_{A1}^T C V_{B1} \Sigma_B^{-1}) U_{B1}^T + V_{A2} Z_1 U_{B1}^T + V_A Z_2 U_{B2}^T, \tag{12}$$

where Z_1, Z_2 are arbitrary matrices. Note that the existence conditions require the matrix C to lie in the left nullspace of A and in the right nullspace of B.

4 Control Design Using LMIs

Let the system be described by the state equations

$$\begin{bmatrix} \dot{x}_p \\ y \\ z \end{bmatrix} = \begin{bmatrix} A_p & D_p & B_p \\ C_p & D_y & B_y \\ M_p & D_z & 0 \end{bmatrix} \begin{bmatrix} x_p \\ w \\ u \end{bmatrix}, \quad \begin{bmatrix} u \\ \dot{x}_c \end{bmatrix} = \begin{bmatrix} D_c & C_c \\ B_c & A_c \end{bmatrix} \begin{bmatrix} z \\ x_c \end{bmatrix} = G \begin{bmatrix} z \\ x_c \end{bmatrix}, \tag{13}$$

where z is the measurement vector, y is the output to be controlled, u is the control vector, x_p is the plant state vector, x_c is the state of the controller, and w is the external disturbance (we will often treat w as zero-mean white noise in these discussions). We seek to choose the control matrix G to satisfy given upper bounds, $E[yy^T] \leq \bar{Y}$, where E represents the steady state expectation operator in the stochastic case (when w is white noise), and in the deterministic case E represents the infinite integral of the matrix $[yy^T]$. The math we do here is the same, with appropriate interpretations of certain matrices. For a rigorous equivalence of deterministic and stochastic interpretations see Skelton (1988)Defining

$$x = \begin{bmatrix} x_p \\ x_c \end{bmatrix}, \quad \begin{bmatrix} A_{cl} & B_{cl} \\ C_{cl} & D_{cl} \end{bmatrix} = \begin{bmatrix} A & D \\ C & F \end{bmatrix} + \begin{bmatrix} B \\ H \end{bmatrix} G \begin{bmatrix} M & E \end{bmatrix} \tag{14}$$

$$A = \begin{bmatrix} A_p & 0 \\ 0 & 0 \end{bmatrix}, B = \begin{bmatrix} B_p & 0 \\ 0 & I \end{bmatrix}, M = \begin{bmatrix} M_p & \\ 0 & I \end{bmatrix}, D = \begin{bmatrix} D_p \\ 0 \end{bmatrix}, E = \begin{bmatrix} D_z \\ 0 \end{bmatrix} \tag{15}$$

$$C = \begin{bmatrix} C_p & 0 \end{bmatrix}, H = \begin{bmatrix} B_y & 0 \end{bmatrix}, F = D_y, \tag{16}$$

one can write the closed loop system dynamics in the form

$$\begin{bmatrix} \dot{x} \\ y \end{bmatrix} = \begin{bmatrix} A_{cl} & B_{cl} \\ C_{cl} & D_{cl} \end{bmatrix} \begin{bmatrix} x \\ w \end{bmatrix}. \tag{17}$$

We shall often be interested in the set of all controllers that can satisfy various performance bounds, such as, $E[yy^T] \leq \bar{Y}$, $E[uu^T] \leq \bar{U}$.

4.1 State Feedback Stabilization

For only stability requirements our problem reduces to the case where $C_p = M_p = I$, $C_c = 0$, and $\bar{Y}$ has no specified bound. In this case the necessary and sufficient condition for stability of the closed loop system is

$$(A_p + B_p G)X + X(A_p + B_p G)^T < 0. \tag{18}$$

This is satisfied if and only if there exists an arbitrarily small positive semidefinite matrix that can be added to the left-hand side of this inequality. So lets add the matrix $\frac{1}{\gamma^2} X G^T G X$, for a large enough γ, to get

$$(A_p + B_p G)X + X(A_p + B_p G)^T + \frac{1}{\gamma^2} X G^T G X < 0. \tag{19}$$

By completing the square, write this as,

$$(\frac{1}{\gamma}XG^T + \gamma B_p)(\frac{1}{\gamma}XG^T + \gamma B_p)^T < \gamma^2 B_p B_p^T - A_p X - X A_p^T =: \Phi. \quad (20)$$

Now we will use corollary 2.3.6 in Skelton et al. (1998) which proves that the following Quadratic Matrix Inequality

$$(\frac{1}{\gamma}XG^T + \gamma B_p)(\frac{1}{\gamma}XG^T + \gamma B_p)^T < \Phi, \quad (21)$$

has a solution G if and only if $\Phi > 0$, in which case, all such G are given by

$$G = -\gamma^2 B_p^T X^{-1} + \gamma L \Phi^{1/2} X^{-1}, \quad LL^T = I. \quad (22)$$

Defining $P = \gamma^2 X^{-1}$, $Q = \gamma^2 X^{-1} \Phi X^{-1}$, we have

$$G = -B_p^T P + LQ^{1/2} \quad (23)$$

$$PA_p + A_p^T P - PB_p B_p^T P + Q = 0. \quad (24)$$

Since $\bar{Y}$ was not bounded in this stabilization problem, the existence of $X > 0$ already satisfies, for $C_p = I$, $C_p X C_p^T = X = \gamma^2 P^{-1} < \bar{Y}$.

5 Information Architecture in Estimation and Control Problems

In the typical "control problem" that occupies most research literature, the sensors and actuators have already been selected. Yet the selection of sensors and actuators and their locations greatly affect the ability of the control system to do its job efficiently. Perhaps in one location a high precision sensor is needed, and in another location high precision is not needed, and high precision there would be a waste of resources. These decisions are influenced by the control dynamics which are yet to be designed. How do we know where to spend money to improve the system?

Lets consider the problem of selecting the control law jointly with the selection of the precision (inverse of the noise intensity) of each actuator/sensor, subject to the constraint of specified upper bounds on the covariance of output error and control signals, and specified upper bounds on the sensor/actuator cost. (We assume the price of these devices are proportional to their precision). Traditionally, with full order controllers, and *pre specified* sensor/actuator instruments (with specified precisions), this is a well-known solved convex problem and can be converted to an LMI problem. If we enlarge the domain of the optimization to include sensor/actuator precisions it is not obvious whether the problem is convex or not. The following shows that this problem of including the sensor/actuator precisions within the control design problem is indeed convex and therefore completely solved. The proof is provided in Li et al. (2010).

Consider the linear system (1)-(5). There exists a dynamic controller G that satisfies the constraints

$$E[uu^T \leq \bar{U}, \quad E[yy^T] \leq \bar{Y}, \quad trPW^{-1} \leq \bar{\Omega} \tag{25}$$

if and only if there exists Matrices L, F, Q, X, Z, W^{-1} such that

$$trPW^{-1} \leq \bar{\Omega} \tag{26}$$

$$\begin{bmatrix} \bar{Y} & CX & C \\ (CX)^T & X & I \\ C^T & I & Z \end{bmatrix} > 0, \begin{bmatrix} \bar{U} & L & 0 \\ L^T & X & I \\ 0 & I & Z \end{bmatrix} > 0, \begin{bmatrix} \Phi_{11} & \Phi_{21}^T \\ \Phi_{21} & -W^{-1} \end{bmatrix} < 0 \tag{27}$$

$$\Phi_{21} = \begin{bmatrix} D & 0 \\ ZD & F \end{bmatrix}, \quad \phi = \begin{bmatrix} AX + BL & A \\ Q & ZA + FM \end{bmatrix}, \quad \Phi_{11} = \phi + \phi^T. \tag{28}$$

Furthermore, the solution of the problem (13) is given below

Assume a solution for (L,F,Q,X,Z,W) is found from (14)-(16). Then the problem (13) is solved by the controller

$$G = \begin{bmatrix} 0 & I \\ V_l^{-1} & -V_l^{-1}ZB \end{bmatrix} \begin{bmatrix} Q - ZAX & F \\ L & 0 \end{bmatrix} \begin{bmatrix} 0 & V_r^{-1} \\ I & -MXV_r^{-1} \end{bmatrix}, \tag{29}$$

where V_l and V_r are left and right factors of the matrix $I - YX$ (which can be found from the singular value decomposition $I - YX = U\Sigma V^T = (U\Sigma^{1/2})(\Sigma^{1/2}V^T) = V_l V_r$).

To emphasize the theme of this article, to relate optimization to LMIs, we note that three optimization problems present themselves in the above problem with three constraints: control effort $\bar{U}$, output performance $\bar{Y}$, and instrument costs $\bar{\Omega}$. To solve optimization problems, one can fix any two of these pre-specified upper bounds and iteratively reduce the level set value of the third "constraint" until feasibility is lost. This process minimizes the resource expressed by the third constraint, while enforcing the other two constraints.

As an example, if cost is not a concern, one can always set large limits for $\bar{\Omega}$ and discover the best assignment of sensor/actuator precisions for the specified performance requirements. These precisions produced by the algorithm are the values W_{ii}^{-1}, produced from the solution (14)-(16), where the observed rankings $W_{ii}^{-1} > W_{jj}^{-1} > W_{kk}^{-1} >$ ..... indicate which sensors or actuators are most critical to the required performance goals $(\bar{U}, \bar{Y}, \bar{\Omega})$. If any precision $W_n^{-1}n$ is essentially zero, compared to other required precisions, then the math is asserting that the information from this sensor (n) is not important for the control objectives specified, or the control signals through this actuator channel (n) is ineffective in controlling the system to these specifications. This information leads us to a technique for choosing the best sensor actuators and their location.

The previous discussion provides a solution to the precisions required of each sensor and each actuator in the system. Our final application of this theory locates sensors and actuators in a large scale system, by discarding the least effective ones. Suppose we solve any of the above feasibility problems, by starting with as many sensors and actuators as is possible for the system (without regard to cost). For example in a flexible structure control problem we might not know whether to place a rate sensor or a displacement sensors at a given location, so we add both. We might not know whether to use torque or force actuators, so we add both. We fill up the system with all the possibilities we might want to consider, and let the above precision rankings (available after the LMI problem is solved) reveal how much precision is needed at each location and at each sensor/actuator. If there is a large gap in the precisions required (say $W_{11}^{-1} > W_{22}^{-1} > W_{33}^{-1} >> .....W_{nn}^{-1}$), then delete the sensor/actuator n and repeat the LMI problem with one less sensor or actuator. Continue deleting sensors/actuators in this manner until feasibility of the problem is lost. Then this algorithm, stopping at the previous iteration, has selected the best distribution of sensors/actuators for solving the specific problem $(\bar{\Omega}, \bar{U}, \bar{Y})$. In my opinion this is the most important contribution of the algorithm, to enlarge the set of solved linear control problems, from solutions of linear controllers with sensors/actuators pre-specified, to solutions which specify the sensor/actuator requirements jointly with the control solution.

6 Many Control Problems Reduce to the Same LMI

Let the left and right null spaces of B be defined by $U_B^T B = 0$, $U_B^T U_B > 0$, $BV_B = 0$, , $V_B^T V_B > 0$. For any given matrices Θ, Φ, Q, the book Skelton et al. (1998) provides all G which solve

$$\Theta G \Phi + (\Theta G \Phi)^T + Q < 0, \tag{30}$$

and there exists such a matrix G if and only if the following two conditions hold

$$U_\Theta^T Q U_\Theta < 0, \quad \text{or} \quad \Theta\Theta^T > 0, \quad V_\Phi^T Q V_\Phi \quad < 0, \quad \text{or} \quad \Phi^T \Phi > 0. \tag{31}$$

The significance of this theorem is that 17 different control problems (using either state feedback or full order dynamic controllers) all reduce to this same mathematical problem. That is, by defining the appropriate Θ, Φ, Q a very large number of different control problems (including the set of all stabilizing controllers, covariance control, H-infinity control, L-infinity control, LQG control, H_2 control), can be reduced to the *same* matrix inequality (18).

7 Conclusions

LMI techniques provide more powerful tools for designing controllers or estimators for linear systems then minimizing a scalar functional for optimization. LMI techniques extend the range of solvable system design problems beyond just control design. By integrating information architecture and control design, one can decide what precision is required of all sensor/actuators to satisfy the closed loop performance constraints. A most amazing fact is that most of the common linear control design problems all reduce to the exact same LMI problem (18).

Bibliography

S. Boyd, L. El Ghaoui, E. Feron, and V. Balakrishnan. *Linear Matrix Inequalities in System and Control Theory.* SIAM, 1994.

F. Li, R. Skelton, and M. C. de Oliveira. Integrating information architecture and control or estimation. *SICE Journal of Control, Measurement, and System Integration, Springer*, 1:120–128, 2010.

R. Skelton. *Dynamic Systems Control.* John Wiley and Sons, 1988.

R. Skelton, T. Iwasaki, and K. M. Grigoriadis. *A Unified Algebraic Approach to Control Design.* Taylor and Francis, 1998.

G. Zhu and R. Skelton. A two-riccati, feasible algorithm for guaranteeing output l constraints. *Journal of Dynamic Systems, Measurement and Control*, 114:329–338, 1992.

G. Zhu, M.A. Rotea, and R. Skelton. A convergent algorithm for the output covariance constraint control problem. *SIAM Journal on Control and Optimization*, 35:341–61, 1997.

Damping Mechanisms

Adnan Akay* and Antonio Carcaterra†

* Bilkent University, Mechanical Engineering Department, Ankara, Turkey

‡ University of Rome, La Sapienza, Department of Mechanics and Aeronautics, Via Eudossiana, 18, 00184, Rome, Italy.

1 Linear Response and Damping

Introduction

The term *damping* is used to describe the means by which oscillation amplitudes are reduced through irreversible removal of vibratory energy in a mechanical system or a component. Dissipation, on the other hand, refers to the mechanism by which irreversible energy transfer, from vibratory to thermal, takes place. In this sense, damping is a macro-scale manifestation of atomic-scale dissipation.

High damping is desirable to attain low vibration and noise levels whereas low damping is desirable for increased sensitivity in sensors and certain precision instrumentation.

Damping is most obvious at resonance where the stiffness and inertia forces become equal. As a result, damping is a key factor in predicting vibration response of structures.

As we will see in the following sections, there are numerous paths to damping and in a complex structure several means of damping may take place simultaneously at different locations throughout the structure. Accordingly, in determining the response of a vibrating structure, the total effect of all types of damping that may be distributed throughout a structure must be taken into account.

Measurements of damping normally indicate the total damping a system experiences. It is difficult to isolate a component or a subsystem or a material within a system and measure its damping. In describing the various damping mechanisms, we will examine each through its effect on a single-degree-of-freedom (sdof) oscillator.

In this section, we will review the response of a simple oscillator and examine the role of damping on it and review the basic methods of measurement criteria for damping properties of structures. However, we will

P. Hagedorn, G. Spelsberg-Korspeter (Eds.), *Active and Passive Vibration Control of Structures*, CISM International Centre for Mechanical Sciences
DOI 10.1007/ 978-3-7091-1821-4_6

not consider here the role of damping in dynamic behaviors such as chaos, stability, etc.

Dissipation of vibratory energy takes place in both fluid and solid media, initiated by a number of possible macro activities. Accordingly, we will consider damping methods to reflect the media in which dissipation takes place when addressing damping methods in the next section. Models of fundamental dissipation mechanisms that describe energy transfer from ordered energy to disordered or thermalized energy are briefly summarized in the last section.

1.1 Simple Harmonic Oscillator

We employ the simple harmonic oscillator as the platform to describe damping models and measures through its linear response.

Initial Value Problem Equation of motion for free vibrations of an undamped sdof oscillator with mass M and stiffness K can be expressed as

$$M\ddot{\eta} + K\eta = 0$$

or in a simplified form

$$\ddot{\eta} + \omega_0^2\eta = 0$$

where $\omega_0 = \sqrt{K/M}$ is the natural frequency of the oscillator. General solution for displacement η can be expressed as

$$\eta = A\cos(\omega_0 t - \phi).$$

Expressing the initial conditions at $t = 0$ as $\eta(0) = \eta_0$ and $\dot{\eta}(0) = \dot{\eta}_0$, we can write

$$\eta_0 = A\cos(\phi) \qquad \dot{\eta}_0 = \omega_0 A\sin(\phi)$$

and the vibration amplitude in terms of the initial conditions becomes

$$A = \sqrt{\eta_0^2 + (\dot{\eta}_0/\omega_0)^2}.$$

Free Damped Motion When damping is proportional to oscillator velocity, represented by a constant of proportionality C, the equation of motion becomes:

$$M\ddot{\eta} + C\dot{\eta} + K\eta = 0$$

with a corresponding solution for free vibrations:

$$\eta(t) = A\,e^{-\gamma t}\cos(\omega_d t - \phi)$$

where $\omega_d = \omega_0\sqrt{1-\zeta^2} = \sqrt{\omega_0^2 - \gamma^2}$ is the damped natural frequency and $\gamma = C/2M$ is the *decay constant* and is related to the *damping ratio* $\zeta = \gamma/\omega_0 = C/C_c$, which is the ratio of damping constant to its critical value $C_c = 2\sqrt{KM}$.

For *underdamped* cases, $\zeta < 1$, response to initial conditions can be written as

$$\eta(t) = e^{-\zeta\omega_0 t}\left[\frac{\dot{\eta}_0 + \zeta\omega_0\eta_0}{\omega_d}\sin\omega_d t + \eta_0\cos\omega_d t\right].$$

When $\gamma > \omega_0$, or $\zeta > 1$, ω_d becomes complex and oscillations are not possible and the system is referred as *overdamped.*

Between these two cases, when $\gamma = \omega_0$ or $\zeta = 1$, the oscillator is considered *critically damped.*

Forced Motion Response η of a sdof oscillator to a force $F(t)$ can be described with:

$$M\ddot{\eta} + C\dot{\eta} + K\eta = F(t). \tag{1}$$

Fourier transforming the motion equation (1) according to

$$\eta(\omega) = \int_{-\infty}^{\infty}\eta(t)e^{-j\omega t}dt$$

we obtain response equation in the frequency domain:

$$\left[-M\omega^2 - j\omega C + K\right]\eta(\omega) = F(\omega).$$

► **Harmonic Excitation** Response of a simple oscillator to harmonic excitation $F_0e^{j\omega t}$ can be expressed in terms of receptance (or compliance) frequency response function (FRF), $H(\omega)$, of the oscillator

$$\eta(\omega) = H(\omega)F_0e^{-j\omega t}$$

where

$$H(\omega) = \left\{K\left[1 - (\omega/\omega_0)^2 - j(2\zeta\omega/\omega_0)\right]\right\}^{-1}.$$

Frequency response function is the Fourier transform of the impulse response function and is generally a complex quantity:

$$H(\omega) = |H(\omega)|\, e^{j\Phi}$$

where

$$\tan\Phi = \frac{\omega C}{K - \omega^2 M} = \frac{2\zeta(\omega/\omega_0)}{1 - (\omega/\omega_0)^2}.$$

Instead of displacement FRF, if we use the velocity FRF, we can write

$$\dot{\eta} = -j\omega H(\omega) F_0 e^{-j\omega t} = Y(\omega) F_0 e^{-j\omega t}$$

where the mobility relates to the impedance expression as $Y(\omega) = 1/Z(\omega)$. The relationship between mobility and receptance is

$$Y(\omega) = -j\omega H(\omega) = \omega |H(\omega)| e^{j\theta}$$

where $\theta = \Phi - \pi/2$, since $\exp(-j\pi/2) = -j$.

Similarly acceleration and excitation force are related through accelerance (or inertance) $\mathcal{A}(\omega) = (-j\omega)^2 \eta / F$ and

$$\mathcal{A}(\omega) = -j\omega Y(\omega) = \omega |Y(\omega)| e^{-j\alpha} = -\omega^2 H(\omega)$$

and $\alpha = \theta - \pi/2 = \Phi - \pi$.

►**Impulse or Step Excitation** Impulse response of an undamped simple oscillator can be considered as equivalent to response to an initial velocity and described as:

$$h(t) = \frac{1}{M\omega_0} \sin \omega_o t, \quad t > 0$$

and when damping is present:

$$h(t) = \frac{1}{M\omega_d} e^{-\zeta \omega_d t} \sin \omega_d t, \quad t > 0.$$

►**Arbitrary Excitation** Response $\eta(t)$ of a system to an arbitrary excitation can be obtained through a convolution integral of the input $F(t)$ and the impulse response $h(t)$ of the linear system:

$$\eta(t) = \int_{-\infty}^{\infty} F(\tau)\, h(t-\tau) d\tau = \int_{-\infty}^{\infty} F(t-\tau)\, h(\tau) d\tau. \qquad (2)$$

However, for the system to be causal, its impulse response also must be causal:

$$h(t) = 0 \quad \text{for} \quad t < 0.$$

The causality condition states that response must follow the excitation and not anticipate or precede it. Invoking causality, the limits of the convolution expression given in (2) can be modified as:

$$\eta(t) = \int_0^t F(\tau)\, h(t-\tau) d\tau = \int_0^t F(t-\tau)\, h(\tau) d\tau.$$

1.2 Causality

The most significant consequence of causality emerges from the Fourier transform of a causal impulse response function. Causal functions exhibit a strong linkage between the real and imaginary parts of their Fourier transforms, expressed in terms of Hilbert transform pairs or Kramer-Krönig relations.

The spectrum of $h(t)$, namely the frequency response function $H(\omega)$

$$H(\omega) = \int_{-\infty}^{\infty} h(t)e^{-j\omega t}dt$$

must reflect in its spectrum the causal properties of $h(t)$. Accordingly, its inverse transform must have the following properties (Pierce, 2008):

$$\frac{1}{2\pi}\int_{-\infty}^{\infty} H(\omega)\, e^{j\omega t}dt = \begin{cases} 0 & \text{if } t < 0 \\ h(t) & \text{if } t > 0 \end{cases}. \tag{3}$$

The frequency response function $H(\omega)$ that satisfies the causality condition in Eq. (3), can be obtained by Fourier transforming the $h(t)$ by expressing its causal property with a unit step, or Heaviside's, function $\mathcal{U}(t)$,

$$H(\omega) = \mathcal{F}\{h(t)\} = \mathcal{F}\{h(t)\mathcal{U}(t)\}$$

which can be expressed as a convolution of the Fourier Transforms of h and $\mathcal{U}$

$$H(\omega) = \mathcal{F}[h(t)] * \mathcal{F}[\mathcal{U}(t)]$$

where the Fourier transform of the unit step function is

$$\mathcal{F}[\mathcal{U}(t)] = \int_{-\infty}^{\infty} \mathcal{U}(t)e^{-j\omega t}dt = \left[\pi\delta(\omega) - j\left\{\frac{1}{\omega}\right\}\right].$$

It is understood that the second term in the brackets is interpreted as a distribution and when combined with a function its Cauchy principle value (**p.v.**) is taken. Substituting in the convolution equation above

$$H(\omega) = \frac{1}{2\pi}\int_{-\infty}^{\infty} H(\omega')\left[\pi\delta(\omega-\omega') - \left\{\frac{j}{\omega-\omega'}\right\}\right]\, d\omega'$$

$$H(\omega) = \frac{1}{2}H(\omega) - \frac{j}{2\pi}\,\mathbf{p.v.}\int_{-\infty}^{\infty}\left\{\frac{H(\omega')}{\omega-\omega'}\right\}\, d\omega'$$

$$H(\omega) = -\frac{j}{\pi}\ \mathbf{p.v.} \int_{-\infty}^{\infty} \left\{ \frac{H(\omega')}{\omega - \omega'} \right\} d\omega'$$

switching ω and ω' produces Hilbert transform of $H(\omega)$:

$$H(\omega) = \frac{j}{\pi}\ \mathbf{p.v.} \int_{-\infty}^{\infty} \left\{ \frac{H(\omega')}{\omega' - \omega} \right\} d\omega'.$$

Separating the real and imaginary parts of $H(\omega)$ yields:

$$H_R(\omega) = -\frac{1}{\pi}\ \mathbf{p.v.} \int_{-\infty}^{\infty} \frac{H_I(\omega')}{\omega' - \omega} d\omega' \tag{4}$$

$$H_I(\omega) = \frac{1}{\pi}\ \mathbf{p.v.} \int_{-\infty}^{\infty} \frac{H_R(\omega')}{\omega' - \omega} \omega'. \tag{5}$$

These relations between the real and imaginary parts of the frequency response function represent and assure the causality of the impulse response. They also show that when the real part of a causal function is known, the imaginary part can be obtained and vice versa.

Since in vibrations we deal with positive frequencies, the integral in the Hilbert transform pair (4) and (5) can be re-expressed so long as H has the property $H(-\omega) = H^*(\omega)$ that allows us to make the substitutions $H_R(-\omega) = H_R(\omega)$ and $H_I(-\omega) = -H_I(\omega)$ in the following derivations:

$$H_R(\omega) = -\frac{1}{\pi}\ \mathbf{p.v.} \int_{-\infty}^{0} \frac{H_I(\omega')}{\omega' - \omega} d\omega' - \frac{1}{\pi}\ \mathbf{p.v.} \int_{0}^{\infty} \frac{H_I(\omega')}{\omega' - \omega} d\omega'$$

$$H_R(\omega) = \frac{1}{\pi}\ \mathbf{p.v.} \int_{0}^{\infty} \frac{H_I(-\omega')}{\omega' + \omega} d\omega' - \frac{1}{\pi}\ \mathbf{p.v.} \int_{0}^{\infty} \frac{H_I(\omega')}{\omega' - \omega} d\omega'$$

$$H_R(\omega) = -\frac{1}{\pi}\ \mathbf{p.v.} \int_{0}^{\infty} H_I(\omega') \left(\frac{1}{\omega' + \omega} + \frac{1}{\omega' - \omega} \right) d\omega'$$

$$H_R(\omega) = -\frac{2}{\pi}\ \mathbf{p.v.} \int_{0}^{\infty} H_I(\omega') \left(\frac{\omega'}{\omega'^2 - \omega^2} \right) d\omega'. \tag{6}$$

Similarly,

$$H_I(\omega) = \frac{2}{\pi}\ \mathbf{p.v.} \int_{0}^{\infty} H_R(\omega') \left(\frac{\omega}{\omega'^2 - \omega^2} \right) d\omega'. \tag{7}$$

The last two equations (6) and (7) are known as the Kramers-Krönig relations that are used to describe causal impulse response functions (viz., Waters et al., 2005).

1.3 Damping Measurement Criteria

Since damping can only be measured *indirectly* by observing the response of a system, we review below the different means by which damping can be characterized. This section also introduces the terms commonly used in association with damping in vibrating systems. For example, in elastic systems, a measure of damping during time-dependent or cyclic motion is defined as the dissipated part $\triangle W$ of total elastic energy stored, W, during one cycle. Their ratio is called the *specific damping ratio*

$$\Psi = \frac{\triangle W}{W}$$

and the corresponding *loss factor* is defined as

$$\chi = \frac{\Psi}{2\pi} = \frac{1}{2\pi}\frac{\triangle W}{W}.$$

As described later, loss factor is related to the Q-value as:

$$\chi = \frac{1}{Q}.$$

Logarithmic Decrement Logarithmic decrement method is used in conjunction with decaying free vibration response of an oscillator taking advantage of the exponential nature of the response envelope as described in the transient or complementary solution expressions above:

$$\eta(t) = Ae^{-\gamma t}\cos(\omega_d t - \alpha).$$

Again, the quantity $\gamma = C/2M$ is the *decay constant* and its inverse is the *decay time*, τ. Hence, the amplitude variation depends on time exponentially, $\exp(-t/\tau)$.

The relative change of the amplitude in one period is called the *logarithmic decrement*, δ, which describes the rate of decay

$$\delta = \gamma T_d = \gamma 2\pi/\omega_d = (\pi C/\omega_0 M)(\omega_0/\omega_d).$$

Logarithmic decrement is also related to the *damping ratio* or *damping factor* ζ:

$$\delta = \frac{2\pi\zeta}{\sqrt{1-\zeta^2}}$$

and for very small values of $\zeta << 1$

$$\delta \approx 2\pi\zeta.$$

Logarithmic decrement is determined by measuring the response at two maxima on its envelope that are apart by one or more periods:

$$\delta = \frac{1}{n} \ln \frac{\eta^{(0)}}{\eta^{(n)}}$$

where n is the number of periods between the measurement positions with amplitudes $\eta^{(0)}, \eta^{(n)}$ measured at times $t = t_0, t_n$, respectively.

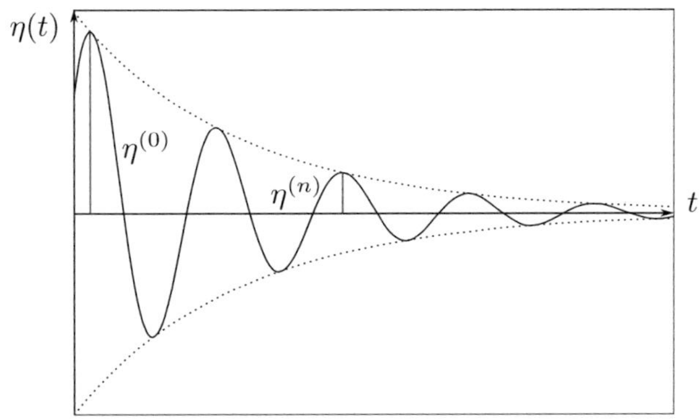

Figure 1. Logarithmic decay of amplitude of damped vibrations of a sdof oscillator.

Quality Factor Damping values can be directly obtained from the frequency response functions. Normalizing the displacement amplitude of a harmonically forced oscillator with the static displacement η_{st} gives:

$$\left|\frac{\eta(\omega)}{\eta_{st}}\right| = \frac{1}{\sqrt{[1-(\omega/\omega_0)^2]^2 + 4\zeta^2(\omega/\omega_0)^2}}$$

where $\eta_{st} = F_0/K$.

From the response measurements, the maximum amplitude is measured at approximately $\omega \approx \omega_0$

$$\left[\frac{|\eta(\omega)|}{\eta_{st}}\right]_{max} \approx \frac{1}{2\zeta} = Q$$

resulting in the *Q-value. Quality Factor* of the system, which is inversely related to the damping factor ζ, can now be directly measured from the FRF.

Using Q-value to determine amplitude is useful when damping is low and the resonant amplitudes are high. In frequency response plots, bandwidth of the resonance at half-power points provide another measurement method.

Frequency Response & Half-Power At half power points in an FRF, we can write

$$\frac{Q}{\sqrt{2}} = \frac{1}{\{[1-(\omega/\omega_0)^2]^2 + 4\zeta^2(\omega/\omega_0)^2\}^{1/2}} = \frac{1}{2\sqrt{2}\zeta}.$$

Expanding the denominator

$$(\omega/\omega_0)^4 - (\omega/\omega_0)^2(2-4\zeta^2) + (1-8\zeta^2) = 0$$

produces the roots:

$$(\omega_1/\omega_0)^2 = 1 - 2\zeta^2 - 2\zeta\sqrt{1+\zeta^2}$$

$$(\omega_2/\omega_0)^2 = 1 - 2\zeta^2 + 2\zeta\sqrt{1+\zeta^2}.$$

For small values of damping ratios, such that $\zeta < 0.05$

$$(\omega_1/\omega_0)^2 \approx 1 - 2\zeta$$

$$(\omega_2/\omega_0)^2 \approx 1 + 2\zeta.$$

Subtracting these equations

$$\omega_2^2 - \omega_1^2 = 4\zeta\omega_0^2$$

$$(\omega_2 - \omega_1)2\omega_0 = 4\zeta\omega_0^2.$$

Then the damping ratio follows as

$$\zeta = \frac{\triangle\omega}{2\omega_0}$$

where $\triangle\omega = \omega_2 - \omega_1$ is the bandwidth at half-power points. Effect of damping on displacement amplitude is illustrated in Fig. (2) with transfer functions for different damping values.

There are numerous other approaches to measure and quantify damping in materials and mechanical systems such as using Nyquist plots and Bode plots that can be found in the literature (viz., Mead, 1998; Nashif et al., 1985).

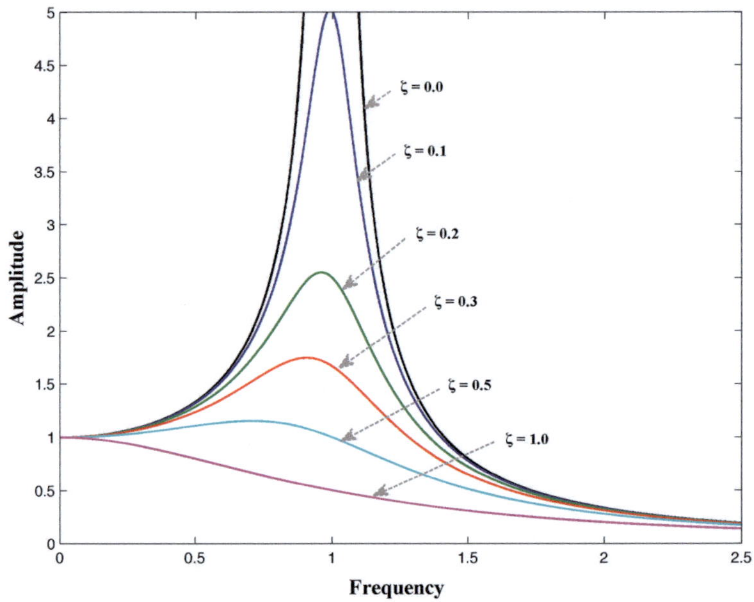

Figure 2. Frequency response function of a SDOF oscillator for different damping values.

2 Damping in Structures

Vibratory energy is dissipated in structures through numerous damping mechanisms. Several such damping mechanisms may take place simultaneously in a complex a mechanical system, or even in a simple component. The total damping that characterizes a system is a combination of energy dissipation effected by different mechanisms throughout the structure.

Damping mechanisms addressed here can be considered largely in three categories according to their fundamental mode of energy dissipation:

- dissipation within a solid,
- dissipation within or to a fluid medium, and
- dissipation at the interfaces between solids or between a solid and a fluid.

Some of the damping mechanisms described below are common to almost all systems and yet others are more specific to operating conditions and even length scales involved, particularly in cases of design and manufacture of MEMS and nano-scale devices.

2.1 Dissipation within Solids: Material or Internal Damping

Internal or material damping refers to inherent energy dissipation during cyclic motion or deformation of a material. The kinetic energy is irreversibly converted to thermal energy through one or more mechanisms. These mechanisms are associated with the internal structure of the material and have different length scales, such as those associated with dislocations, grain boundaries, or atomic motion. Internal damping also refers to those dissipations that arise from thermal, electronic and magnetic fields in the materials. As such, effectiveness of internal damping mechanisms range over different temperatures and frequencies.

Internal damping properties of materials can be enhanced by changing the molecular structure of the materials, or by, using alloys and viscoelastic materials. For example, carbon in cast iron is known to increase its damping properties. In the case of composites, however, macroscopic modification of the material structure provides for increased damping properties. Various fiber enhancements and foam-type structures are other examples.

In engineering, internal damping is generally characterized by a single value and, where appropriate, with frequency and temperature dependence. Internal damping is, in fact, an aggregate of energy dissipation due to numerous microscopic sources and mechanisms in a material.

This section attempts to delineate and explain some of these fundamental mechanisms of energy dissipation in types of solids that are of interest in structural vibrations, namely anelastic and viscoelastic materials. We exclude here nonlinear elasticity and plasticity and consider linear elasticity to explain the anelastic properties.

Ideal (linear) elasticity assumes an instantaneous relationship between stress and strain with a unique equilibrium value and a perfectly reversible deformation, i.e., with a complete recovery. On the other hand, *anelasticity* exhibits the same properties of recoverability and linearity but without the instantaneous response. In an anelastic material, a unique equilibrium value of strain corresponds to every stress (and vice versa), but the equilibrium is reached after a finite time rather than instantaneously. In *viscoelasticity*, in addition to time dependence, the initial equilibrium is not completely recoverable.

Anelastic solids, also described as thermodynamic solids, reach a thermodynamic equilibrium in response to a change in applied external forces. Through self-adjustment the solid reaches the new equilibrium through a process called as *anelastic relaxation*, which takes place a over period of *relaxation time*. For instance, when a constant stress is applied, anelastic relaxation manifests itself as a time-dependent (or frequency-dependent) equilibriation of strain and vice versa. This external manifestation of anelasticity reflects the thermodynamic equilibriation of internal variables in the solid. In this manner, for each stress level, a strain relaxation develops in conjunction with a new internal equilibrium of the solid. In anelastic solids the stress and strain relationship has different moduli corresponding to the initial and new equilibria and such a change in the modulus requires a transport process, for example, of atomic migration, dislocation displacements, grain boundary sliding, and phase transformations as well as thermal relaxation all of which lead to anelastic behavior.

Thermodynamic damping is the most fundamental internal damping mechanism as it can develop without the presence of material inhomogeneities. In the presence of material inhomogeneities, other peaks with respective Lorentz distributions appear. Among these, Zener peaks refer to dislocation relaxation, Bordoni peaks describe grain boundary relaxation, and Snoek peaks refer to defect pair reorientation. Presence of multiple such relaxation times may lead to multiple or broadened Debye peaks (Lifshitz and Roukes, 2000).

A three-parameter solid is used commonly to describe thermoelastic damping or thermoelastic relaxation of materials, which is referred as Zener model or standard linear solid. The Zener model consists of either a Maxwell model (spring and damping elements in series) parallel to another spring or a Voigt element (spring and damping elements in series) in series with another spring. Such models are found to effectively represent anelastic material behavior.

In what follows, we summarize the derivation of relaxation relations with Zener's anelastic solid model based on the thermodynamic model mentioned above. The resulting expressions describe dissipation due to inhomogeneities in a material, representing *thermoelastic damping*.

The thermoelastic damping expression has the form of a Lorentz distribution, with a maximum sometimes called the Debye peak, magnitude of which is the *relaxation strength* and the peak frequency corresponds to the inverse of relaxation. Different anelastic relaxations have different frequencies and temperature dependence. When multiple peaks develop, their effects are superposed.

2.2 Zener's Anelasticity Model

A homogeneous material subject to homogeneous stress undergoes thermal relaxation through heat exchange with its environment. However, if the stress or strain field is inhomogeneous, the resulting temperature gradients can lead to thermal relaxation through internal flow or "thermal currents" to reach new equilibrium from one part of the material to another (Nowick and Berry, 1972). Such coupling between stress fields and thermal fields in a solid gives rise to thermoelastic damping. *Thermoelastic coupling* is quantified by *thermal expansion coefficient*, α, as the coupling constant. Thermal expansion, change in strain due to change in externally applied temperature and the converse, and thermoelastic effect that describes small changes in temperature due to isentropic changes in dilatational stress are examples of thermoelastic coupling.

Thermoelastic effect can be considered by combining strain induced by temperature change with that obtained under pure elastic conditions, by considering strain to depend only on stress, σ, and temperature, T,

$$\epsilon = \frac{\sigma}{E_R} + \alpha \triangle T \tag{8}$$

where E_R is the *relaxed* or *isothermal modulus*, α represents the linear thermal expansion coefficient, and $\triangle T$ is the deviation from standard temperature (Zener, 1948).

The variation in temperature, $\triangle T$, is caused by either diffusion or change in strain. Temperature change caused by diffusion (or equalization or relaxation) of thermal fluctuations, can be approximated by

$$\left(\frac{d}{dt}\triangle T\right)_{diffusion} = -\frac{1}{\tau}\triangle T \tag{9}$$

where τ is the *relaxation* time. Relaxation time may have different values depending on the stress and strain restrictions imposed. For example, τ_ϵ represents the relaxation time for stress relaxation and temperature relaxation under constant strain. Analogously, a relaxation time τ_σ is defined for strain relaxation and temperature relaxation under constant stress. Just as an increase in temperature leads to an increase in length, an *adiabatic* increase in length leads to a decrease the temperature:

$$\left(\frac{d}{dt}\triangle T\right)_{adiabatic} = -\gamma\dot{\epsilon} \tag{10}$$

with $\gamma = (\partial T/\partial\epsilon)_{adiabatic}$. Combining the two mechanisms of temperature change in (9) and (10), we have:

$$\frac{d}{dt}\triangle T = -\frac{1}{\tau_\epsilon}\triangle T - \gamma\dot{\epsilon}. \tag{11}$$

Isolating $\triangle T$ from the coupled equation (8) and substituting it and its derivative in equation (11) above eliminates $\triangle T$ and leaves us with the stress-strain relationship:

$$E_R\,\epsilon + E_U\,\tau_\epsilon\,\dot{\epsilon} = \sigma + \tau_\epsilon\,\dot{\sigma} \tag{12}$$

where the unrelaxed modulus is

$$E_U = (1 + \alpha\,\gamma)E_R.$$

Rewriting Eq. (12) leads to an expression:

$$\sigma + \tau_\epsilon\dot{\sigma} = E_R(\epsilon + \tau_\sigma\dot{\epsilon}) \tag{13}$$

which describes the deviation from an elastic (Hook) solid into an anelastic, or standard, solid described by Zener (1948).

The relationship between τ_σ and τ_ϵ can be obtained by integrating both sides of (13) over a very small time δt, which yields

$$\int_0^{\delta t}(\sigma + \tau_\epsilon\dot{\sigma})dt = E_R\int_0^{\delta t}(\epsilon + \tau_\sigma\dot{\epsilon})dt. \tag{14}$$

The first term on each side of (14) vanishes as $\delta t \to 0$ leaving

$$\tau_\epsilon\,\triangle\sigma = E_R\,\tau_\sigma\,\triangle\epsilon.$$

In this case, the relation between the changes in stress and strain take place over such a short time that there is no time for relaxation to take effect and the relationship between them, an adiabatic process, is through an *unrelaxed elastic modulus*, E_U as

$$\triangle\sigma = E_U \, \triangle\epsilon$$

and

$$\frac{\tau_\sigma}{\tau_\epsilon} = \frac{E_U}{E_R}. \tag{15}$$

Deviation of the ratio in (15) from unity indicates relaxation of stress or strain.

Assuming $\epsilon = \dot{\epsilon} = 0$, equation (13) becomes:

$$\sigma + \tau_\epsilon \dot{\sigma} = 0$$

with the solution

$$\sigma(t) = \sigma_0 e^{-t/\tau_\epsilon}$$

where $\sigma_0 = \sigma(0)$. When an initial strain ϵ_0 is suddenly applied at $t = 0$, the relaxation of stress follows

$$\sigma(t) = E_R\epsilon_0 + (\sigma_0 - E_R\epsilon_0)e^{-t/\tau_\epsilon}.$$

After the relaxation is completed, the relationship is simply $\sigma(t) = E_R\epsilon_0$ and, hence, the modulus is the relaxed elastic modulus. Analogously, for a suddenly applied stress σ_0, the corresponding strain time history becomes (Beltzer, 1988)

$$\epsilon(t) = \frac{1}{E_R}\sigma_0 + \left(\epsilon_0 - \frac{1}{E_R}\sigma_0\right) e^{-t/\tau_\sigma}.$$

Substituting harmonic excitation and response expressions

$$\sigma(t) = \sigma_0 e^{-j\omega t} \text{ and } \epsilon(t) = \epsilon_0 e^{-j(\omega t - \phi)}$$

in the anelastic solid expression (13) yields:

$$\sigma_0 = \frac{1 - j\omega\tau_\sigma}{1 - j\omega\tau_\epsilon} E_R \, \epsilon_0 = \mathcal{E}\epsilon_0. \tag{16}$$

The angle by which strain lags behind stress is a measure of internal friction:

$$\tan\phi = Q^{-1} = \frac{\Im\{\mathcal{E}\}}{\Re\{\mathcal{E}\}} = \frac{\omega(\tau_\epsilon - \tau_\sigma)}{1 + (\omega\tau_\epsilon)(\omega\tau_\sigma)}$$

which can be simplified as a Lorentzian distribution:

$$\tan\phi = \frac{E_R - E_U}{\bar{E}} \frac{\omega\bar{\tau}}{1 + (\omega\bar{\tau})^2} \tag{17}$$

with the geometric means: $\bar{\tau} = \sqrt{\tau_\sigma \tau_\epsilon}$ and $\bar{E} = \sqrt{E_R E_U}$. The first part of the internal loss expression (17) represents the relative difference in relaxed and unrelaxed elastic moduli and the second part represents its frequency dependence. The relative difference of the moduli is defined as the *relaxation strength*

$$\triangle_E = \frac{E_R - E_U}{\bar{E}}.$$

The loss in Eq.(17) has a maximum value at $\omega\bar{\tau} = 1$:

$$(\tan\phi)_{max} = Q^{-1}_{max} = \frac{1}{2}\triangle_E.$$

The magnitude of the complex modulus $\mathcal{E}$ in (16), the ratio of stress to strain, is expressed as:

$$|\mathcal{E}| = \frac{1+\omega^2\tau_\sigma^2}{1+\omega^2\tau_\sigma\tau_\epsilon} E_R = E_U - \frac{E_U - E_R}{1+\omega^2\bar{\tau}^2}$$

for which approximate values are expressed for low and high frequencies as (Beltzer, 1988):

$$\left.\begin{aligned} E_U - |\mathcal{E}| &= \frac{E_U - E_R}{(\omega\bar{\tau})^2} \\ \tan\phi &= \frac{E_U - E_R}{\bar{E}\,(\omega\bar{\tau})} \end{aligned}\right\} \qquad \omega\bar{\tau} \gg 1$$

$$\left.\begin{aligned} |\mathcal{E}| - E_R &= (E_U - E_R)\,(\omega\bar{\tau})^2 \\ \tan\phi &= \frac{E_U - E_R}{\bar{E}}\,(\omega\bar{\tau}) \end{aligned}\right\} \qquad \omega\bar{\tau} \ll 1$$

which can be further simplified as:

$$|\mathcal{E}| = \begin{cases} E_U & \omega\bar{\tau} \gg 1 \\ E_R & \omega\bar{\tau} \ll 1 \end{cases}.$$

2.3 Thermoelastic Damping

It is known that a homogeneous material under homogeneous stress can undergo thermal relaxation only by heat exchange with its surroundings since there is no other heat flux path. However, if the stress field changes periodically, as it does during vibration, it gives rise to periodic changes in temperature, even if the material is homogeneous, resulting in temperature gradients. Heat flux due to temperature gradients lead to increase in entropy indicating an increase of internal energy by reducing mechanical energy (Zener, 1940).

Thermoelastic damping that develops during vibration of a homogeneous system has been calculated for transverse and longitudinal waves by computing the values for relaxation strength and relaxation time. An example of thermoelastic damping is given for a beam of thickness h that vibrates in flexure at a frequency f (Zener, 1937, 1938)

$$Q^{-1}(f,T) = \triangle_T \frac{f \cdot f_0}{f^2 + f_0^2}$$

which has the same functional form as a Lorentz distribution and a Debye peak as a function of frequency given by:

$$f_0 = \frac{\pi k_T}{2h^2 \rho C_p}$$

where ρ is the density and C_p is the specific heat capacity under constant pressure or stress and the relaxation strength is defined as

$$\triangle_T = \alpha^2 \frac{E_U T}{\rho C_p}.$$

The relaxation time, in terms of the Debye peak frequency, f_0, is

$$\tau_T = 1/2\pi f_0 = h^2/\pi^2 D_{th}$$

with the thermal diffusivity $D_{th} = k_T/\rho C_p$ and k_T is the thermal conductivity. These relations are used in various forms to determine thermoelastic damping in mechanical systems, particularly in micro- and nano-mechanical systems where it can have a significant influence.

2.4 Viscoelastic Damping

The basic relations between stress and strain no longer hold through a simple proportionality of a modulus for linear viscoelastic materials under a time-dependent stress and strain. Compared with a completely recoverable behavior of an *anelastic* solid, *viscous* and *viscoplastic* properties describe nonrecoverable behavior and *viscoelasticity* falls in between and may have both recoverable and nonrecoverable parts. Such a partially recoverable behavior can be modeled using a four-parameter model that consists of a Voigt model in series with a spring and a damping element, which is commonly used to describe viscoelastic behavior. Origin of such models may be explained through a partial differential equation of arbitrary order (Nowick and Berry, 1972).

$$a_0\sigma + a_1\frac{\partial\sigma}{\partial t} + a_2\frac{\partial^2\sigma}{\partial t^2} + \cdots + a_n\frac{\partial^n\sigma}{\partial t^n}\cdots = b_0\epsilon + b_1\frac{\partial\epsilon}{\partial t} + b_2\frac{\partial^2\epsilon}{\partial t^2} + \cdots + b_n\frac{\partial^n\epsilon}{\partial t^n}\cdots$$

This differential stress-strain equation can be used to express most of the complex relationships for a viscoelastic material.

For example, by keeping only a_0 and b_0, it describes an elastic solid:

$$\sigma = E\epsilon$$

or, keeping a_0 and b_1 and setting other coefficients to zero produces Newton's law of viscosity with the coefficient of viscosity η_0:

$$\sigma = \eta_0\left(\frac{\partial}{\partial t}\right)\epsilon.$$

The Voigt model that is commonly used to describe viscoelastic solids has a spring and a dashpot with a_0, b_0, b_1

$$\sigma = \left[E + \eta_0\left(\frac{\partial}{\partial t}\right)\right]\epsilon = E\epsilon + \eta_0\,\dot{\epsilon}.$$

The anelastic solid described earlier has an additional spring term that makes it a Voigt model in series with a spring:

$$\sigma + \tau_\epsilon\,\dot{\sigma} = E_R\,\epsilon + E_U\,\tau_\epsilon\,\dot{\epsilon}.$$

where the relaxation time τ_ϵ and the relaxed and unrelaxed elastic moduli E_R and E_U are those described earlier.

2.5 Friction Damping

Miscroslip vs. Sliding Friction or contact damping refers to conversion of the kinetic energy associated with the relative motion of two surfaces in contact to thermal energy. Contact stresses generate inhomogeneous stress-strain fields on and near the surfaces leading to temperature gradients and, thus, transport of thermal energy from the contact areas.

Contacts that generate friction damping can be characterized by the relative motion between the surfaces: microslip and sliding. When contact is between nominally conforming surfaces that do not have a relative rigid-body motion, contact behavior is sometimes described as micromotion or microslip, and may not reach slip or sliding conditions. Friction remains more or less in the "static" range and is associated with the tangential stiffness of the contact zone. Examples of microslip can be found in bolted or riveted joints, braided wire ropes, and inserted gas turbine blades.

When the contacting surfaces have a relative whole-body motion as in the case of brakes, damper rings in gears, and Lanchester dampers, full slip can develop between the surfaces. In such cases, friction damping has a preferred range of normal contact force within which it becomes most effective. Below such an optimum normal force, excess relative motion at the interface develops without significant energy dissipation. Above it, excess pressure inhibits the relative sliding motion for friction to act as an effective damper (Akay, 2002).

The simplest case to illustrate the effects of friction on vibration is an oscillator with a friction damper as represented in the following equation of motion (Den Hartog, 2013):

$$m\ddot{x} + kx = F_0 \cos\omega t - \mu N \mathbf{sgn}(\dot{x}).$$

Action of a friction damper on a simple oscillator is analogous to the fluid damper where the drag force acts against the direction of motion of the oscillator. There is rich literature on the dynamic response of systems in the presence of sliding friction including on the damping effects of friction (viz., Dowell and Schwartz, 1983b,a; Dowell, 1986, 1983).

Focusing on microslip or quasi-static contact damping, knowledge of friction characteristics is necessary to predict its damping effects, much like any other dynamic problem that involves friction. However, this knowledge is normally obtained through measurements, since we do not yet have an acceptable model of friction that is based on first principles.

Friction damping that develops during microslip, by necessity, is associated with the tangential stiffness between the contacting surfaces. The tangential forces that develop not only resist relative motion but effectively change the boundary conditions and the resonant frequencies in the structure and thus lead to nonlinearities in the response.

Considering a simple oscillator with friction force resisting its motion, its motion can be described by:

$$m\ddot{x} + c\dot{x} + kx = f_0 \cos\omega t - f_\mu(t)$$

where the nonlinear friction force is approximated by a spring and a viscous damper acting in the direction of motion

$$f_\mu(t) \approx k_e x + c_e \dot{x}$$

where k_e and c_e represent the effective values of stiffness and damping, respectively, and are found from measurements (Filippi et al., 2004).

Granular Damping Granular damping, or particle damping, in vibrations refers to absorption of waves or oscillations through the use of a collection of particles or granular materials. Granular materials are known to provide effective means of dissipation of vibratory energy largely through inelastic collisions and friction among the granules.

Physical properties of individual granular materials very much differ from their ensemble properties in terms of elasticity and dissipation. The collective behavior is governed not only by the physical properties of individual grains but also through their interactions with each other and the manner by which the ensemble is contained. These parameters include friction between the particles, filling factor or packing force, shape of the device in which they are contained and frequency and amplitude of the vibrations to which they are subjected. Based on these factors, the collective behavior of granular particles may be a plug-like solid, a fluid, or a gas, each phase having different dissipation characteristics. Elastic particles, such as ball bearings, absorb vibration energy effectively only when exposed to a vibration field collectively, similar to the so-called beanbag absorbers. Although each ball bearing may rebound upon impact on an elastic solid, when collected in a flexible container such as a bag, they behave inelastically, due to friction among the particles and due to diffusion of their energy (Jaeger et al., 1996).

Packing force of granular materials in a rigid container determines their density and, thus, directly affects their collective behavior. Very high packing forces can severely limit relative motion among the grains and thus reduce damping effectiveness.

2.6 Damping in Fluid Media

Radiation Damping Radiation damping describes a broad range of damping effects such as those associated with gravitational fields, quantum mechanics and optics, but we focus here on radiation damping effects on engineering structures. Radiation damping is a term also used in connection with civil engineering piles and footings to describe soil-structure interaction.

Radiation damping of a structure refers to energy lost from a vibrating surface through sound radiation. As a simple demonstration, we consider a sdof oscillator in a waveguide with the usual properties K, M excited by a harmonic force $F_0 e^{i\omega t}$ such that $2\pi c/\omega > D$, where c is the speed of sound and D is the diameter of the duct, so that radiation can be assumed to be a plane wave. Since the pressure acts on the mass against its surface, the forces arising from radiation have opposite signs and add together to form the total acoustic radiation force and can be expressed as

$$F_a = 2\pi(D^2/4)\rho c\,\dot{\eta}$$

where ρ is the density of the fluid medium in the duct, and η represents the displacement of the oscillator. The equation of motion for the oscillator then becomes

$$M\frac{d^2\eta}{dt^2} + (\pi D^2/2)\rho c\frac{d\eta}{dt} + K\eta = F_0 e^{i\omega t}.$$

The harmonic solution yields a compliance expression:

$$\left|\frac{\eta}{F}\right| = \frac{1}{\sqrt{(K - M\omega^2)^2 + (\pi D^2 \rho c\omega/2)^2}}.$$

We note here that radiation damping is frequency dependent, which becomes significant in cases of radiation damping that involves higher-order systems, such as plates and beams that have their own modal and critical frequencies.

The rather straightforward approach and the expression described above takes a more complicated form in the case of more complex structures. Since radiation damping results from loss of energy radiated into the adjacent medium, its prediction requires modeling vibrations of the structure coupled with the medium. In media such as air, radiation damping from heavy machinery may be negligible whereas in water or soil, it is more significant. The fluid-structure coupling determines both the frequencies and magnitude of damping due to radiation. Although approximate expressions are available, more complete expressions for damping require solution of the coupled equations.

By defining a damping factor for radiation damping as the ratio of acoustic energy radiated to the maximum kinetic energy stored per cycle of vibration in such a structure, an approximate value can be given for the first mode of a rectangular plate with fixed edges as (Mangiarotty, 2005; Mead, 1998):

$$\delta_{ac} = 1.155X10^{-5}\frac{\rho_0}{\rho_m}\sqrt{\frac{E}{\rho_m}}\left(N + \frac{1}{N}\right)$$

where $N = a/b$ is the ratio of its length to width and ρ_m and E are the material elastic properties and ρ_0 is the density of the fluid medium. For a simply-supported panel a similar result is given as:

$$\delta_{ac} = 1.155X10^{-5}\frac{\rho_0}{\rho_m}\sqrt{\frac{E}{\rho_m}}\left(N^2 + \frac{2}{3} + \frac{1}{N^2}\right)^{1/2}.$$

We note that acoustic damping factor of a given material in these expressions depends on the density of the medium and the ratio of its length to width of the plate.

Fluid Damping - Drag Different from radiation damping, fluid damping refers to energy lost through drag forces on a solid body moving in a fluid. The fundamental mechanisms may involve vortex shedding, but not always.

Drag force F_D is proportional to the square of the relative velocity and, thus, is dominant at high relative velocities and may be negligible at low velocities. Drag force is also proportional to the density of the fluid medium and to the cross-sectional area of the body facing the flow:

$$F_D = -c_D S \left(\frac{1}{2} \rho v^2 \right).$$

The drag coefficient c_D depends on the shape of the body and usually found empirically. Because drag force acts against the relative motion, it is usually accompanied by a sign switching function that depends on the direction of velocity v. For a unidirectional motion, drag force is:

$$F_D = -c_D S \left(\frac{1}{2} \rho v^2 \right) \mathbf{sgn}(v).$$

The corresponding equation of motion for a simple oscillator becomes:

$$m\ddot{x} + \frac{1}{2} c_D \rho S \, \dot{x}^2 \, \mathbf{sgn}(\dot{x}) + kx = 0.$$

Analogous to the dry friction problems, the above equation can be expressed to obtain the phase plane for $(x, \dot{x})$ by a first-order differential equation for $\dot{x}^2$:

$$\frac{d\dot{x}^2}{dx} + \frac{1}{2} \frac{c_D \rho S}{m} \dot{x}^2 \, \mathbf{sgn}(\dot{x}) = -2 \frac{k}{m} x.$$

Examples of its solution can be found elsewhere (Kneubühl, 1997).

Squeeze-Film or Fluid-Film Damping

►**Air-Film Dampers** It is known that when two plates are brought in close contact with each other (for example, through spot-welding), the vibrational damping rate and the attenuation of radiated sound from the plates are markedly enhanced. This method of vibration damping and noise reduction can be used effectively in several applications where standard viscoelastic layer damping cannot be used, such as in hostile chemical environments and/or at high temperatures. For example, vibrations of combustion chambers can be reduced by applying patches of metal plates at strategic locations on the chambers. Other applications include damping of vibrations in centrifugal separators and circular saw blades. In the latter case, damping can be obtained by means of the thin air layer between the blade and a rigid block placed in close proximity of the blade (Meins, 1963; Allen, 1977). They are particularly useful to reduce vibration of moving devices, such rotating shafts.

The principle of operation is based on the motion of fluid in a narrow gap between a moving vibrating surface and a fixed one. As the moving surfaces oscillate toward and away from the fixed surface, pressure in the fluid layer periodically increases and decreases, squeezing the fluid out and pulling it back in, respectively. When the layer thickness is small, fluid motion is largely normal to the oscillation direction and parallel to the surfaces. The energy required to pump the fluid either through its edges or from a region of compression to a region of rarefaction under a vibrating plate is supplied by the motion of the plate and results in the damping of its motion.

The flow impedance per unit length of a thin film of fluid with density ρ_l can be expressed as (Morse, 1986; Ingard and Akay, 1987)

$$Z = R + jX = -j\omega\rho_l/(1 - F)$$

with

$$F(x) = x(1 + j)\tanh[(1 - j)/2x]$$

where $x = d_\nu/D$ is the ratio of viscous boundary layer and fluid layer thickness. Approximate expressions of F are given as:

$$F(x) \approx \begin{cases} (1 + j)x & x << 1 \\ 1 + (j/6x^2) - (1/30x^4) & x >> 1 \end{cases}$$

with these values the flow impedance becomes:

$$Z = \begin{cases} \rho_l\omega\,[(d_\nu/D) - j] & d_\nu/D << 1 \\ \rho_l\omega\,\left[12\mu/D^2 - j(1.2\omega)\right] & d_\nu/D >> 1 \end{cases}$$

where μ is the viscosity of the fluid.

At very low frequencies, where traditional damping techniques are not as effective, flow resistance per unit length becomes $R = 12\mu/D^2$ and the corresponding damping factor for bending vibrations of a plate takes the form:

$$\delta \sim (\lambda/D)(\lambda\mu/D^2)(1/\omega_0 m)$$

where λ represents the bending wavelength of the plate with mass m per unit length vibrating at frequency ω_0.

Q-value of the plate due to fluid layer damping is obtained as

$$Q = \omega_0 m D^3/\mu\lambda^2.$$

Further damping can develop from acoustic streaming that develops in a squeeze film between two flat surfaces. Beyond a certain oscillation frequency and amplitude combination, in addition to the periodic flow in and out of the layer edges, continuous streams develop changing the flow pattern in the layer (Akay and Xu, 1998).

2.7 Other Fluid Damping Mechanisms

It is worth mentioning other fluid damping mechanisms details of which can be found in literature.

►**Couette Flow Damping** Similar to squeeze film damping but the surfaces move parallel to each other developing a unidirectional flow. Its effects on MEMS devices continue to be of interest.

►**Damping in Porous Materials: Biot Damping** Named after M. A. Biot who developed the theory of mechanics for porous media, Biot damping describes dissipative effects of a material with fluid-filled pores (Biot and Tolstoy, 1992). Damping in porous materials results from a combination of the damping in the solid porous structure, the fluid that saturates it, and the relative motion at the interface of the two (Göransson, 2006). The skeleton of the porous solid carries the stress waves as the acoustic pressure waves propagate through the fluid medium. The configuration and properties of the porous material and the fluid pressure influence the energy balance between the two media and thus the resulting dissipation. Biot damping is particularly significant in sound transmission and soil consolidation problems in geophysics.

►Aerodynamic & Hydrodynamic Damping Aerodynamic damping is commonly used in connection with wind effects on structures such as buildings, airplanes, and cables. The forces generated by wind may attenuate vibrations but more commonly induce vibrations and instability. When the wind forces cause instability, they are described as negative damping forces. In aerodynamics, terms pitch-, yaw-, and roll-damping refer to moments due to differential forces rather than dissipation of energy.

Similarly, hydrodynamic damping used in describing motion of marine vessels, typically consist of any combination of radiation damping into the fluid medium, viscous damping and its contributions to turbulent boundary layer, the so-called wave drift damping that describes added resistance due to waves in sea, and damping that results from vortex shedding (Kareem and Gurley, 1996).

►Damping with Magnetorheological & Ferro Fluids Effective viscous and stiffness properties of a fluid, such as oil, can be adjusted as desired when mixed with ferrous particles and subjected to magnetic field. Magnetorheological (MR) fluids refer to liquids with micron-size particles unlike ferro fluids (FF) which contain nano-sized particles. The ability to control their physical characteristics allows their use as lubricants, seals and dampers, for example.

Damping Materials and Devices Choice of damping treatment is governed by the amount of vibration energy to be dissipated and by the operating conditions under which a damping treatment is used. Environmental conditions with hazardous chemicals or extreme temperatures may prohibit use of some of the treatments.

The basic mechanisms of dissipation are utilized in many different forms as mentioned above. In addition, use of tuned dampers, impact dampers, pendulum dampers also dissipate energy largely through momentum exchange between moving parts and the dissipation mechanisms within them. One can consider sloshing fluids in the same category.

3 Dissipation Mechanisms

Dissipation refers to conversion of kinetic energy associated with motion, oscillations or waves to heat, which is described as the thermalized oscillations of molecules of the medium in which the propagation takes place. Whether in a solid or a fluid, the conversion process leads to an increase in the kinetic energy of the molecules in the medium. In this sense, dissipation describes the conversion of ordered kinetic energy at macroscopic scale to disordered kinetic energy at the molecular scale. As described in the previous section, damping models contain empirical constants that describe dissipation through, for example, viscosity, coefficient of friction, or material losses in solids.

Modeling dissipation based on first principles, requires knowledge of the heat capacity of the medium, which describes the increase in its internal energy per unit temperature. The various damping mechanisms described earlier involve some form of external excitation that leads to an increase in internal energy of the medium. Since internal energy is proportional to the average kinetic energy of its molecules, understanding the dynamic response of molecules and accurate prediction of their behavior constitutes a key to modeling dissipation.

Dissipation is fundamentally a nonlinear process during which energy transfer takes place irreversibly (viz., Çelik and Akay, 2000). The first-principles based quantitative models that can accurately predict dissipation await further advances in molecular dynamics simulations. However, qualitative relationships exist for thermal energy at the molecular scale in terms of average kinetic energy of molecules. Simulation studies that consider solids as a set of oscillators in a lattice have been continuing since the introduction of the FPU problem (Fermi, Pasta, and Ulam, 1955; Ford, 1992). Simulations invariably assume that the vibrations of molecules are in thermal equilibrium, or "thermalized," such that all states of the molecules have equal probability of having equal energy. As a result, investigations of thermalization process look for conditions leading to energy equipartitioning. Chaotic response of, say, molecules in a lattice describing a solid, is considered by some as the indicator of thermalization. Later the KAM theorem explained the role of nonlinearities in such models indicating that not all nonlinearities lead to thermalization of their oscillations (Kolmogorov, 1979; Arnol'd, 1963; Moser, 1962; Salamon, 1986).

Notwithstanding the nonlinearity of a dissipation process at molecular level, linear dissipative media have long been modeled with a set of independent linear oscillators. In most of these approaches, a dissipative property emerges from the collective behavior of an ensemble of independent linear oscillators. Such models have been developed to describe, for example, Landau damping, dissipation in Langevin equation, and virtual damping due to phase diffusion.

A final point to note about these models is that the linear oscillators describing a dissipative environment must have a continuous distribution and, by implication, the number of oscillators is very large, reaching infinity. Models based on such requirements fall short of describing conditions where the oscillator numbers are not necessarily very large. Special cases when a finite number of linear oscillators can be used to describe dissipation of a medium with near irreversibility is also described later in this section.

3.1 Linear Modeling of Dissipative Systems

In its most fundamental form, classical or quantum dissipation can be considered as an interaction of a particle with its environment. An approach, first introduced by Feynman and Vernon (1963), is to model the environment as a continuous set of non-interacting, linear independent oscillators into which energy flows from the particle of interest. Presumption of infinite number of linear oscillators in the environment permits irreversible energy flow into it, which then acts as an energy sink as briefly summarized below.

Motion for a harmonically bound particle of mass M and spring constant K, with N oscillators attached to it bilinearly, can be described by a set of coupled equations:

$$M\ddot{x}(t) + Kx(t) = \sum_n m\omega_n^2(x - q_n) + F(t) \tag{18}$$

$$\ddot{q}_n(t) + \omega_n^2 q_n(t) = \omega_n^2 x(t) \qquad n = 1.2.3\ldots N \tag{19}$$

where $F(t)$ is an external force and the summation terms represent the force by the oscillators, each with displacement $q_n(t)$, describing the environment. The force by the oscillators consists of a dissipative, or systematic, part and a fluctuating, or random, part described below.

To determine the force exerted by the oscillators on the bound particle, response of each oscillator is expressed as:

$$q_n(t) = -\int_0^t \cos\omega_n(t-\tau)\,\dot{x}(\tau)\,d\tau + q_n(t_0)\cos\omega_n(t'-t_0) + \frac{\dot{q}_n(t_0)}{\omega_n}\sin\omega_n(t'-t_0)$$

where $q_n(t_0)$ and $\dot{q}_n(t_0)$ are values of $q_n(t)$ and $\dot{q}_n(t)$ at $t' = t_0$. Substituting for $q_n(t)$ from above in equation (18), with $t = t' - t_0$:

$$M\ddot{x}(t)+\left(M\Omega_0^2+\sum_{n=1}^{N} m\omega_n^2\right)x(t)+\int_0^t \sum_{n=1}^{N} m\omega_n^2 \cos\omega_n(t-\tau)\dot{x}(\tau)d\tau$$
$$=\sum_{n=1}^{N} m\omega_n^2\left[q_n(0)\cos\omega_n t+\frac{\dot{q}_n(0)}{\omega_n}\sin\omega_n t\right]+F(t). \tag{20}$$

We now set

$$\Gamma'(t)=\sum_{n=1}^{N} m\omega_n^2\cos(\omega_n\tau),\ \Pi'(t)=\sum_{n=1}^{N} m\omega_n^2\left[q_n(0)\cos\omega_n t+\frac{\dot{q}_n(0)}{\omega_n}\sin\omega_n t\right] \tag{21}$$

and the system frequency

$$\Omega^2=\Omega_0^2+\sum_{n=1}^{N}\frac{m}{M}\omega_n^2$$

where the prime indicates the discrete summations for Γ and Π. With these notations the equation of motion (22) takes the form:

$$M\ddot{x}(t)+M\Omega^2(t)+\int_0^t \Gamma'(\tau)\dot{x}(t-\tau)d\tau=\Pi'(t)+F(t). \tag{22}$$

Coupled equations of motion for an unforced ($F=0$) particle of mass M with attached oscillators that are initially quiescent has the form:

$$M\ddot{x}(t)+M\Omega^2 x(t)+\int_0^t \Gamma'(\tau)\dot{x}(t-\tau)d\tau=0$$

$$\ddot{q}_n(t)+\omega_n^2 q_n(t)=\omega_n^2 x(t) \qquad n=1.2.3\ldots N.$$

The solution of these equations show recurrence: energy is exchanged between the particle and the attached oscillators. However, when the attached oscillators are distributed continuously with a particular frequency distribution, energy is irreversibly absorbed by the attached particles even in the absence of a dissipation source in the classical sense.

Feynman and Vernon (1963) showed how a continuous distribution of oscillators describe loss using a perfectly linear, undamped set of oscillators. A simple demonstration of their approach is made using the distribution introduced by Caldeira and Leggett (1983) for the oscillators as $G(\omega)=2M\gamma_0/\pi m\omega^2$ in making the summation Γ' an integral Γ:

$$\sum_n \to \int_0^\infty G(\omega)\, d\omega.$$

Expression $\Gamma(\tau)$ can be written and evaluated as

$$\Gamma(\tau) = \int_0^\infty G(\omega) m\omega^2 \cos(\omega\tau) d\omega = 2M\gamma_0\, \delta(\tau)$$

where the constant γ_0 is analogous to a velocity-dependent damping constant that introduces the familiar viscous damping into the equation of motion:

$$\ddot{x}(t) + \gamma_0 \dot{x}(t) + \Omega^2 x(t) = 0.$$

As shown later in this section, in cases where the summation can not be substituted by an integral, special frequency distributions are shown to closely imitate the dissipative behavior of a continuously distributed, i.e., infinite number of oscillators (Carcaterra and Akay, 2004, 2007, 2011; Carcaterra et al., 2006; Koç et al., 2005; Akay et al., 2005).

►**Dynamics of an Ensemble of Oscillators** It is instructive to examine the behavior of the attached oscillators independently of the particle to which they are attached.

Consider an ensemble of simple oscillators each with a slightly different frequency, ω. If the entire set of oscillators start with the same initial conditions, their motions become out of phase with time and the average displacement of the ensemble decays, displaying a process known as *decoherence* (Ng, 2006, 2010; Carcaterra and Akay, 2004).

Motion of an oscillator in the ensemble with a displacement y and angular frequency ω is described by

$$\ddot{y} + \omega^2 y = 0$$

with the general solution

$$y(t) = y_0 \cos \omega t + \dot{y}_0\, \frac{\sin \omega t}{\omega}$$

where y_0 and $\dot{y}_0$ represent the initial values of y and $\dot{y}$, respectively.

If all the oscillators start with the same initial conditions, the average displacement of the ensemble in time becomes

$$< y(t) >= y_0 \mathcal{H}(t) \int \rho(\omega) \cos \omega t\, d\omega, \qquad \dot{y}_0 = 0$$

alternatively, the average velocity is

$$< \dot{y}(t) >= \dot{y}_0 \mathcal{H}(t) \int \rho(\omega) \cos \omega t \, d\omega, \qquad y_0 = 0$$

where $\mathcal{H}$ represents the Heaviside step function, $\rho(\omega)$ is the distribution of the oscillation frequency in the ensemble, with the constraint

$$\int_{-\infty}^{\infty} \rho(\omega) \, d\omega = 1.$$

The impulse response of the ensemble is always real and vanishes for $t < 0$.

$$G(t) = \mathcal{H}(t) \int \rho(\omega) \cos \omega t \, d\omega, \qquad \dot{y}_0 = 0.$$

For ensemble frequencies having a Lorentz distribution with half-width-at-half-maximum $\triangle\omega$, we have

$$\rho(\omega) = \frac{1}{\pi} \frac{\triangle\omega}{(\omega - \bar{\omega})^2 + (\triangle\omega)^2}$$

where $\bar{\omega}$ is the mean angular frequency of the ensemble. For this distribution, the impulse response is

$$G(t) = \mathcal{H}(t) \Re e^{j(\bar{\omega} + j\triangle\omega)t} = \mathcal{H}(t) e^{-(\triangle\omega)t} \cos \bar{\omega} t$$

which shows that the average displacement of the oscillators decay with a relaxation time $\tau^{-1} \approx \triangle\omega$. The process described above is also called kinematic decoherence (Ng, 2006, 2010).

Phase Damping Instead of frequency distribution, if all oscillators are assumed to have the same frequency but each with a time-dependent phase, their collective response again exhibits dissipation. Since the response decay does not involve any dissipation in a real sense, sometimes phase damping is also called virtual damping when used in the context of phase noise in electronics (Ham and Hajimiri, 2003). Describing, as before, the response of a single oscillator as:

$$y(t) = y_0 \cos[\omega_0 t + \phi(t)]$$

where y_0, ω_0 and $\phi(t)$ are initial displacement, oscillation frequency and the phase variation, respectively.

Assuming that the phase distribution $\phi(t)$ can be characterized as a Wiener process, that is a zero-mean Gaussian random process, it can be shown that: $< \phi^2(t) >= 2Dt$ where D is the diffusion constant.

For a Gaussian distribution of $\phi(t)$ for all oscillators at any given time, t:

$$<\cos\phi>= e^{-<\phi^2>/2} \qquad <\sin\phi>= 0$$

and, therefore, the average response

$$<y(t)>= y_0\, e^{-\phi^2(t)/2}\,\cos\omega_0 t$$

and the autocorrelation

$$<y(t)y(t+\tau)>= \frac{1}{2}\, y_0^2\, e^{-Dt}\,\cos\omega_0\tau$$

show the decaying behavior of the collective response.

The corresponding power spectral density of the oscillators is a Lorentzian distribution.

$$S_{y,y}(f) = y_0^2\,\frac{D}{(\omega-\omega_0)^2+D^2}$$

where the diffusion constant D represents the damping rate.

While the concept of virtual damping is used in connection with phase noise in electronics, as seen above it has a direct bearing on dissipation models in vibrations.

Forced Oscillations - Landau Damping When the object, to which the oscillators are attached, is subjected to a harmonic force, response of each oscillator is described by:

$$\ddot{y}+\omega^2 y = F_0\cos\Omega t$$

with the general solution

$$y(t) = y_0\cos\omega t + \dot{y}_0\frac{\sin\omega t}{\omega} + \frac{F_0}{\omega^2-\Omega^2}[\cos\Omega t - \cos\omega t].$$

Considering only the forced response, displacement of the center-of-mass of an ensemble with a distribution $\rho(\omega)$ can be expressed as:

$$<y(t)>= F_0\int_{-\infty}^{\infty}\frac{\cos\Omega t-\cos\omega t}{\omega^2-\Omega^2}\,\rho(\omega)\,d\omega. \qquad (23)$$

For distribution $\rho(\omega)$ that is narrow and centered at $\bar{\omega}$ and without other peaks, when the ensemble is driven at a frequency close to the mean frequency, $\Omega\approx\bar{\omega}$, using the expansion $\omega=\Omega+(\omega-\Omega)$, $<y(t)>$ can be expressed as (Ng, 2006):

$$< y(t) >= \frac{F_0}{2\bar{\omega}} \left[\cos \Omega t \int_{-\infty}^{\infty} \frac{1 - \cos(\omega - \Omega)t}{\omega - \Omega} \rho(\omega) d\omega \right.$$
$$\left. + \sin \Omega t \int_{-\infty}^{\infty} \frac{\sin(\omega - \Omega)t}{\omega - \Omega} \rho(\omega) d\omega \right].$$

The ensemble response has a fast oscillating component at frequency Ω and a slow-oscillating component $\omega - \Omega$ that acts like an envelope.

Noting that $\triangle\omega$ is a measure of the width of the frequency distribution $\rho(\omega)$, all transients vanish in the asymptotic behavior of the ensemble average displacement for $t \gg 1/\triangle\omega$ leaving:

$$< y(t) >= \frac{F_0}{2\bar{\omega}} \left[\cos \Omega t \ \textbf{p.v.} \int_{-\infty}^{\infty} \frac{\rho(\omega)}{\omega - \Omega} d\omega + \pi \rho(\Omega) \sin \Omega t \right].$$

For a narrow frequency distribution $\rho(\omega)$ centered around $\bar{\omega}$, the assembly average given in Eq. (23) can be approximated as (Ng, 2006, 2010)

$$< y(t) >= \frac{F_0 \sin \bar{\omega} t}{\bar{\omega}} \int_{-\infty}^{\infty} \frac{\sin \frac{1}{2}(\omega - \Omega)t}{\omega - \Omega}.$$

Considering the amplitude, A, of an oscillator with frequency ω as

$$\mathrm{A}(\omega) = \frac{F_0}{\bar{\omega}} \frac{\sin \frac{1}{2}(\omega - \Omega)t}{\omega - \Omega}$$

all oscillators with frequency ω are excited at $t = 0$ and reach a maximum at $t \approx \pi/(\omega - \Omega)$ and vanish again at $t \approx 2\pi/(\omega - \Omega)$. Such absorption and return of energy to the ensemble is a familiar display of recurrence. For frequencies ω closer to Ω, amplitude is larger and the return time is longer. For oscillators with $\omega = \Omega$, amplitude grows linearly with time and the absorption continues indefinitely. This process of vanishing amplitude growth except for a few oscillators near Ω is called the *Landau Damping* (Ng, 2006).

3.2 Energy Sinks

Energy sink is a concept based on the linear models of dissipation described above but adapted for use with a finite number of oscillators. As described above, linear oscillators when attached to a primary structure can irreversibly absorb its vibration energy provided their frequencies follow certain frequency distributions continuously. Since it is not practical in engineering applications to attach oscillators with a continuous frequency distribution or, equivalently, to use an infinite number of them, the models are adapted for a finite set of oscillators (Carcaterra and Akay, 2004, 2007, 2011; Carcaterra et al., 2006; Koç et al., 2005; Akay et al., 2005).

The energy sinks described here rely on the use of a set of undamped, linear oscillators with a particular frequency distribution that enables it to rapidly absorb vibration energy from a structure and retain nearly all of it.

3.3 Dissipation in Finite Systems

Returning to the equation of motion (20) of an harmonically excited particle with a set of linear oscillators attached to it:

$$M\ddot{x}(t) + M\Omega^2 x(t) + \int_0^t \Gamma(\tau)\dot{x}(t-\tau)d\tau = F(t)$$

$$\ddot{q}_n(t) + \omega_n^2 q_n(t) = \omega_n^2 x(t) \qquad n = 1.2.3\ldots$$

where we have assumed quiescent initial conditions, $q_n(t_0) = 0$, $\dot{q}_n(t_0) = 0$, for the attached oscillators. The kernel of the dissipative term is

$$\Gamma(\tau) = \sum_n m_n \omega_n^2 \cos(\omega_n \tau).$$

In general, harmonic series such as $\Gamma(\tau)$ have properties similar to those of almost-periodic-functions and exhibit recurrence (Bohr, 1947), precluding the use of independent linear oscillators to model irreversible absorption of vibratory energy, except for early times of observation (Carcaterra and Akay, 2004).

However, it has been reported that when the attached oscillators follow particular frequency distributions, they very closely mimic the dissipation characteristics of an infinite number of oscillators, reducing and nearly eliminating the effects of recurrence (Carcaterra and Akay, 2004, 2007, 2011; Carcaterra et al., 2006; Koç et al., 2005; Akay et al., 2005).

►**Example** The prototypical system under consideration consists of a rigid primary structure with a substructure comprised of a set of linear oscillators of equal mass m_n attached to it, as described in Fig. 3. The system does not possess any mechanism to dissipate energy in the classical sense, thus stiffness alone characterizes the connections between the substructure and the primary structure. The total mass, m ($m = \sum m_n$), of the attachments is assumed to be fraction of the primary mass, M, say, $m/M \leq 0.1$. Under these conditions, the significance of the stiffness or frequency distribution is demonstrated for two different cases.

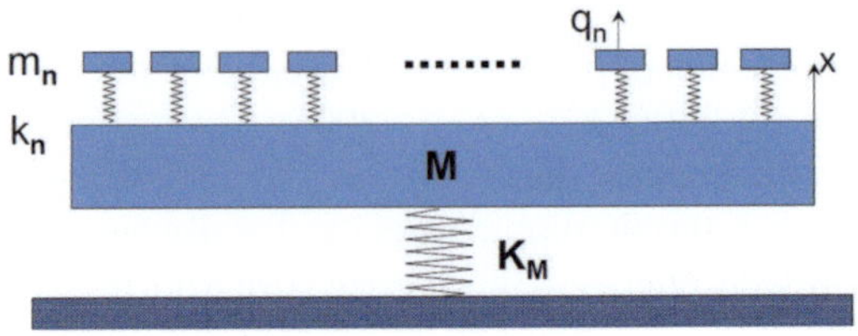

Figure 3. Schematic description of a SDOF structure with a set of linear oscillators attached to it.

For example, Figure 4 shows a typical impulse response of the primary structure with the attached set having a linear frequency distribution as shown in Fig. 4 (a). The oscillators have a constant frequency difference between the neighboring frequencies. As expected of linear oscillators with a linear frequency distribution, the response shows a recurrence (b); and as shown in Fig. 4(c), energy periodically returns to the primary structure when the number of oscillators is finite, in this case $N = 100$. The return time corresponds to the constant frequency difference $t^* = 2\pi/\Delta\omega$ (Koç et al., 2005).

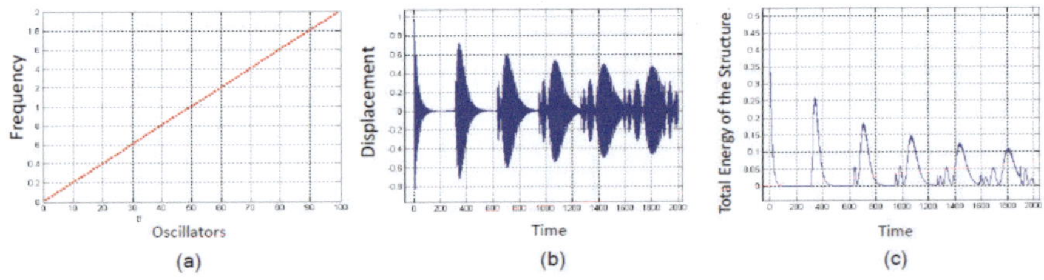

Figure 4. Response of a SDOF structure with attached linear oscillators as shown in Fig.(3): (a) Attached oscillators have a linear frequency distribution, (b) displacement response of the structure with periodic increases, (c) energy of the structure displaying recurrence.

It has been shown that there exist optimum frequency distributions for such finite sets that increase the period of recurrence, effectively absorbing vibratory energy in a nearly irreversible manner (Koç et al., 2005; Carcaterra and Akay, 2004).

In the example shown in Figure 5, for the particular frequency distribution shown in (a), recurrence nearly disappears in the response. A comparison of the energy of the primary oscillator in Figure 5 (c) with that in Figure 4 (c) also shows the reduction in the energy of the structure as a result of the frequency distribution of the attached oscillators. Experimental verifications of such systems have also been reported (Akay et al., 2005).

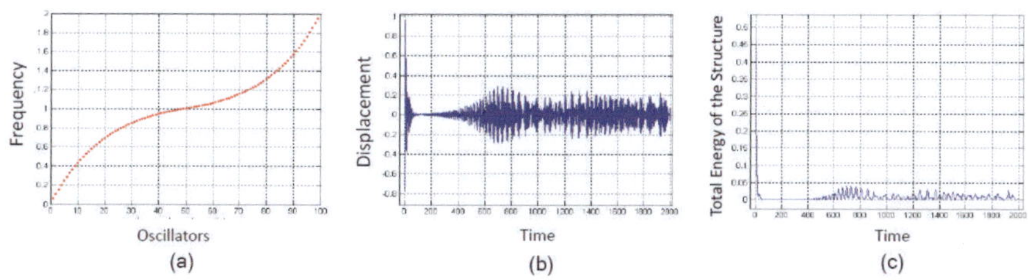

Figure 5. Response of the SDOF structure as in Fig.(3): (a) Frequency distribution among the attached oscillators, (b) displacement response of the structure is no longer periodic, (c) energy of the structure has been absorbed by the oscillators.

Compared with the conventional methods of vibration reduction, the concept of energy sinks presents a unique and viable alternative for cases where the classical vibration absorption or damping techniques have limited applicability, particularly at low frequencies and under transient conditions. Energy sinks described here consist of a set of oscillators that collectively absorb and retain energy when attached to a vibrating structure. An energy sink admits flow of energy from a transiently excited structure and retains it in the collective phase space of the oscillators.

3.4 Damping and Dissipation in an Ideal Gas

Damped motion of a piston in a cylinder presents another example of dissipation at the molecular level where the piston energy is irreversibly converted to thermal energy in a gas. Piston motion is similar to that of a particle motion in a thermal environment discussed earlier where its dissipation is modeled with independent linear oscillators. In this case, however, dissipation modeling is nonlinear and arises from individual impacts of particles on the piston.

Considering a prototypical system illustrated in Fig. 6 in which a piston

P with a cross-sectional surface area S moves freely without wall friction. It moves along the cylinder axis x at speed V_P, separating the two reservoirs of the piston, each maintained at temperatures such that $T_H > T_C$.

Gas is modeled as a set of freely moving particles with a velocity distribution, rather than as attached oscillators. Pressures in volumes H and C that act on the piston in opposite directions are calculated by summing the impulses from particles impinging on the two surfaces of the piston.

Considering first the pressure in the H reservoir, a single gas particle has speed components v_{Hx} and v'_{Hx} along x, before and after its impact on the piston surface, respectively, and are related through the piston velocity:

$$v'_{Hx} = -v_{Hx} + 2V_P.$$

Note that this relationship is valid for cases where the particle mass m is very small compared with the mass M of the piston. The momentum variation of a particle associated with the impact is $2m(V_P - v_{Hx})$. The force exerted on the piston surface H is then calculated as a sum of the x-components of the pulses:

$$F_H(V_P, t) = 2m \sum_i (v_{Hx_i} - V_P)\, \delta(t - t_{iH}) \qquad (24)$$

where F_H depends both on time t and the piston speed V_P. This force evolves through a sequence of random impulses $\delta(t - t_{iH})$ of random amplitude and random time delay. Analogous to the Fluctuation-Dissipation

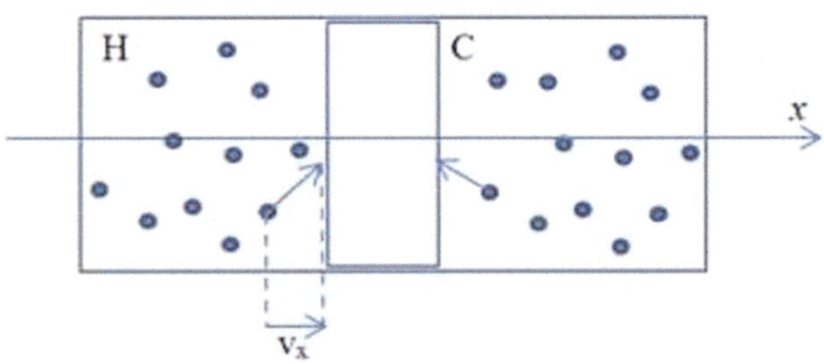

Figure 6. Piston in a cylinder.

Theorem, we separate the mean and fluctuating parts of F_H in Eq. (24). Designating by $n(v_{Hx})\, dv_{Hx}$ the number of particles per unit volume of gas that have the speed component v_{Hx}, the number dN of particles that hit the piston surface within the time interval dt can be expressed as:

$$dN = \frac{1}{2}(v_{Hx} - V_P)\, n(v_{Hx})\; S\, dv_{Hx}\, dt$$

the factor 1/2 comes from the consideration of particles traveling in only one direction along x. Then, the force produced on the piston by dN number of impinging particles in the reservoir H becomes:

$$dF_H(v_{Hx}) = mS\,(v_{Hx} - V_P)^2\, n(v_{Hx})\, dv_{Hx}.$$

Introduction of a velocity distribution $f_{Hx}(v_{Hx}) = n(v_{Hx})/N$ results in the corresponding average total force expression:

$$\bar{F}_H = mNS \int_0^\infty (V_P - v_{Hx})^2\, f_{Hx}\, dv_{Hx} \tag{25}$$

where N is the number of particles per unit volume. For piston velocities that are small compared with the particle velocity in the gas, we can assume the standard Maxwell distribution for f_{Hx}. The influence of piston motion can be evaluated by modifying the Maxwell distribution to include a drift velocity $V_{GH} = -V_P/2$ equivalent to the velocity of the center of mass of the gas in reservoir H, with respect to the piston reference frame:

$$f_{Hx} = \sqrt{\frac{m}{2\pi k T_H}}\; e^{-\frac{m}{2kT_H}(v_{Hx} - V_{GH})^2}. \tag{26}$$

The distribution in (26) is a valid approximation for small drift velocities compared to the gas particle velocity, $V_{GH} \ll \sqrt{\overline{v^2_{Hx}}}$, where the upper bar represents the average value calculated with the standard Maxwell distribution, in the absence of any drift effects ($V_P = 0$). Integrating equation (25) using (26) produces an average force on the piston facing reservoir H:

$$\bar{F}_H(V_P) = mNS\left[\bar{v}^2_{Hx} - 3V_P\bar{v}_{Hx} + \frac{9}{8}V_P^2\right]$$

and similarly for its opposite side facing C:

$$\bar{F}_C(V_P) = mNS\left[\bar{v}^2_{Cx} + 3V_P\bar{v}_{Cx} + \frac{9}{8}V_P^2\right].$$

These expressions differ only in the sign of the linear piston velocity terms. Independent of its direction piston motion induces damping. When both reservoirs are at the same temperature, equivalently when the piston is under pressure equilibrium, the average net force on the piston reduces to:

$$\bar{F}(V_P) = \bar{F}_H(V_P) - \bar{F}_C(V_P) = -6mNSV_P\,\bar{v}_x \tag{27}$$

which reveals the viscous nature of the average damping force. The instantaneous net force on the piston follows from equation (24):

$$F(V_P, t) = 2m \left[\sum_i (v_{xi} + V_P)\, \delta(t - t_{iH}) - \sum_i (v_{xi} - V_P)\, \delta(t - t_{iC}) \right]. \tag{28}$$

The force (28) on the piston can be expressed as a combination of its average value and a fluctuating part:

$$F(V_P, t) = \bar{F}(V_P) + f(t) \tag{29}$$

where the mean value of force $\bar{F}(V_P)$ from Eq. (27) represents the damping force and $f(t)$ represents the fluctuating force that evolves due to impulses from the particles in the gas.

An explicit expression for the fluctuation can be obtained simply by considering the case for a stationary piston, and setting $V_P = 0$, in Eq. (29):

$$F(0, t) = \bar{F}(0) + f(t)$$

resulting in the expression for fluctuation:

$$f(t) = 2m \sum_i v_{xi} \delta(t - t_i).$$

Finally the general expression for the interaction force between the piston and the gas can be expressed as:

$$F(V_P, t) \approx -6mNSV_P\bar{v}_x + 2m \sum_i v_{xi} \delta(t - t_i).$$

This shows how the interaction with the particles in the reservoirs generates a viscous damping, with damping coefficients $C_D = 6mNS\bar{v}_x$, and a random force on the right-hand side. Moreover, it appears the two terms are not independent, since they appear to be both driven by the speed v_x of the particles in the gas. This fact again manifests the fluctuation-dissipation duality. Finally, note how the damping coefficient depends on the average speed $\bar{v}_x$ of the particles contained in the reservoirs, that it is proportional to the root of the temperature of the gas since $T \propto \overline{v_x^2} \propto \bar{v}_x^2$ and therefore $C_D \propto \sqrt{T}$.

Acknowledgement

Adnan Akay would like to express his gratitude to the Alexander von Humboldt Foundation for supporting this work through a Research Award.

Bibliography

A. Akay. Acoustics of friction. *The Journal of the Acoustical Society of America*, 111(4):1525–1548, 2002.

A. Akay and Zh. Xu. Experiments on acoustic streaming in a fluid layer between vibrating surfaces and amplitude-dependent damping of vibrations. *The Journal of the Acoustical Society of America*, 103(2):865–871, 1998.

A. Akay, Zh. Xu, A. Carcaterra, and I. M. Koç. Experiments on vibration absorption using energy sinks. *The Journal of the Acoustical Society of America*, 118(5):3043–3049, 2005.

C. H. Allen. Vibration damper for circular saws. In *INTER-NOISE and NOISE-CON Congress and Conference Proceedings*, volume 1977, pages 569–572. Institute of Noise Control Engineering, 1977.

V. I. Arnol'd. Small denominators and problems of stability of motion in classical and celestial mechanics. *Russian Mathematical Surveys*, 18(6): 85–191, 1963.

A. I. Beltzer. *Acoustics of solids.* Springer, 1988.

M. A. Biot and I. Tolstoy. *Acoustics, elasticity, and thermodynamics of porous media: twenty-one papers.* Acoustical Society of Amer, 1992.

H. A. Bohr. *Almost periodic functions.* American Mathematical Soc., 1947.

A. O. Caldeira and A. J. Leggett. Path integral approach to quantum brownian motion. *Physica A: Statistical Mechanics and its Applications*, 121(3):587–616, 1983.

A. Carcaterra and A. Akay. Transient energy exchange between a primary structure and a set of oscillators: Return time and apparent damping. *The Journal of the Acoustical Society of America*, 115(2):683–696, 2004.

A. Carcaterra and A. Akay. Theoretical foundations of apparent-damping phenomena and nearly irreversible energy exchange in linear conservative systems. *The Journal of the Acoustical Society of America*, 121(4):1971–1982, 2007.

A. Carcaterra and A. Akay. Dissipation in a finite-size bath. *Physical Review E*, 84(1):011121, 2011.

A. Carcaterra, A. Akay, and I.M. Koç. Near-irreversibility in a conservative linear structure with singularity points in its modal density. *The Journal of the Acoustical Society of America*, 119(4):2141–2149, 2006.

C. E. Çelik and A. Akay. Dissipation in solids: Thermal oscillations of atoms. *The Journal of the Acoustical Society of America*, 108(1):184–191, 2000.

J. P. Den Hartog. *Mechanical vibrations*. Courier Dover Publications, 2013.

E. H. Dowell. The behavior of a linear, damped modal system with a non-linear spring-mass-dry friction damper system attached. *Journal of Sound and Vibration*, 89(1):65–84, 1983.

E. H. Dowell. Damping in beams and plates due to slipping at the support boundaries. *Journal of Sound and Vibration*, 105(2):243–253, 1986.

E. H. Dowell and H. B. Schwartz. Forced response of a cantilever beam with a dry friction damper attached, part ii: Experiment. *Journal of Sound and Vibration*, 91(2):269–291, 1983a.

E.H. Dowell and H.B. Schwartz. Forced response of a cantilever beam with a dry friction damper attached, part i: Theory. *Journal of Sound and Vibration*, 91(2):255–267, 1983b.

E. Fermi, J. Pasta, and S. Ulam. Studies of nonlinear problems. Technical report, I, Los Alamos Scientific Laboratory Report No. LA-1940, 1955.

R. P. Feynman and F. L. Vernon Jr. The theory of a general quantum system interacting with a linear dissipative system. *Annals of physics*, 24:118–173, 1963.

S. Filippi, M. M. Gola, and A. Akay. Measurement of tangential contact hysteresis during microslip. *Journal of tribology*, 126(3):482–489, 2004.

J. Ford. The fermi-pasta-ulam problem: paradox turns discovery. *Physics Reports*, 213(5):271–310, 1992.

P. Göransson. Acoustic and vibrational damping in porous solids. *Philosophical Transactions of the Royal Society A: Mathematical, Physical and Engineering Sciences*, 364(1838):89–108, 2006.

D. Ham and A. Hajimiri. Virtual damping and einstein relation in oscillators. *Solid-State Circuits, IEEE Journal of*, 38(3):407–418, 2003.

K. U. Ingard and A. Akay. On the vibration damping of a plate by means of a viscous fluid layer. *Journal of Vibration, Acoustics Stress and Reliability in Design*, 109(2):178–184, 1987.

H. M. Jaeger, S. R. Nagel, and R. P. Behringer. Granular solids, liquids, and gases. *Reviews of Modern Physics*, 68(4):1259–1273, 1996.

A. Kareem and K. Gurley. Damping in structures: its evaluation and treatment of uncertainty. *Journal of Wind Engineering and Industrial Aerodynamics*, 59(2):131–157, 1996.

F. K. Kneubühl. *Oscillations and waves*. Springer, 1997.

I. M. Koç, A. Carcaterra, Zh. Xu, and A. Akay. Energy sinks: vibration absorption by an optimal set of undamped oscillators. *The Journal of the Acoustical Society of America*, 118(5):3031–3042, 2005.

A. N. Kolmogorov. Preservation of conditionally periodic movements with small change in the hamilton function. In *Stochastic Behavior in Classical and Quantum Hamiltonian Systems*, pages 51–56. Springer, 1979.

R. Lifshitz and M. L. Roukes. Thermoelastic damping in micro-and nanomechanical systems. *Physical review B*, 61(8):5600, 2000.

R. A. Mangiarotty. Acoustic radiation damping of vibrating structures. *The Journal of the Acoustical Society of America*, 35(3):369–377, 2005.

D. J. Mead. *Passive vibration control.* Wiley Chichester, 1998.

W. Meins. *Geräuschuntersuchungen an Kreissägemaschinen.* PhD thesis, TH Braunschweig, 1963.

P. M. Morse. *Theoretical acoustics.* Princeton University Press, 1986.

J. Moser. *On Invariant Curves of Area-preserving Mappings of an Annulus...* Vandenhoeck; ruprecht, 1962.

A. D. Nashif, D. I. G. Jones, and J. P. Henderson. *Vibration Damping.* Wiley New York et al., 1985.

K.-Y. Ng. *Physics of intensity dependent beam instabilities.* World Scientific, 2006.

K.-Y. Ng. Landau damping. Technical report, Fermi National Accelerator Laboratory (FNAL), Batavia, IL, 2010.

A. S. Nowick and B. S. Berry. *Anelastic relaxation in crystalline solids.* Academic Press, 1972.

A. D. Pierce. Causality and mathematical models in vibration and acoustics, a realistic perspective. In *Proceedings of Meetings on Acoustics*, volume 5, page 065001. Acoustical Society of America, 2008.

D. Salamon. The kolmogorov-arnold-moser theorem. *Zürich preprint*, 1986.

K.R. Waters, J. Mobley, and J.G. Miller. Causality-imposed (kramers-krönig) relationships between attenuation and dispersion. *Ultrasonics, Ferroelectrics and Frequency Control, IEEE Transactions on*, 52(5):822–823, 2005.

C. Zener. Internal friction in solids. i. theory of internal friction in reeds. *Physical review*, 52(3):230, 1937.

C. Zener. Internal friction in solids ii. general theory of thermoelastic internal friction. *Physical Review*, 53(1):90, 1938.

C. Zener. Internal friction in solids. *Proceedings of the Physical Society*, 52 (1):152, 1940.

C. Zener. *Elasticity and anelasticity of metals.* University of Chicago press, 1948.

Vibration Control and Failure Diagnosis in Rotating Machinery by means of Active Magnetic Bearings

R. Nordmann
TU Darmstadt, Darmstadt, Germany

Abstract Mechatronic components are getting more and more common in mechanical systems. As an example Active Magnetic Bearings (AMB) are often used in Rotating Machinery. Besides their function of an oil-, contact- and frictionless levitation of the rotor, they are best suited to be used as an exciter and measurement instrument to extract more information from the system under observation. In this paper it is shown, how Active Magnetic Bearings can be used for identification, diagnosis and optimization purposes.

1 Introduction

In the field of rotating machinery the number of applications using mechatronic components is increasing. In comparison to conventional systems such mechatronic products, consisting of mechanical, electrical and electronical components have the ability to pick up changes in their environment by sensors and react to the system or process by means of actuators after an appropriate information processing, carried out in a microprocessor (Aenis (2002)).
Nowadays, rotors running with active magnetic bearings or with other mechatronic bearings or components already offer a variety of advantages. Some of them are the tuning possibilities for stiffness and damping, the absence of friction and wear, the high running speeds, the vibration isolation, the active vibration damping and possible unbalance compensation. However there is much more potential in such systems with respect to a smart behaviour. In rotating machinery with mechatronic components, consisting of built in control, sensors, microprocessors, actuators and last but not least integrated software, different novel features like identification, diagnosis and correction can be realized. In this way it is possible to design new machines with higher performance, higher reliability and longer lifetime.
This paper particularly describes rotating machines with mechatronic components and concentrates on the mentioned smart features of identification,

P. Hagedorn, G. Spelsberg-Korspeter (Eds.), *Active and Passive Vibration Control of Structures*, CISM International Centre for Mechanical Sciences
DOI 10.1007/ 978-3-7091-1821-4_7

process diagnosis and correction and process optimization as well. After a short introduction to Mechatronics, technical applications of rotating systems with mechatronic components will be presented.

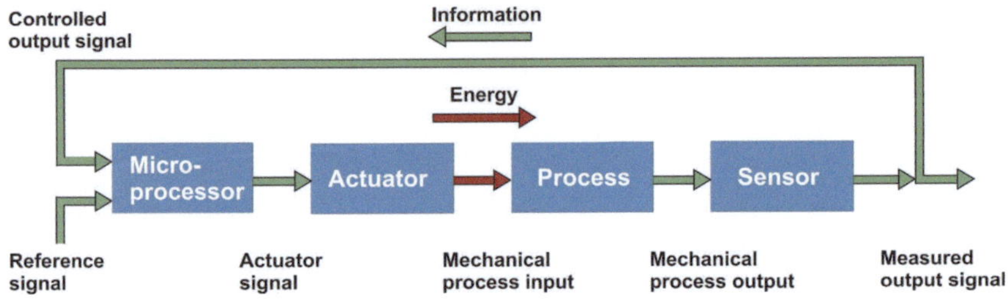

Figure 1. Block diagram of a Mechatronic system

2 What is Mechatronics

Mechatronics is a multidisciplinary area of engineering including mechanical and electrical engineering together with information technology. In mechatronic systems signals are measured by sensors and processed in embedded microcontrollers. Actuator forces and/or moments then act on the process, controlled by the microprocessor functions. In Figure 1 the block diagram shows the different components of a mechatronic system (microprocessor, actuator, process and sensor) and their connections for the flow of energy and information. However, a diagram like this can also be applied for classical mechanical-electrical control systems. In comparison modern mechatronic systems have much more potential. They are characterized by two types of integration: the hardware integration and the integration of functions. The hardware integration or integration of components results from designing the mechatronic system as an overall system embedding the sensors, actuators and microcontrollers into the mechanical process. This spatial integration may be limited to the process and sensor or the process and actuator. The microcomputers can be integrated with the actuator, the process or sensor, or be arranged at several places.

Integration by information processing (software integration) is mostly based on advanced control functions. Besides a basic feed forward and feedback control, an additional influence may take place through the process knowledge and corresponding on-line information processing. This includes the solution of tasks like supervision with fault diagnosis, identification, correction and process optimization. In this paper applications of such integrated functions will be presented especially for rotating machinery.

3 Rotating Machinery: Configuration and Technical Applications

In various technical areas rotating machinery systems are in operation, like turbines, pumps, compressors, motors and generators etc.

Figure 2. Configuration of Mechanical Components in Rotating Machinery

An example of a steam turbine rotor is shown in figure 2 with its main mechanical components. The rotating shaft with the blading system (energy transfer from thermal to mechanical) is supported in oil film bearings. The rotating parts are arranged in the turbine housing and sealed to the environment by means of labyrinths. Users expect, that their machines are running safe and reliable and have a high efficiency and availability as well. In order to satisfy these requirements mechanical problems especially vibrations have to be considered very careful. Particularly the dynamic behaviour of the rotating components and their different interactions with the stator are of great importance for the durability and lifetime of a machine. Therefore the machine designer already starts in an early stage of the development to investigate and to predict the machines dynamic behaviour and the corresponding internal forces and stresses, respectively Ω. Figure 3 shows as another example a blower system consisting of the rotating shaft, supported in two roller bearings. The impeller in the center of the shaft delivers a specified gas flow for the process in the plant. In order to investigate the vibration problem of forced vibrations including the resonance problem a relatively simple model can be used with a flexible shaft, rigid bearings and the impeller with its inertia characteristics.

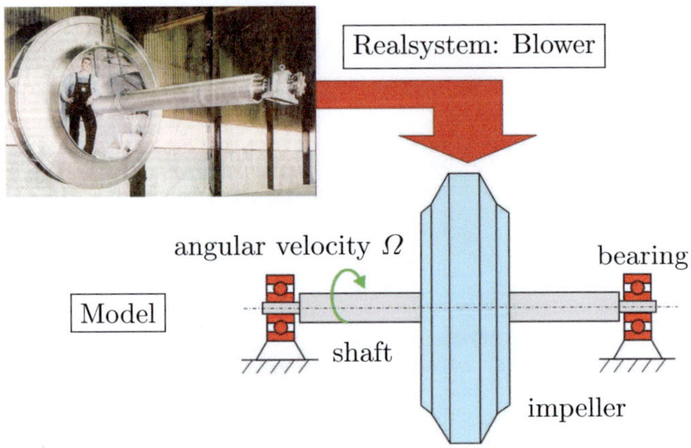

Figure 3. Blower system with model

The natural frequency f (Hz) of the first bending of the rotor system can easily be calculated by

$$f = \frac{1}{2\pi}\omega = \frac{1}{2\pi}\sqrt{\frac{c}{m}}. \tag{1}$$

With this natural frequency f or the circular natural frequency $\omega = 2\pi f$ the critical angular of velocity is known. c is the shaft stiffness in N/m and m is the mass of the impeller in kg.
Due to the fact, that the center of rotation is not equal to the center of gravity a mass eccentricity e has to be considered. This leads to an excitation by unbalance forces, with the rotational angular of velocity. The resulting forced unbalance vibration

$$\frac{x(\Omega)}{e} = \frac{m\Omega^2}{\sqrt{(c - m\Omega^2)^2 + (d\Omega)^2}} \tag{2}$$

is shown in figure 4.

The diagram shows the relative amplitude of vibration x (related to the mass eccentricity) versus the running speed (related to the natural frequency

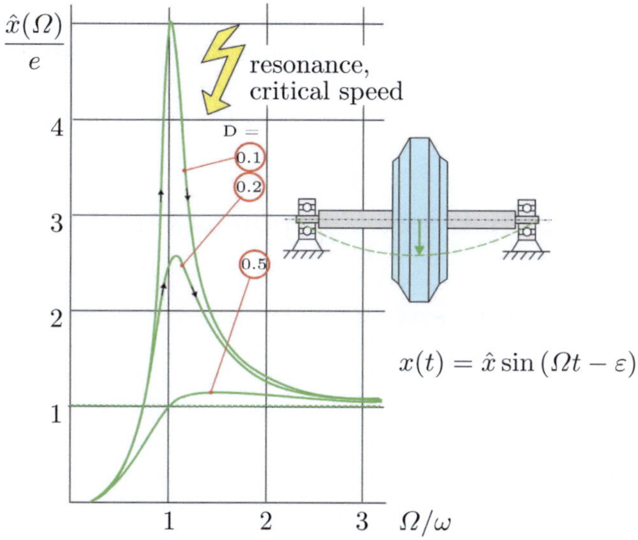

Figure 4. Run up curve for blower

ω). The different curves belong to different damping values. It can clearly be seen, that the resonance effect occurs (critical speed), when the rotational frequency is equal to the natural frequency ($\Omega = \omega$). Damping (passive or active) helps to decrease the vibrations in the resonance. The example of figure 3 is relatively easy to solve. More general rotordynamic tools for computer simulations are available nowadays, usually based on the Finite Element method. These routines allow to include all important components like shafts, impellers, bearings, seals etc. and take into consideration the corresponding effects like inertia, damping, stiffness, gyroscopics, unbalance and fluid structure interaction forces. They predict modal parameters like natural frequencies, damping values, mode shapes and unbalance and transient vibrations as well. While these powerful tools itself usually work without difficulties, problems more often occur in finding the correct input data. Particularly not all of the physical parameters are available from theoretical derivations. This is especially true for the rotordynamic coefficients, describing various fluid structure interactions. In such cases the required data have to be taken from former experience or have to be determined experimentally via identification procedures. The identification of the dynamic characteristics of a rotating system by means of mechatronic components (Active Magnetic Bearings) offers further possibilities besides identification. The developed procedures are an important base for failure detection and failure diagnosis during operation. By using the actuators as elements to introduce static and dynamic forces into the system process optimization and correction can also be performed. This will be shown in the next section.

4 Use of Mechatronic Components in Rotating Machinery

Identification techniques have already been used in different applications in order to find rotordynamic coefficients (stiffness, damping, inertia) e.g. in bearings and seals. One of the main problems to work with identification techniques in rotordynamics is the excitation of a rotating structure during operation. On the one side it is not easy to have access to the rotor and on the other side the force measurement is difficult, especially when a machine is running with full power and speed and the signal to noise ratio is bad. In some recent investigations Active Magnetic Bearings (AMB's) have been used in order to solve this difficult task. These new techniques seem to be very promising, because AMB's do not only support the rotor, but act as excitation and force measurement equipment as well. In cases, where active magnetic bearings arc designed as bearing elements for turbomachinery systems, it seems helpful to use them also as excitation and force measurement tool. In such applications identification of the dynamic behaviour of the rotating machinery system would be possible during normal operation.

4.1 Principle of active magnetic bearings

Unlike conventional bearing systems, a rotor in magnetic bearings is carried by a magnetic field. This means, that sensors and controllers are necessary to stabilize the unstable suspense state of the rotor. Therefore, essential dynamic characteristics like stiffness and damping properties of the whole system can be influenced by the controller (DSP-Board). For the following applications the rotor can be moved on almost arbitrarily chosen trajectories independently of the rotation, e.g. harmonic motions in one plane, forward or backward whirls. Additionally, an imbalance compensation can be performed. A magnetic bearing system consists of four basic components: magnetic actuator, controller, power amplifier, and shaft position sensor. To keep the rotor in the bearing center, the position sensor signal is used as input for a control circuit to adequately adjust the coil currents. The bearing configuration is composed of four horseshoe- shaped magnets (8 magnetic poles) and is operated in the so-called differential driving mode, where one magnet is driven with the sum of bias current and control current, and the other one with the difference (Figure 5).

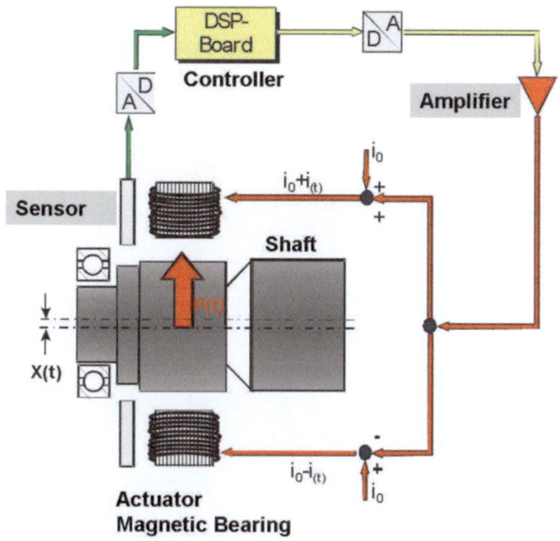

Figure 5. Principle of an Magnetic Bearing

4.2 Failure diagnosis of a centrifugal pump

Today, monitoring systems are normally not an integral component of turbomachines. With these failure detection systems, the relative and/or absolute motions of the rotor are measured as output signals. After signal processing, certain features (threshold values, orbits, frequency spectra etc.) are created from the measured data. With the deviations of these features from a faultless initial state, the diagnosis attempts to recognize possible faults. The difficulty with these procedures is that the causes of the modifications of the output signals can not be detected clearly. The reason might either be a change of the process or a modification of the system itself. An improvement of the existing diagnostic techniques can be achieved by using AMBs. They are well suited to operate contactless as actuator and sensor elements in rotating machinery. Consequently, frequency response functions (stiffness or compliance frequency responses) can be determined from the measured input and output signals, of which the physical parameters or modal parameters (natural frequencies, eigenmodes, modal damping) of the system can be identified. A magnetically suspended centrifugal pump is used to validate and to demonstrate the performance of

the developed model based diagnosis methods.
Figure 6 shows the single stage pump with the pipe system and the driving motor. Two radial and axial magnetic bearings support the pump shaft and serve in addition as actuators and sensors for the determination of the frequency response functions (compliance functions).

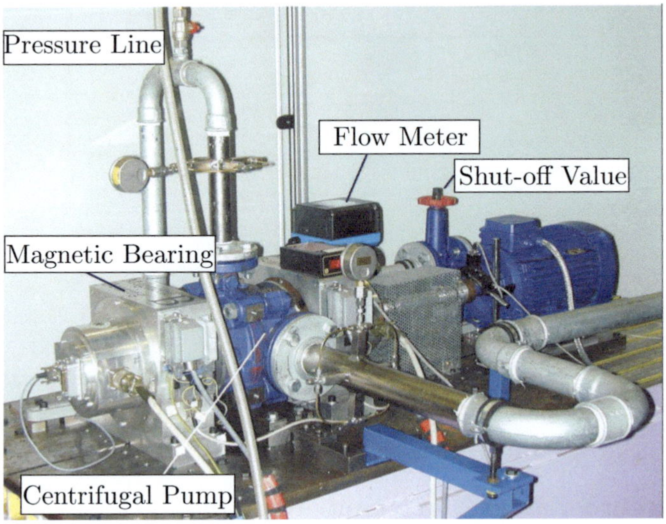

Figure 6. Centrifugal Pump with Magnetic Bearings

As an example, figure 7 represents a measured frequency response function (reference) without any failure. If a failure occurs, e.g. dry run of the pump system, the frequency response function will change its behaviour, e.g. to the curve Dry Run (measured). Due to the dry run the fluid structure interactions in narrow seals are no longer working and this changes the pump dynamic behaviour completely (stiffness and damping coefficients of the fluid). From the frequency response change it can be concluded that the failure dry run has occurred. The task is to find out, when a fault occurs and to determine the faults type, location and extend. A model of the pump system can help to solve this problem (model based diagnosis).
In figure 7 the simulation for the dry run system leads to another curve, which fits very well to the measurements of the faulty rotor. The corresponding pump parameters for this simulation belong to the dry run situation.

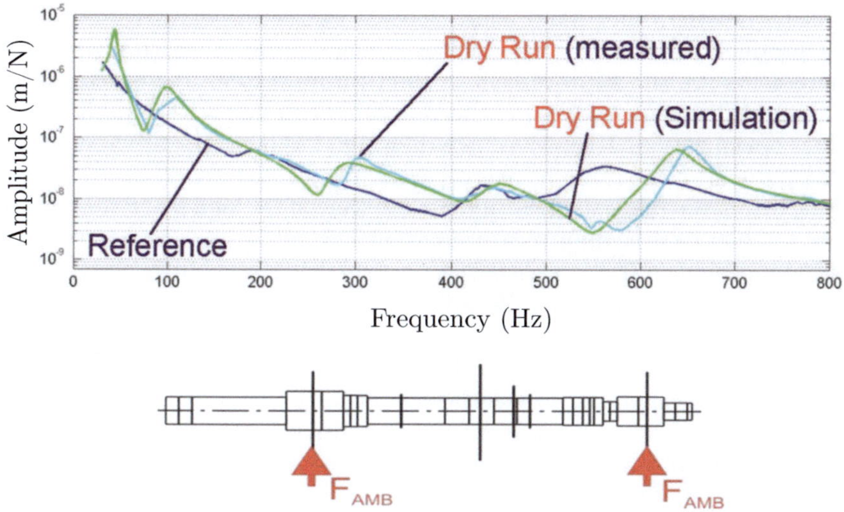

Figure 7. Frequency Response functions for the centrifugal pump

4.3 Diagnosis and process optimization of a high speed grinding process

Internal grinding is applied on products such as outer rings of ball bearings or injection parts of combustion engines. The requirements for this process are very high and contradictory. On the one hand, very high shape and size accuracies as well as surface quality of the work-pieces are de-

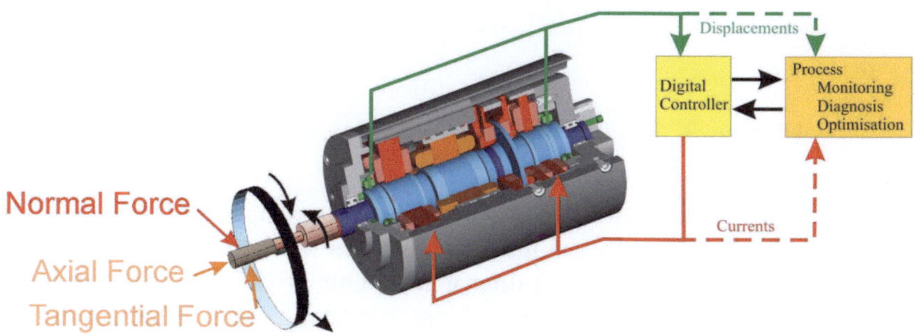

Figure 8. Grinding Spindle with Magnetic Bearings

manded. On the other hand short process cycles are wanted due to the mass production. The mean measures which can be taken to meet these requirements are the improvement of the bearing system, the process monitoring and diagnosis as well as the process correction with regard of the process optimization. High speed grinding spindles in AMBs are best suited in order to carry out these measures(figure 8).

Regarding the magnetic bearing system, they allow high rotational speeds up to 180.000 rpm leading to a decreasing normal process force. The not existing mechanical friction within the bearings permits a long bearing life. Furthermore a high static spindle stiffness can be reached by an appropriate controller design. Using the measured displacement signals and the given control currents supplied to the AMB system the AMB-forces are obtained and out of it the normal, tangential and axial process forces respectively. The normal process force is the essential quantity for the process diagnosis since it permits a direct conclusion to the process state (figure 9).

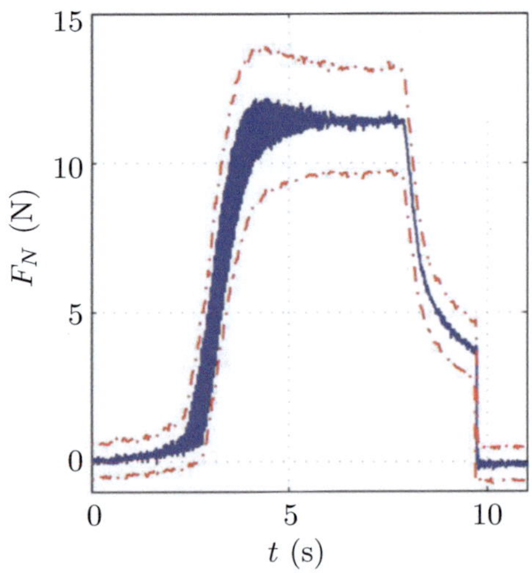

Figure 9. Process diagnosis with normal grinding force

Based on the measured and calculated quantities the diagnosis algorithms are carried out to evaluate the process state, for instants with regard of chatter or a broken grinding wheel. Out of the diagnosis procedure or directly out of the signals correction procedures take place in order to optimize the grinding process. In this phase the AMBs are used as actuators to move the spindle to a required position. One correction procedure is shown in figure 10. Due to its flexible property the work- tool is bent during grinding leading to an undesirable conical bore shape. By tilting the spindle within the bearing this effect does not take place. The described concept is developed and tested on a AMB high speed grinding spindle test rig. Another process optimization is possible by oscillation of the spindle in axial direction with a frequency of about 30 Hz and small amplitudes of about 20μm. This motion, generated by the axial magnetic bearing, leads to an improvement of the workpiece surface.

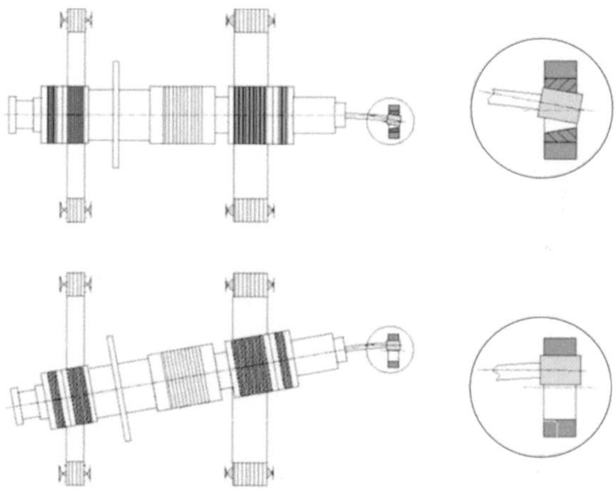

Figure 10. Static tool bending compensation

5 Conclusions

The use of mechatronic components in rotating machinery may contribute to a better performance of this type of machines. As an example Active Magnetic Bearings can also be used as sensor, actuator and exciter. Accurate measurements of forces and displacements and the generation of axial and radial shaft motions are possible. In this way AMBs offer new applications in turbomachinery, e.g. improved identification, diagnosis and optimization techniques. This technology supports the development of new products in the field of rotating systems.

Bibliography

M. Aenis. *Einsatz aktiver Magnetlager zur modellbasierten Fehlerdiagnose in einer Kreiselpumpe.* Dissertation, TU Darmstadt, 2002.

MIX
Papier aus verantwortungsvollen Quellen
Paper from responsible sources
FSC® C105338

If you have any concerns about our products,
you can contact us on
ProductSafety@springernature.com

In case Publisher is established outside the EU,
the EU authorized representative is:
Springer Nature Customer Service Center GmbH
Europaplatz 3, 69115 Heidelberg, Germany

Printed by Libri Plureos GmbH
in Hamburg, Germany